Computational Earthquake Physics: Simulations, Analysis and Infrastructure, Part I

Edited by
Xiang-chu Yin
Peter Mora
Andrea Donnellan
Mitsuhiro Matsu'ura

2006

Birkhäuser Verlag
Basel · Boston · Berlin

Reprint from Pure and Applied Geophysics
(PAGEOPH), Volume 163 (2006) No. 9

Editors

Xiang-chu Yin
Institute of Earthquake Science
Chinese Earthquake Administration
Beijing 10036
China

e-mail: xcyin@public.bta.net.cn

Peter Mora
Earth Systems Science
Computational Centre
The University of Queensland
Brisbane, 4072 Qld
Australia
e-mail: director@esscc.uq.edu.au

Andrea Donnellan
Jet Propulsion Laboratory
4800 Oak Grove
Drive-Mail Stop 183-335
Pasadena, CA 91109
USA

e-mail: andrea.donnellan@jpl.nasa.gov

Mitsuhiro Matsu'ura
Dept. of Earth and Planetary Science
The University of Tokyo
7-3-1 Hongo
Bunkyo-Ku
Tokyo 113-0033
Japan
e-mail: matsuura@eps.s.u-tokyo.ac.jp

Library of Congress Control Number: 2006935063

Bibliographic information published by Die Deutsche Bibliothek:
Die Deutsche Bibliothek lists this publication in the Deutsche Nationalbibliographie; detailed bibliographic data is available in the internet at <http://dnb.ddb.de>

ISBN 3-7643-7991-X Birkhäuser Verlag, Basel – Boston – Berlin

Part of Springer Science+Business Media
Printed on acid-free paper produced from chlorine-free pulp
Printed in Germany

ISBN 10: 3-7643-7991-X
ISBN 13: 978-3-7643-7991-9

e-ISBN 10: 3-7643-7992-8
e-ISBN 13: 978-3-7643-7992-6

9 8 7 6 5 4 3 2 1

Contents

Pure appl. geophys. 163 (2006) 1737–1740
0033–4553/06/091737–4
DOI 10.1007/s00024-006-0112-4

Pure and Applied Geophysics

Computational Earthquake Physics
PART I: Introduction

Xiang-chu Yin,[1,2] Peter Mora,[3] Andrea Donnellan,[4] and Mitsuhiro Matsuura[5]

Large earthquakes are terrible natural disasters which usually cause massive casualties and huge property loss. In the beginning of the new century, large earthquakes violently struck the world, especially the Asia-Pacific region. Nearly 300,000 people were killed by the magnitude 9.0 Northern Sumatra earthquake and tsunami, and the magnitude 7.8 Pakistan earthquake of October 8th, 2005 resulted in 90,000 deaths. In the meantime, there has been great progress in computational earthquake physics. New understanding of earthquake processes, numerous ideas on earthquake dynamics and complexity, new numerical models and methods, higher performance super-computers, and new data and analysis methods are emerging. These include the LSM (Lattice Solid particle simulation Model) Australian Computational Earth Systems Simulator (ACcESS), Japan's Earth Simulator, GeoFEM, GeoFEST, QuakeSim, SERVO grid, iSERVO, LURR (Load-Unload Response Ratio), PI (Pattern Informatics), Critical Sensitivity, earthquake Critical Point Hypothesis, the friction law and seismicity, tremor, the Virtual California model, interaction between faults and the conversation of earthquakes, ROC (Relative Operating Characteristic), MFEM (Multiscale Finite-Element Model), etc. Most of these are the outcome of ACES-related research and activities, and will be presented in this volume.

The APEC Cooperation for Earthquake Simulation (**ACES**) [1], endorsed by APEC (Asia-Pacific Economic Cooperation) in 1997, capitalizes on this new

[1] Institute of Earthquake Science, Chinese Earthquake Administration, Beijing 100036. E-mail: xcyin@public.bta.net.cn
[2] State Key Laboratory of Nonlinear Mechanics, Institute of Mechanics, Chinese Academy Sciences, Beijing 100080. E-mail: yin@LMN.imech.ac.an
[3] Earth Systems Science Computational Centre, The University of Queensland, Brisbane, 4072 Qld, Australia. E-mail: director@esscc.uq.edu.au
[4] Jet Propulsion Laboratary, 4800 Oak Grove Drive-Mail Stop 183-335, Pasadena, California 91109. E-mail: andrea.donnellan@jpl.nasa.gov
[5] Department of Earth and Planetary Science, The University of Tokyo, 7-3-1 Hongo, Bunkyo-Ku, Tokyo 113-0033. E-mail: matsuura@eps.s.u-tokyo.ac.jp

opportunity and the complementary strengths of the earthquake research programs of individual APEC member economies via collaboration towards development of such models, the necessary research infrastructure to enable large-scale simulations, and to assimilate data into the models.

The inaugural workshop, the second and the third workshop of ACES were held in 1999, 2000 and 2002, respectively [2–5]. During the week of July 9–14, 2004, China hosted the 4th ACES workshop [6] in Beijing. The 4th ACES Workshop was a milestone for ACES as unanimous agreement was reached for the follow-on to ACES, the ACES-iSERVO [7] International Institute (International Solid Earth Research Virtual Observatory Institute). A colloquium on iSERVO was held at the 4th ACES Workshop leading to broad endorsement for establishment of the iSERVO Institute by the international group of over 100 scientists in attendance, and subsequent signing of a formal agreement – "The Beijing Declaration" [8] – to establish the institute which will be a frontier international research institute on simulating the solid earth. A special issue on Earth Systems Modelling overviewed contributions to the development of the iSERVO institute [7] by its key participants. The institute will consist of a node in each participating economy, and will build on complementary national programs, centers and facilities for solid earth simulation. The institute's focus will be development of predictive capabilities for solid earth phenomena via simulation and breakthrough science using the computational simulation capabilities aimed at understanding solid earth system complexity.

This special issue is divided into two parts. The first part (part I) incorporates Micro-Scale Simulation, Macro-Scale Simulation and Scaling Physics. Topics covered range from numerical developments, rupture and gouge studies of the particle model, Liquefied Cracks and Rayleigh Wave Physics, studies of catastrophic failure and critical sensitivity, numerical and theoretical studies of crack propagation, development in finite-difference methods for modeling faults, long time scale simulation of interacting fault systems, modeling of crustal deformation, through to mantle convection. The second part (Part II, PAGEOPH Vol 163, No 11/12 (2006)) incorporates Computational Environment and Algorithms, Data Assimilation and Understanding, Model Applications and iSERVO.

The 4th ACES workshop (2004) was planned by the ISB (International Science Board) of ACES, consisting of Peter Mora, Xiang-chu Yin, Mitsuhiro Matsuura, Andrea Donnellan and Jean-Bernard Minster, and was hosted by the Institute of Earthquake Science, China Earthquake Administration and LNM (State Key Laboratory of Nonlinear Mechanics), Institute of Mechanics, Chinese Academy of Sciences.

We appreciate our sponsors including China Earthquake Administration, Chinese Academy of Sciences, Chinese National Natural Science Foundation, Chinese Ministry of Science and Technology, Chinese Ministry of Finance, Australia-China Fund, Australian Research Council, The University of Queensland and Earth Systems Science Computational Centre, Australian Computational Earth Systems Simulator Major National Research Facility, Japan Society for the Promotion of Science, Research Organization for Information Sciences and Technology (RIST), the National Aeronautics and Space Administration (NASA) and the US National Science Foundation (NSF) and the United States Geological Survey (USGS).

References

[1] http://www.esscc.uq.edu.au/ACES.

[2] Microscopic and Macroscopic Simulation: Towards Predictive Modelling of the Earthquake Process (2000), eds. Mora, P., Matsu'ura, M., Madariaga, R., and Minster, J-B. Pure Appl. Geophys. *157*, 11/12, 1817–2383.

[3] 2nd ACES Workshop Proceedings, eds. Mitsuhiro Matsuura, Kengo Nakajima, and Peter Mora (2001), (ACES, Brisbane, Australia, ISBN 186499 510 6, GoPRINT).

[4] Computational Earthquake Science Part I (2004), eds. Andrea Donnellan, Peter Mora, Mitsuhiro Matsu'ura, and Xiang-chu Yin, Pure Appl. Geophys. *161*, 9/10, 1823–2118.

[5] Computational Earthquake Science Part II, eds. Andrea Donnellan, Peter Mora, Mitsuhiro Matsu'ura, and Xiang-chu Yin, Pure Appl. Geophys. *161*, 11/12, 2119–2460.

[6] http://www.aces-workshop-2004.ac.cn/.

[7] Special issue on "Earth System Modeling" (2005), guest editor: John B. Rundle, authors: Peter Mora, Xiang-chu Yin, Mitsuhiro Matsuura, Lisa B. Grant and Andrea Donnellan *et al.*, Computing in Science & Engineering *7*, 24–66.

[8] Preface, 4th ACES Workshop Proc., eds. Xiang-chu Yin, Peter Mora, Andrea Donnellan and Mitsuhiro Matsuura (2006) (ACES, Brisbane, Australia, GoPRINT).

Exciting developments in earthquake science have benefited from new observations, improved computational technologies, and improved modeling capabilities. Designing a realistic supercomputer simulation model for the complete earthquake generation process is a grand scientific challenge due to the complexity of phenomena and range of scales involved from microscopic to global. The APEC Cooperation for Earthquake Simulation (ACES) aims to develop such models. Since 1997 four ACES Workshops have been held in Brisbane and Noosa in Australia, Tokyo and Haoken in Japan, Maui, Hawaii in USA and Beijing, China on July 10–14, 2004, respectively. The book mainly contains the results presented in the 4th ACES Workshop in Beijing and the new outcomes from 4th ACES Workshop to the present. The book covers: Microscopic simulation of earthquake, scaling physics, macroscopic simulation, computational environment and algorithms, data assimilation and understanding, model applications and iSERVO (International Solid Earth Research Virtual Observation).

Part I of the book focuses on microscopic from numerical and physical simulation, scaling physics, dynamic rapture and wave propagation, earthquake generation, cycle and seismic pattern.

To access this journal online:
http://www.birkhauser.ch

Pure appl. geophys. 163 (2006) 1741–1750
0033–4553/06/091741–10
DOI 10.1007/s00024-006-0094-2

Pure and Applied Geophysics

Fracture of a Liquefied Crack and the Physics of Rayleigh Waves

L. Knopoff[1,2] and J. A. Landoni[1]

Abstract—The standard free-surface boundary conditions for in-plane crack dynamics are shown to be identical to the conditions for crack dynamics on a liquefied crack. The surfaces of both the free and liquefied cracks do not separate during faulting and hence the static normal stress is not relaxed by the faulting. A crack with either free or liquid boundary conditions deforms in the transverse direction during slip. It follows that both the free and liquefied cracks may represent solutions to the heat-flow paradox. As an application of the proof, we derive a physical understanding of the properties of harmonic Rayleigh waves on a uniform elastic half-space without solving a cubic equation.

Key words: Fracture, faulting, boundary conditions, Rayleigh waves.

Introduction

The conventional mathematical model of the dynamics of slip on an in-plane shear fracture on a planar fracture surface with normal in the z-direction, involves the solution to the elastic wave equation for the slip under the condition of a sudden decrease in the shear traction. The slip is given by the solution to the elastic wave equation under the free-crack conditions

$$\tau_{zz}^{(1)} = \tau_{xz}^{(1)} = 0, \quad \tau_{zz}^{(2)} = \tau_{xz}^{(2)} = 0 \tag{1}$$

at $z = 0$; the superscripts refer to each half-space. (We restrict discussion to the two-dimensional case of crack growth without loss of generality.) Since the conditions (1) are also those for stress-free surfaces, one interpretation of the condition $\tau_{zz} = 0$ has been that faulting develops under a crack-opening process in which slip takes place as though the two surfaces of the fault were separated sufficiently that they are no longer in contact. To account for initiation and evolution of slip by crack opening, hopping and Schallamach modes of deformation have been proposed (see Brune,

[1] Institute of Geophysics and Planetary Physics, University of California, Los Angeles, CA 90095, U.S.A. E-mail: knopoff@physics.ucla.edu
[2] Department of Physics and Astronomy, University of California, Los Angeles, CA 90095, U.S.A.

et al., 1993). However, it is unlikely that an in-plane shear fracture could be accompanied by a decrease in the component τ_{zz} even at modest depths in the earth where the hydrostatic stress is nonzero. One proposal to bypass this difficulty has been to assume that fluids are present and that faulting takes place when pore pressures match the normal stress τ_{zz}. Another proposal suggests that slip without crack opening takes place under conditions of significant contrast in elastic properties on the two sides of a fault, a condition under which head waves of slip propagate along the fault (Andrews and Ben-Zion, 1997). The requirement that there be crack opening will be shown to be the consequence of a misinterpretation of the boundary condition $\tau_{zz} = 0$. The proposal for fluidization is a vital step, completely consistent with the conditions (1), and neither crack opening nor material contrast across the fault are needed to find a solution to the problem of in-plane faulting dynamics.

Slip on a Liquefied Crack

Consider the problem of antisymmetric slip on a fault imbedded in a homogeneous, elastic medium; the surfaces slip in opposite directions equally. Near the advancing edge of a fracture there must be a gradient of the slip with increasing distance from the edge. Thus the region of gradient whose direction of slip is toward the edge, must be under horizontal dynamical compressional strain $\epsilon_{xx} < 0$, and the opposing surface must be under horizontal dynamical tensional strain $\epsilon_{xx} > 0$ (Fig. 1). The gradients of horizontal deformation of the surface must be accompanied by a deformation in the z-direction. Since one surface is in horizontal compression and the other in tension, the vertical deformation of the two opposing surfaces must be in the same direction, i.e., one must deform inward as an indentation into one half-space and the other must deform outward as a bulge into the other half-space. (We defer discussion of the sign of the vertical component of motion to the next section of this paper.) For perfectly antisymmetric slip, the amplitudes of the two vertical components of deformation are equal. Hence u_z is continuous across the boundary $z = 0$ and the two slipping surfaces are in contact throughout the slip history; u_z is symmetric in coordinate z, although it is antisymmetric with respect to the outward drawn normals to the two surfaces. It follows that the stress τ_{zz} is also continuous across the boundary. Thus the appropriate formulation of the boundary conditions for slip on a crack in an infinite medium is

$$\tau_{zz}^{(1)} - \tau_{zz}^{(2)} = 0, \quad u_z^{(1)} - u_z^{(2)} = 0, \quad \tau_{xz}^{(1)} = \tau_{xz}^{(2)} = 0 \tag{2}$$

at $z = 0$. The whole-space conditions (2) are similar to the half-space conditions (1) except a) there is a new condition describing the continuity of u_z, and b) the

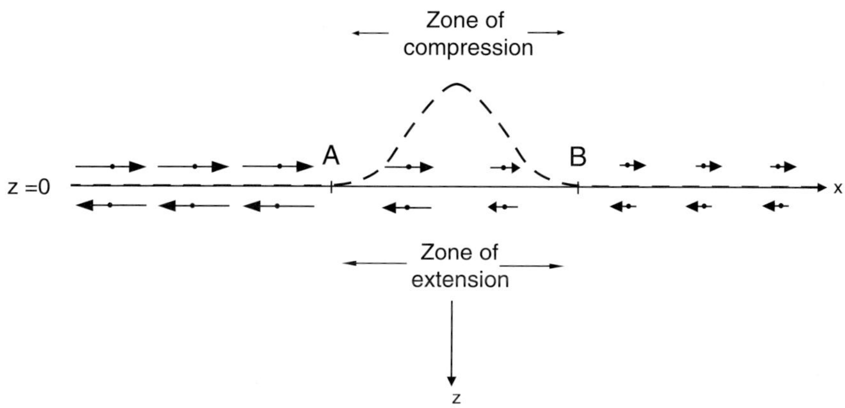

Figure 1
Antisymmetric in-plane slip on a fault surface $z = 0$ with a gradient in the slip in the interval between sites A and B. Where the surface is compressed horizontally at the gradient there is an indentation in the surface indicated by the dashed curve; where the surface is stretched, it bulges outward. Hence the z-component of the displacement is a symmetric function of z.

condition $\tau_{zz} = 0$ in the half-space case is relaxed to the less stringent continuity condition on τ_{zz}.

Since the quantity u_x is an antisymmetric function and u_z a symmetric function of coordinate z across $z = 0$, it follows that τ_{zz} is antisymmetric and τ_{xz} is symmetric. From the antisymmetry with respect to $z = 0$, $\tau_{zz}^{(1)} = -\tau_{zz}^{(2)}$ while from the condition of continuity in (2), $\tau_{zz}^{(1)} = \tau_{zz}^{(2)}$. Thus

$$\tau_{zz}^{(1)} = \tau_{zz}^{(2)} = 0,$$

on $z = 0$, which is the result obtained by DAS and AKI (1977). Thus a set of boundary conditions on $z = 0$ equivalent to (2) is

$$\tau_{zz}^{(1)} = \tau_{zz}^{(2)} = 0, \quad u_z^{(1)} - u_z^{(2)} = 0, \quad \tau_{xz}^{(1)} = \tau_{xz}^{(2)} = 0. \tag{2'}$$

The condition $\tau_{zz} = 0$ in (1) is a property of conditions (2) as well.

The correspondence to the usual conditions (1) is now complete. The continuity condition $u_z^{(1)} = u_z^{(2)}$ must be added as in (2′) to conditions (1). In-plane antisymmetric slip on a liquid interface is accompanied by a transverse motion of the fault. The fallacy in assuming that there is crack-opening during faulting is in the assumption of an inappropriate symmetry: the surfaces do not move apart; instead they remain in contact at all times. Thus neither crack-opening and its variations, nor exotic combinations of properties are needed to understand faulting.

Since the stress τ is derived from the elastic wave equation and is therefore associated with the *dynamics of faulting*, the dynamic constraint $\tau_{zz}^{(1)} = \tau_{zz}^{(2)} = 0$ on

$z = 0$ is a condition that the static stress before fracture T_{zz} be unchanged during the dynamic rupture. Hence faulting can take place in the presence of significant hydrostatic stress without reduction of the static prestress. The conditions $(2) \equiv (2')$ require liquidity of the interface. The formulation of the free-crack problem (1) on each of the two half-spaces abutting the crack is identical to the liquid interface formulation for faulting in an infinite medium for antisymmetric slip, $(1) \equiv (2')$. From the equivalence $(1) \equiv (2)$, the conditions (1) are also the conditions for liquidity of the interface.

Rayleigh wave motions with the usual phase velocity are not only a property of motions on a half-space with a stress-free surface (1), but as long as faulting is antisymmetric, they are also a property of a whole space with an internal fault surface with liquid boundary conditions $(2')$, since the conditions $\tau_{xz} = \tau_{zz} = 0$ on $z = 0$ appear in both. The addition of the continuity condition on the normal displacement u_z to the usual conditions (1) does not change our expectation of an important Rayleigh wave component to the slip on the surface of the fracture.

The infinite medium Green's function for dynamic fracture on a liquefied crack is easily obtained from the solution by RICHARDS (1979) where the terms of oppositely directed tangential forces are selected, corresponding to the application of a traction T_{xz} at each surface. The single-couple nature of this source function is not in conflict with its double-couple body force equivalence in an unfaulted medium.

The generation of frictional heat during faulting should be minimal during fracture on a crack liquefied with an ideal frictionless liquid. It follows from the demonstration above, that the same result should hold for a crack with free surface conditions. These observations may represent a resolution of the heat flow paradox.

The Physics of Rayleigh Waves

Questions concerning the physics of the propagation of a Rayleigh wave on an elastic homogeneous half-space are often answered with the irrelevant mathematical response that the Rayleigh wave is the solution to a familiar cubic equation. The cubic equation arises from the mathematics of wave propagation with boundary conditions at the surface $\tau_{nn} = \tau_{nt} = 0$ where n is the normal to the surface of the half-space and t is the tangential direction, taken to be the direction of wave propagation. The reliance on mathematics is in direct contrast to the immediacy and transparency with which compression and shear waves can be understood in terms of a simple physical picture of deformation of an elastic solid. The minimal physical argument of the preceding section can be used as the basis for understanding the simple physics of deformation in Rayleigh wave propagation at the surface of a homogeneous half-space. It will be shown without difficult mathematics that the wave velocity is less than that of S waves, that the two components of motion are Hilbert transforms of one another, that the motions will be retrograde elliptical on the surface and prograde elliptical at depth for sinusoidal Rayleigh wave motions, as well as other familiar properties. Ours will be a qualitative theory. We amend arguments presented elsewhere (KNOPOFF, 2001).

Consider an elastic half-space deformed dynamically by horizontal displacements at the planar surface $z = 0$ (Fig. 2); the deformation moves to the right with Rayleigh wave velocity. To the left of point A, the surface is displaced uniformly to the right, while to the right of point B there is no displacement of the surface. Since there is no compression or extension of the undeformed regions at the far left and right, there is no vertical component of the displacement in these regions. Vertical displacement of the surface in the central region AB is associated with the gradient of the horizontal component of the displacement as in the first part of this paper.

Again we choose the coordinates (x, z) to correspond to the directions (t, n). The sign of the vertical component of the displacement u_z in the transition interval AB between the two undeformed regions follows from the condition that the shear stress

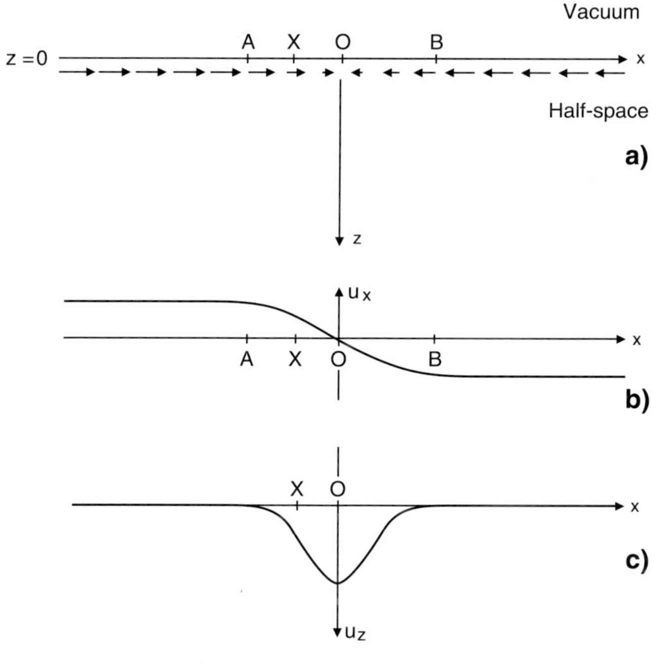

Figure 2
a) Wave of horizontal displacements at the surface of an elastic half-space. The displacements are positive and uniform to the left of A and positive and uniform to the right of B. b) $u_x(x)$ at $z = 0$. c) $u_z(x)$ at $z = 0$.

τ_{xz} vanishes at the surface. We assume that the horizontal component of the displacements of the elastic medium resulting from the deformation of the surface must decrease with increasing depth, and vanishes at infinite depth. For $u_x > 0$ in AB, as in the figure, then $\partial u_x/\partial z < 0$. Since the boundary condition for the shear stress at the surface is $\tau_{xz} = \mu(\partial u_x/\partial z + \partial u_z/\partial x) = 0$, then $\partial u_z/\partial x > 0$. Hence the horizontal compression at sites in the transition zone in the region between A and B must have a corresponding indentation of the surface. Conversely a traveling horizontal extension must have an outward bulge in the vertical direction. An intuitive expectation that the surface under compression would be likely to bulge outward would be correct if the half-space were incompressible, but under the condition of zero shear at the surface, it has the opposite sign. We return to this point below.

There is also a geometrical view of the deflection of the surface. Consider an element of fault area at the surface at X in the transition zone. As above, $\partial u_x/\partial u_z < 0$ and $\partial u_z/\partial u_x > 0$, the two quantities being equal and opposite to each other. These gradients are shown at the left in Figure 3a. After the mean motion, which is rightward and downward, is subtracted (Fig. 3b), there remains a residual net

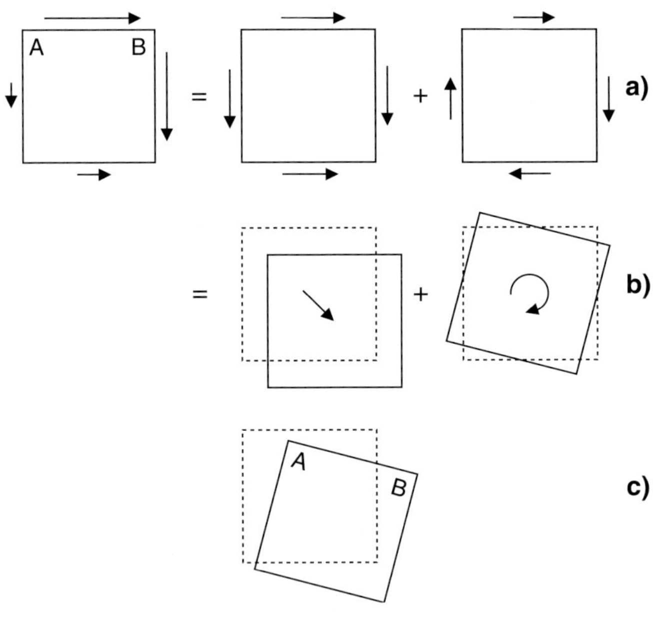

Figure 3
Deformation of an element of area at point X (see Fig. 2) at the surface of the half-space, under a condition of vanishing shear stress. a) Deformation has a negative vertical gradient of u_x and an equal positive horizontal gradient of u_z. These can be decomposed into a translation and a rotation. b) Vectorial decomposition as in a). c) Total motion of element showing depression and rotation of surface line element AB.

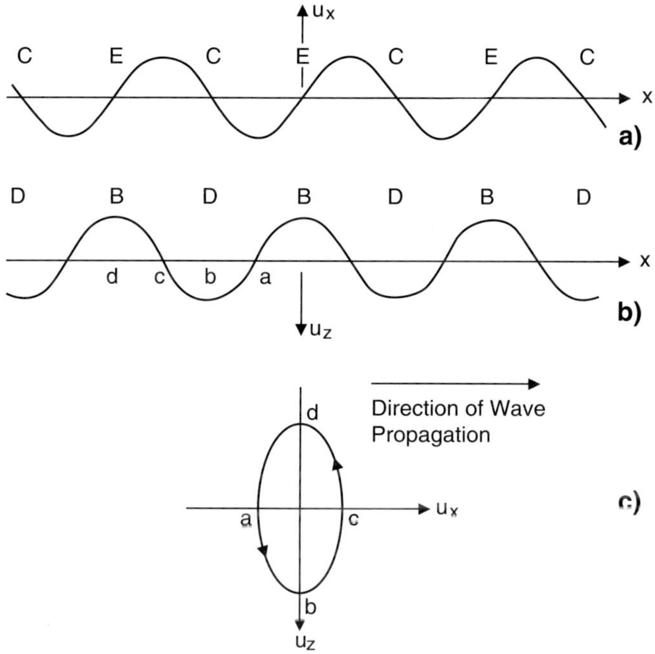

Figure 4
a) Sinusoidal deformation of the surface by horizontal components of motion. C and E are sites of maximum Compression and Extension. b) Schematic vertical component of motion in correspondence with a). D and B denote sites of maximum Depression (indentation) and Bulge. These correspond one-to-one with sites C and E in a). c) Motion is retrograde elliptically polarized. Successive times a, b, c, d correspond to those in b) for a wave moving to the right.

rotation of the elemental area in the clockwise direction (Fig. 3c); the surface is indented as before.

Consider the deformation of the half-space by spatially sinusoidal horizontal displacements at the surface (Fig. 4a) in the moving coordinate system. The vertical component of the displacement is also sinusoidal since it is proportional to the gradient of the surface motion; the maxima of the bulges B and the minima of the indentations D in the vertical component of the displacement at the surface must correspond to the extreme values of extension E and compression C in the gradient of the horizontal component of the displacement, respectively (Fig. 4b). A zero value of the vertical component of the displacement corresponds to a zero value of the gradient of the horizontal component. The two components of the displacement at the surface have a 90° phase difference, and hence the motion has an elliptical trajectory. Thus the two components of motion are Hilbert transforms of one another for any function of $(ct - x)$. From the bulge/extension and indentation/compression phase correlations at times (a,b,c,d), the elliptical motion is retrograde at the surface (Fig. 4c) for either direction of propagation.

To determine that the Rayleigh wave velocity is less than that of S-waves, consider the wave properties of the deformation, characterized by the wave equation. The solutions to the elastic wave equation for horizontally traveling waves are

$$\exp\{i\omega(x/c - t) - \omega\eta_{P,S}z\},$$

where c is the velocity of Rayleigh waves. The vertical spatial decay rates η_S and η_P for the shear and compression wave components of the motion satisfy

$$\frac{1}{c^2} = \frac{1}{(\alpha^2, \beta^2)} + \eta^2_{(P,S)},$$

where $\alpha > \beta$ are the P- and S-wave velocities. By inspection, $c < \beta$, a direct consequence of the vanishing of the motion at infinite depth.

The P-wave component of the motion decays more rapidly with depth than the S-wave component, $\eta_P > \eta_S$. The displacement vectors and stress and strain components, all propagate parallel to the surface with phase velocity c, and all have amplitudes that vary with depth as the sum

$$(Ae^{-\omega\eta_P z} + Be^{-\omega\eta_S z})e^{i\omega(x/c-t)}.$$

The invariant volumetric strain $\epsilon_{kk} = \epsilon_{xx} + \epsilon_{zz}$ is an exception to this rule since it can have no dependence on the shear properties of the system. The volumetric strain must fall off with depth as $e^{-\omega\eta_P z}$, i.e., $B = 0$. There is a 90° phase shift between the components of the pairs (u_x, u_z), $(\epsilon_{xx}, \epsilon_{xz})$, etc.; the vector displacement and strain are elliptically polarized.

The normal stress $\tau_{zz} = (\lambda + 2\mu)\partial u_z/\partial z + \lambda\partial u_x/\partial x = 0$ at the surface. Hence

$$\partial u_z/\partial z = -\frac{\lambda}{\lambda + 2\mu}\partial u_x/\partial x$$

at $z = 0$. A site of maximum horizontal compression at the surface is also a site of maximum amplitude of the vertical component of displacement at the surface. At this site $\partial u_x/\partial x < 0$, and thus $\partial u_z/\partial z > 0$. Thus the vertical component of the displacement at the surface increases with depth. At greater depth the vertical component of displacement must decrease to zero steadily with depth. Hence $\partial u_z/\partial z < 0$ at great depth below a surface site of maximum horizontal compression.

Although both the volumetric and shear strains vanish at depth, at very large but finite depth the volumetric strain is much smaller when compared with the shear strain since $\eta_P \geq \eta_S$. We approximate $\epsilon_{kk} = 0$ at large but finite depth, and obtain $\partial u_x/\partial x > 0$, i.e., there is a reversal of the horizontal component of the displacement

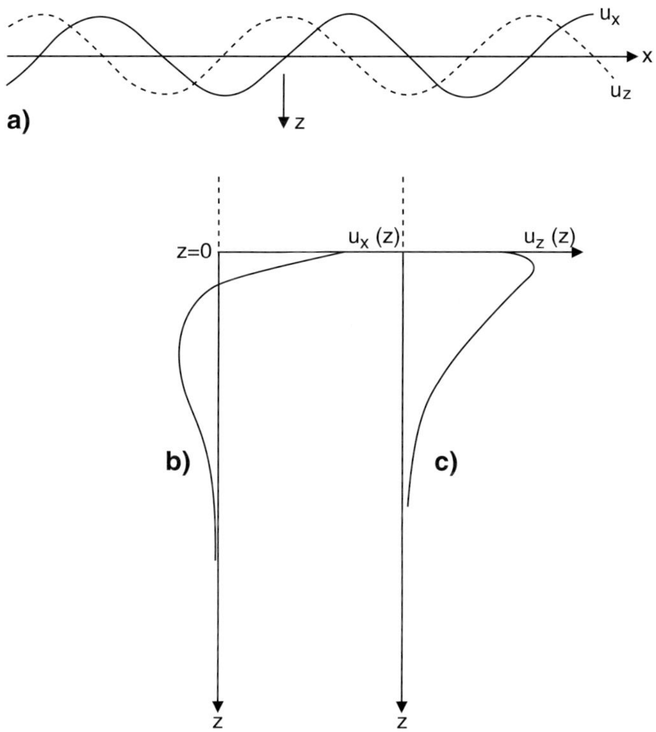

Figure 5
a) Schematic horizontal and vertical components of the motion at $z = 0$, as in Figure 3. b) Schematic amplitude $u_x(z)$ beneath a maximum of u_x, showing reversal of sign at depth. c) Schematic amplitude $u_z(z)$ beneath a maximum of u_z, showing positive gradient near the surface.

at depth relative to that at the surface. Thus the elliptical polarization at depth is prograde. As a consequence, there must be a crossover depth at which the horizontal component of sinusoidal displacements is zero. Near the surface, the motion is controlled by the vanishing of the shear strain while at depth the motion is dominated by the vanishing of the volumetric strain. The general shape of the depth dependence in both components of the motion can now be sketched as in Figure 5.

References

ANDREWS, D.J. and BEN-ZION, Y. (1997), *Wrinkle-like slip pulse on a fault between different materials*, J. Geophys. Res. *102*, 553–571.

BRUNE, J.N., BROWN, S., and JOHNSON, P.A. (1993), *Rupture mechanism and interface separation in foam rubber model of earthquakes: A possible solution to the heat flow paradox and the paradox of large overthrusts*, Tectonophysics *218*, 58–67.

RICHARDS, P.G. (1979), *Elementary solutions to Lamb's problem for a point source and their relevance to three-dimensional studies of spontaneous crack propagation*, Bull. Seismol. Soc. Am. *69*, 947–956.

DAS, S. and AKI, K. (1977), *A numerical study of two-dimensional spontaneous rupture propagation*, Geophys. J. Roy. Astron. Soc. *50*, 643–688.

KNOPOFF, L. (2001), *Rayleigh waves without cubic equations*, Comput. Seismol. *32*, 33–37.

(Received December 30, 2004, revised October 6, 2005, accepted November 7, 2005)

Pure appl. geophys. 163 (2006) 1751–1767
0033–4553/06/091751–17
DOI 10.1007/s00024-006-0095-1

Pure and Applied Geophysics

Experimental Evidence of Critical Sensitivity in Catastrophe

Xianghong Xu,[1] Mengfen Xia,[1,2] Fujiu Ke,[1,3] and Yilong Bai[1]

Abstract—The paper presents an experimental study on critical sensitivity in rocks. Critical sensitivity means that the response of a system to external controlling variable may become significantly sensitive as the system approaches its catastrophic rupture point. It is found that the sensitivities measured by responses on three scales (sample scale, locally macroscopic scales and mesoscopic scale) display increase prior to catastrophic transition point. These experimental results do support the concept that critical sensitivity might be a common precursory feature of catastrophe. Furthermore, our previous theoretical model is extended to explore the fluctuations in critical sensitivity in the rock tests.

Key words: Critical sensitivity, catastrophe, damage fraction, acoustic emission, digital speckle correlation method, fluctuation.

1. Introduction

Many observational evidences show that there are some precursors prior to a major earthquake, such as accelerating moment release (AMR) or power-law increase in the number of intermediate-size events (Jaumé *et al.*, 1999; Bowman *et al.*, 1998; Rundle *et al.*, 2000a), anomalously high values of LURR (Yin *et al.*, 2000), etc.

Recently, based on statistical mesoscopic damage mechanics, Xia *et al.* (Xia *et al.*, 2002; Zhang *et al.*, 2004) suggested that the sensitivity of response to controlling variable might be an effective variable to characterize a system with macroscopic uncertainty. In particular, a properly defined sensitivity of the system may display significant increase as the system approaches its catastrophic point, i.e., the transition point from damage accumulation to catastrophic rupture. Such a behavior is called critical sensitivity. The underlying mechanism behind critical sensitivity is the

[1] State Key Laboratory of Nonlinear Mechanics (LNM), Institute of Mechanics, Chinese Academy of Sciences, Beijing 100080, PR China.
E-mail: xxh@lnm.imech.ac.cn; baiyl@lnm.imech.ac.cn
[2] Department of Physics, Peking University, Beijing 100871, PR China.
E-mail: xiam@lnm.imech.ac.cn
[3] Department of Applied Physics, BeiHang University, Beijing 100088, PR China.
E-mail: kefj@lnm.imech.ac.cn

coupling effect between disordered heterogeneity on multiple scales and dynamical nonlinearity due to damage-induced stress redistribution (STEIN, 1999; XIA *et al.*, 2000, 2002). Critical sensitivity may provide a clue to prediction of catastrophic transition, such as material failure or great earthquakes, provided the sensitivity of the system is measurable or can be monitored.

To validate the concept of critical sensitivity, a series of experiments have been conducted. 167 gabbro samples were tested under uniaxial compression. The deformation and damage processes were observed with acoustic emission (AE) and white Digital Speckle Correlation Method (DSCM) (PETER *et al.*, 1981; MA *et al.*, 2004) synchronously. Then, the sensitivities characterizing the evolution of damage, surface displacement pattern and AE energy induced by boundary displacement can be obtained. The experimental results show that these three kinds of responses of the samples become significantly sensitive to the controlling variable, i.e., boundary displacement, as the samples approach catastrophe. This clearly indicates that the experimental results do support the concept of critical sensitivity reasonably.

In addition, it can be seen that the sensitivities observed in the experiments display fluctuations, while the sensitivity obtained from the previous theoretical approximation (ZHANG *et al.*, 2004) is monotonic and smooth. In order to understand the mechanism governing the fluctuations in sensitivity, a model with multi-peak structure in the distribution function of mesoscopic units' threshold is introduced. It is found that the multi-peak structure might be responsible for the fluctuations shown in critical sensitivities.

2. Observations of Critical Sensitivity in Rock Failure

2.1 Experimental Method

In our tests, rectangular gabbro samples with dimensions of $5 \times 5 \times 13$ mm^3, were compressed uniaxially with a MTS810 testing machine. The loading mode is boundary-displacement control with velocity of 0.02 mm/min. The displacement was measured by an extensometer with resolution of 3 μm and an offset of 1 kN load.

The surface of the specimen was illuminated by a white luminescence and the speckle images were captured and transferred to a computer by a CCD camera. After the experiment, the speckle images were analyzed with DSCM, thus, both displacement and strain fields during the loading process were obtained.

Moreover, two AE sensors were fixed on two sides of a sample with a specially designed clamp. The resonant frequencies of the sensors are 140 kHz and 250 kHz, respectively. The AE signals were recorded and processed by an AE21C system produced by the Institute of Computer Technology of Shenyang, China. As well known, AE is an effective method to detect damage process of rock, so the AE series, such as AE energy, can provide statistical information on damage evolution.

2.2 Definitions of Critical Sensitivity Adopted in Rock Experiments

To deal with experimental data, a dimensionless boundary displacement U is adopted, i.e., $U = U^*/13$, where U^* is the actual boundary displacement. (Hereafter, symbols without superscript * will represent either dimensionless (like stress, displacement, length and energy, etc.) or normalized (like strain) variables, while those with * mean dimensional or non-normalized ones), and factor 13 is the length of the sample along the loading axis with unit of millimeter. Figure 1 shows the processed curves of actual nominal stress σ^{*P} versus dimensionless boundary displacement U^P for 151 gabbro samples (in the following text, symbols with superscript P demonstrate the processed experimental variables. See the processing method in Section 2.2.1). The catastrophe appears at the end of each curve. Clearly, it is hard to forecast when catastrophe will occur beforehand. In particular, the maximum and the failure stresses show large diversity. This macroscopic uncertainty results in great difficulty in rupture prediction.

In order to measure the sensitivity of a rock sample to external controlling variable, we define sensitivity S as

$$S = \frac{\Delta}{\Delta U}\left(\frac{\Delta R}{\Delta U}\right), \tag{1}$$

where the dimensionless boundary displacement U is the external controlling variable and R is the response of the rock sample. Moreover, in the theoretical model, if the second-order derivative of R, d^2R/dU^2, exists, Eq. (1) can be written as

$$S = \frac{d^2R}{dU^2}. \tag{1a}$$

In the present paper, R is defined as the accumulative response of the rock sample. Hence, the first-order derivative of R demonstrates the changing rate of the response

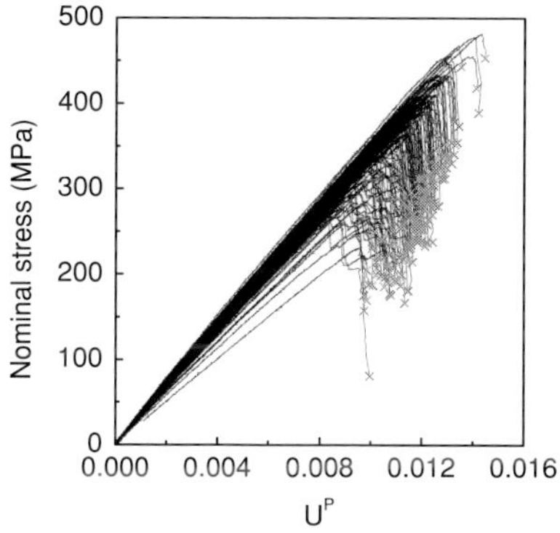

Figure 1
Processed curves of experimental nominal stress σ^{*P} versus dimensionless boundary displacement U^P for 151 gabbro samples under uniaxial compression. The symbols $\times$ indicate catastrophe points.

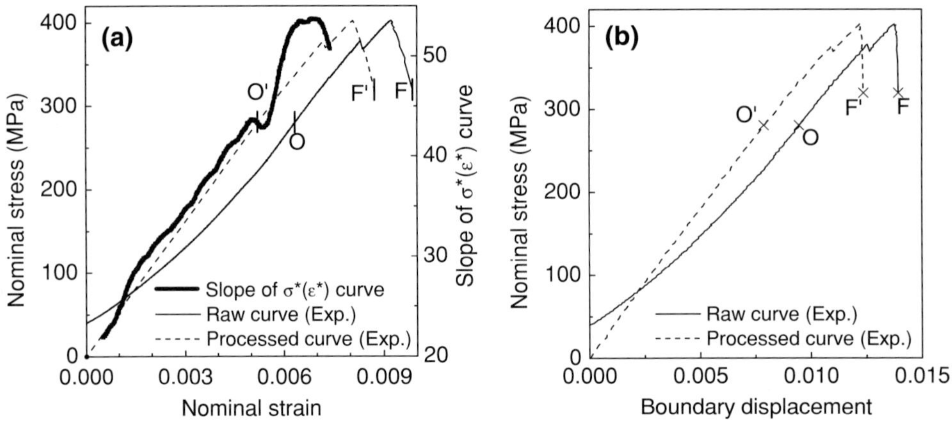

Figure 2

(a) The slope of $\sigma^*(\varepsilon^*)$ curve, namely $\Delta\sigma^*/\Delta\varepsilon^*$ (bulk solid line –). $\sigma^*(\varepsilon^*)$ (solid line –) and $\sigma^{*P}(\varepsilon^{*P})$ (dashed line ---), are the raw and processed experimental nominal stress-strain curves respectively. (b) The curves, $\sigma^*(U)$ (solid line –) and $\sigma^{*P}(U^P)$ (dashed line ---) are the raw and processed experimental nominal stress versus dimensionless boundary displacement curves, respectively. Points O and O′ are the points corresponding to maximum modulus, F and F′ are the catastrophe points.

with respect to the external controlling variable. In order to measure the sensitivity of the changing rate of the response to the external controlling variable, the second-order derivative of R is adopted as the definition of sensitivity, because it is considerably more sensitive to the external controlling variable than the first-order derivative.

Importantly, R could be chosen from different kinds of responses. In this paper, the responses from the behaviors at different scales is adopted, i.e., the mean damage fraction D at global scale, the distance ΔH^* between successive patterns of surface displacement related to the behavior at locally macroscopic scale and the AE energy Θ^* contributed from the events at mesoscopic scale. In data processing, D can be calculated from the experimental nominal stress σ^* and strain ε^* curve, ΔH^* can be calculated from the surface displacement patterns, and Θ^* can be obtained directly from the AE system, as discussed later in detail.

Critical sensitivity means that the response to the controlling variable, i.e., boundary displacement U, may become significantly sensitive, i.e., $S \gg 1$, prior to the catastrophe. Now, we focus on whether critical sensitivity is a common precursor to final rupture in rock experiments.

2.2.1 Sensitivity calculated from damage evolution

At the initial part of the raw experimental nominal stress σ^* and strain ε^* curve (Fig. 2(a), solid line), the slope of $\sigma^*(\varepsilon^*)$ curve, $\Delta\sigma^*/\Delta\varepsilon^*$ (Fig. 2(a), bulk solid line), increases with increasing ε^* due to the closure of micro-cracks. For simplicity, without regard to healing process of micro-cracks, only the weakness induced by damage is considered based on damage mechanics (JAYATILAKA, 1979). Then, the global damage fraction D can be calculated from the processed σ^{*P} (ε^{*P}) curve obtained by the following steps (XU *et al.*, 2004): (1) Calculate the slope of $\sigma^*(\varepsilon^*)$ curve, $\Delta\sigma^*/\Delta\varepsilon^*$ (Fig. 2(a), bulk solid line), and suppose E_0^*, which equals to the maximum value of $\Delta\sigma^*/\Delta\varepsilon^*$ (at point O in Fig. 2(a)), as the initial elastic modulus of the rock sample. (2) Draw a straight line with slope E_0^* from the origin ($\sigma^{*P} = \varepsilon^{*P} = 0$) to O′ (O′ and O locate at the same nominal stress). (3) Parallelly shift the O-F-part of the raw stress-strain curve $\sigma^*(\varepsilon^*)$ to point O′. Then, the entire processed experimental nominal stress-strain curve $\sigma^{*P}(\varepsilon^{*P})$ (dashed line in Fig. 2(a)) is obtained. Similarly, parallelly draw the processed curve of nominal stress σ^{*P} versus boundary displacement U (solid line in Fig. 2(b)) from the origin ($\sigma^{*P} = U^P = 0$). Then, the processed experimental nominal stress σ^{*P} versus dimensionless boundary displacement U^P curve (dashed line in Fig. 2(b)) can be obtained.

Theoretically, global damage fraction D, a macroscopic variable characterizing damage evolution, can be obtained from the processed experimental nominal stress σ^{*P} and strain ε^{*P} curve based on mean field (MF) approximation. According to damage mechanics and MF approximation, the constitutive relation can be

$$\sigma^{*P} = \varepsilon^{*P} E_0^*(1 - D). \tag{2}$$

Then, global damage fraction D can be calculated by

$$D = 1 - \frac{\sigma^{*P}}{\varepsilon^{*P} E_0^*}. \tag{3}$$

The global damage fraction D of rock sample calculated from $\sigma^{*P}(\varepsilon^{*P})$ curve is shown in Figure 3(a). When choosing the response of R to be damage D, the sensitivity is denoted by S_D. Figure 3(b) shows the curve of S_D versus the boundary displacement U for a sample.

In order to compare the sensitivity series of different samples, a normalized boundary displacement U_0 is adopted,

$$U_0 = U/U_c, \tag{4}$$

where U_c is the dimensionless boundary displacement of a sample at its catastrophic point. Figure 3(c) shows the curves of S_D versus U_0 for 151 samples.

2.2.2 Sensitivity calculated from the distance between successive patterns of surface displacement

Since the length-pixel ratio of the imaging system is about 0.028 mm/pixel in DSCM system, the obtained displacement can be understood as an average over the area of $28 \times 28\ \mu m^2$. The surface strain pattern can be calculated from the surface displacement pattern. Owing to damage evolution, the surface strain pattern becomes inhomogeneous on multi-scales, and later strain localization appears. Then, new

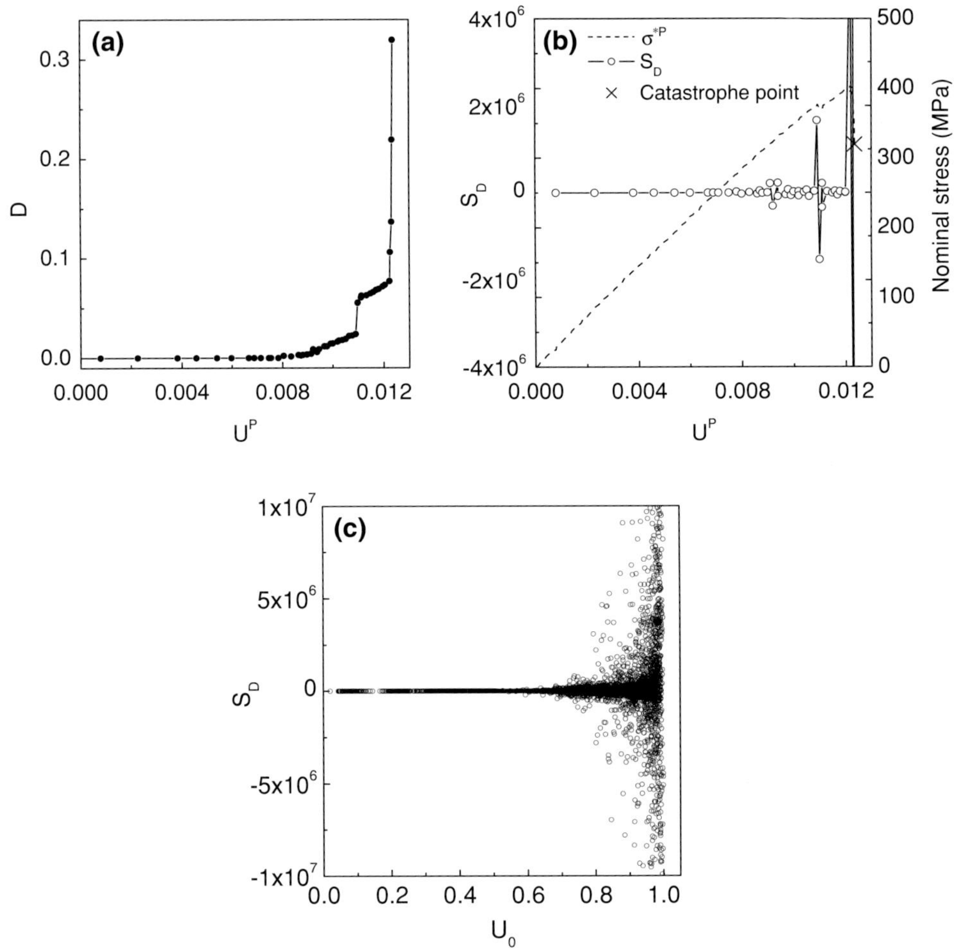

Figure 3
(a) Damage fraction D, obtained from experimental nominal stress-strain curves, versus processed boundary displacement U^P. (b) Sensitivity S_D calculated from damage evolution versus U^P for a sample. (c) S_D versus normalized boundary displacement U_0 for 151 samples.

scales between the pixel scale and the sample scale emerge. These emerging scales are called locally macroscopic scales in this paper. In order to describe the change of response of the system at the locally macroscopic scales, distance between the successive patterns of surface displacement is introduced. The distance between two patterns of surface displacement is defined as

$$\Delta H_i^* = \frac{1}{N_{\text{eff}}} \sum_{x=1}^{131} \sum_{y=1}^{401} \left| \Delta u_i^*(x, y) \right|, \quad i = 1, 2, \tag{5}$$

where $\Delta u_i{*}(x,y)$ is the increment of surface displacement vector $u_i{*}$ at point (x,y) along axis i ($i = 2$ indicates loading direction whereas 1 the direction vertical to loading), 131 and 401 are the number of points in the surface displacement pattern along axis 1 and axis 2, respectively, and N_{eff} is the number of points effective for DSCM calculation.

The curve of ΔH_1^* versus U^{P} is shown in Figure 4(a), and the sensitivity $S_{H_1^*}$ calculated from the distance between surface displacement patterns is shown in Figure 4(b) for a single sample and Figure 4(c) for 143 samples.

2.2.3 Sensitivity calculated from AE energy

Acoustic emission, resulted from mesoscopic damage evolution, is a response of rock sample to boundary displacement on the mesoscopic scale. AE energy is a

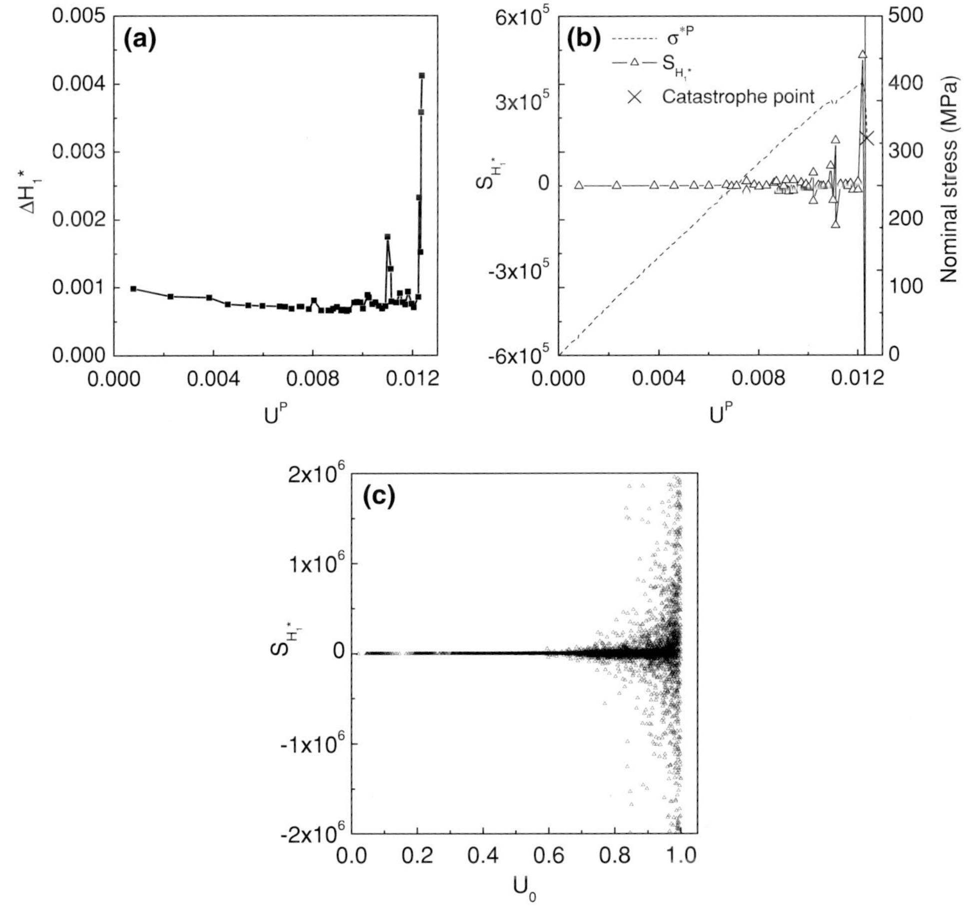

Figure 4

(a) The distance between surface displacement patterns ΔH_1^* versus U^{P}. (b) Sensitivity $S_{H_1^*}$ calculated from distance between surface displacement patterns versus U^{P} for a sample. (c) $S_{H_1^*}$ versus U_0 for 143 samples.

proper parameter characterizing damage and the accumulated AE energy Θ^* can be obtained directly from AE recording, see Figure 5(a).

Then, taking response $R = \Theta^*$, the sensitivity S_{Θ^*} calculated from AE energy is shown in Figure 5(b) for a single sample and Figure 5(c) for 131 samples.

2.3 Critical Sensitivity in Rock Experiments

According to the above-mentioned method, the experimental results of sensitivity reflecting the responses on three scales are obtained, namely S_D, $S_{H_1^*}$ and S_{Θ^*} from the responses on sample scale, locally macroscopic scales and mesoscopic scale, respectively.

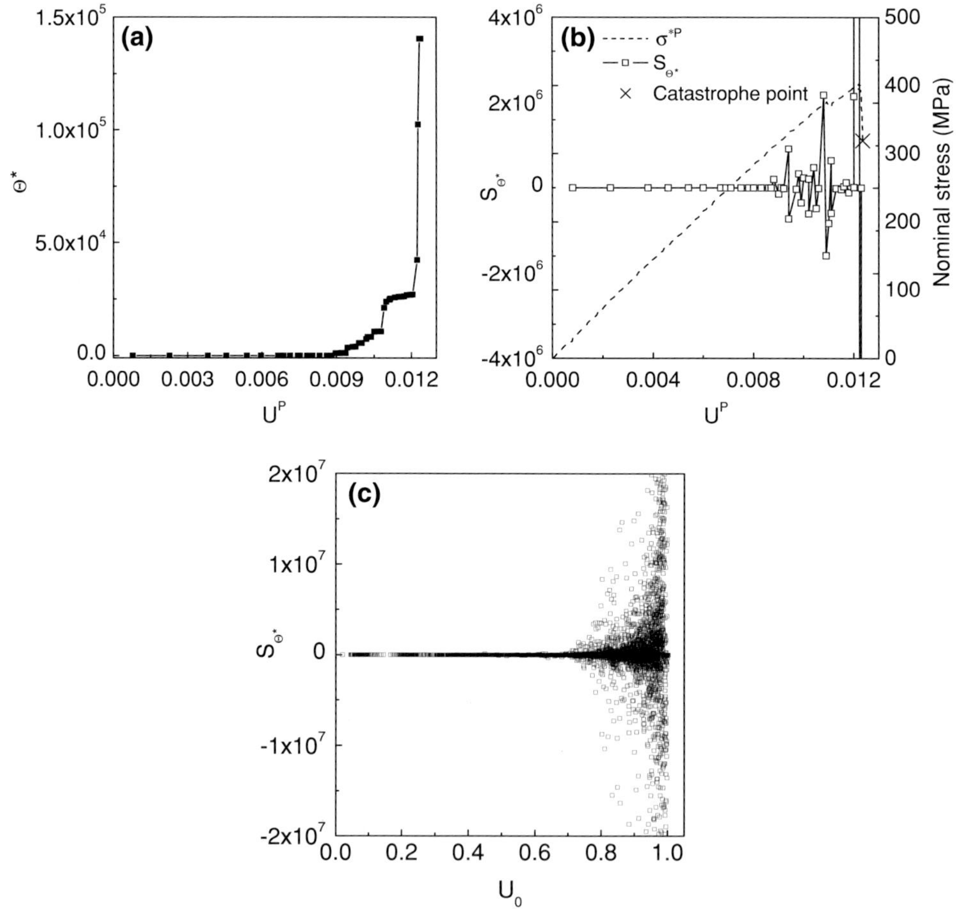

Figure 5
(a) Accumulative AE energy Θ^* versus U^P. (b) Sensitivity S_{Θ^*} calculated from AE energy versus U^P for a sample. (c) S_{Θ^*} versus U_0 for 131 samples.

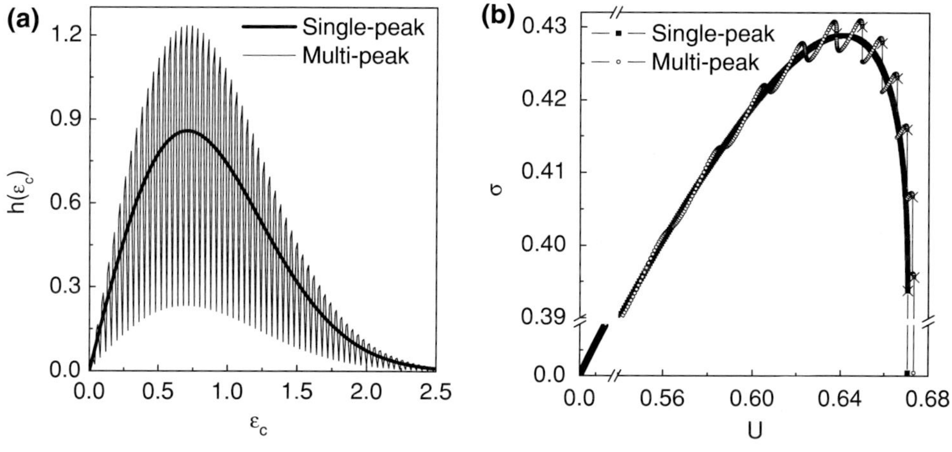

Figure 6
(a) Distribution function of mesoscopic strength ε_c, single-peak function with $m = 2$ in Eq. (29) (bulk solid line –) and multi-peak function with $m = 2$, $N = 100$, and $\Delta = 0.04$ in Eqs. (33)–(37) (solid line –). (b) Nominal stress σ versus boundary displacement U with $k = 0.3$ in Eq. (14), single-peak function (■ and corresponding line) and multi-peak function (○ and corresponding line).

Figures 3(b), 4(b), and 5(b) show the sensitivity S at the three scales for a single sample. At the initial stage, S_{Θ^*} is equal to zero since the events of mesoscopic damage are too small to be detected by AE sensors, $S_{H_1^*}$ and S_D are also equal to zero since the change of the surface displacement patterns is nearly zero and macroscopic damage can be neglected at the initial stage. In other words, at the initial stage, the responses of the system at all scales are not sensitive to the external controlling variable, i.e., the boundary displacement U. As the boundary displacement increases, the three sensitivities remain in low level. This means that the system is in a state with low sensitivity. However, the three sensitivities increase significantly prior to the catastrophic transition point. This implies that the system becomes highly sensitive prior to catastrophic point, from mesoscopic scale to macroscopic scale.

Figures 3(c), 4(c), and 5(c) show the three sensitivities for more than 100 samples. Noticeably, the series of sensitivities are different from sample to sample. That is to say, the catastrophic rupture demonstrates sample specificity. But, more importantly, there is a common trend in sensitivity for all samples, namely significantly increasing sensitivity near the catastrophic transition $U_0 = 1$. This is a strong experimental evidence of critical sensitivity prior to catastrophic rupture in heterogeneous rock.

Since the value of sensitivity of different samples displays large diversity, only a qualitative analysis can be given in present paper. In our later research, more quantitative work will be done.

3. Theoretical Analysis of Critical Sensitivity

However, it is noticeable that the sensitivity shows severe fluctuations in experiments (Figs. 3–5). In order to explore the mechanism underlying the fluctuations of sensitivity in rock experiments, a theoretical model is proposed to explain the phenomenon. Suppose the rock sample and the MTS tester, be two parts in series and driven by boundary displacement quasi-statically. According to MF approximation, the boundary displacement equals to

$$U^* = L^*\varepsilon^* + L_e^*\varepsilon_e^*, \tag{6}$$

where ε^* and ε_e^* are nominal strains of the rock sample and the elastic tester respectively, while L^* and L_e^* are the corresponding initial length along the loading axis of the two parts. Under uniaxial monotonic loading, the equilibrium condition between the rock sample and the elastic part can be written as

$$\sigma^* = \varepsilon_e^* E_e^* = \varepsilon^* E_0^*(1 - D), \tag{7}$$

where σ^* is the nominal stress, E_e^* is the elastic modulus of the tester, and E_0^* is the initial elastic modulus and D is the damage of the rock sample. According to damage mechanics, the true stress σ_s^* and true strain ε_s^* of the rock sample are respectively given by

$$\sigma_s^* = \frac{\sigma^*}{1 - D} \quad \text{and} \quad \varepsilon_s^* = \varepsilon^*. \tag{8}$$

After taking dimensionless stress σ and normalized strain ε,

$$\sigma = \sigma^*/\eta^*, \quad \varepsilon = E_0^*\varepsilon^*/\eta^* \text{ and } \varepsilon_e = E_e^*\varepsilon_e^*/\eta^*, \tag{9}$$

where η^* is the position factor of Weibull distribution (WEIBULL, 1951) (see Eq. (28)). According to Eqs. (7)–(9), the relations between the nominal and true variables, i.e., stress, strain and damage of the damage part, are

$$\sigma = \varepsilon(1 - D), \tag{10}$$

$$\sigma = \sigma_s(1 - D) \quad \text{and} \quad \varepsilon = \varepsilon_s, \tag{11}$$

$$\sigma_s = \varepsilon_s. \tag{12}$$

According to Eqs. (7) and (9), the relations between dimensionless nominal stress and normalized strain ε_e of the elastic part can be

$$\sigma = \varepsilon_e. \tag{13}$$

According to Eqs. (6) and (9), the dimensionless boundary displacement U can be derived as

$$U = \frac{U^*}{\eta^*\left(\frac{1}{E_0^*/L^*} + \frac{1}{E_e^*/L_e^*}\right)} = \frac{1}{k+1}(k\varepsilon + \sigma) = \frac{k+1-D}{k+1}\varepsilon, \tag{14}$$

where k is the ratio between the rigidity of the elastic part and the initial rigidity of the rock sample

$$k = \frac{E_e^*/L_e^*}{E_0^*/L^*}. \tag{15}$$

As soon as damage occurs in heterogeneous elastic-brittle medium, some stored energy will be released. Since the elastic energy of the elastic-brittle (without residual deformation) model under MF approximation

$$\Theta_{el}(\varepsilon) = \frac{1}{2}\sigma\varepsilon = \frac{1}{2}(1-D)\varepsilon^2, \tag{16}$$

and the increment of external work on rock sample is

$$\Delta W = \sigma\Delta\varepsilon = (1-D)\varepsilon\Delta\varepsilon. \tag{17}$$

Then, the increment of energy release can be

$$\Delta\Theta(\varepsilon) = \Delta W(\varepsilon) - \Delta\Theta_{el}(\varepsilon) = \frac{\varepsilon^2}{2}\Delta D. \tag{18}$$

Suppose the rock sample can be simplified as a driven, nonlinear threshold system (RUNDLE *et al.*, 2000b; ZHANG *et al.*, 2004), comprising numerous interacting and nonlinear mesoscopic units, which fails when the force acting on it reaches a predefined threshold. In the present model, it is assumed that all units have the same elastic modulus E_0^* but different breaking stress threshold σ_c^*. Hence, firstly, each unit remains elastic until its own σ_c^*, i.e.,

$$\sigma_s^* = \varepsilon_s^* E_0^*, \tag{19}$$

where σ_s^* and ε_s^* are mesoscopic stress and strain of each unit, i.e., true stress and true strain as mentioned before, respectively. According to Eq. (19), the stress threshold σ_c^* and strain threshold ε_c^* of the mesoscopic unit should also follow

$$\sigma_c^* = \varepsilon_c^* E_0^*. \tag{20}$$

According to definitions (9), the relations between the dimensionless stress threshold σ_c and normalized strain threshold ε_c can be

$$\sigma_c = \varepsilon_c. \tag{20a}$$

Secondly, as soon as σ_s^* reaches σ_c^* on a unit, the unit will be broken and can never support load, i.e.,

$$\sigma_s^* = \sigma_s = 0, \tag{21}$$

when the unit is broken.

Suppose that at the initial state the mesoscopic strength ε_c of units follows a certain type of distribution function $h(\varepsilon_c)$, which should be normalized as

$$\int_0^\infty h(\varepsilon_c)\mathrm{d}\varepsilon_c = 1. \tag{22}$$

For quasi-static loading, i.e., the characteristic time of damage relaxation is much shorter than that of loading, damage D in Eq. (7) can be determined by

$$D = \int_0^\varepsilon h(\varepsilon_c)\mathrm{d}\varepsilon_c. \tag{23}$$

According to Eqs. (1a), (10), (14), (17), and (23), the sensitivities of damage, nominal strain and released energy can be respectively derived as

$$S_D = \frac{\mathrm{d}^2 D}{\mathrm{d}U^2} = \frac{(k+1)^2\left[(k+1-D)h'(\varepsilon) + 2(h(\varepsilon))^2\right]}{(k+1-D-\varepsilon h(\varepsilon))^3}, \tag{24}$$

$$S_\varepsilon = \frac{\mathrm{d}^2 \varepsilon}{\mathrm{d}U^2} = \frac{(k+1)^2[2h(\varepsilon) + \varepsilon h'(\varepsilon)]}{(k+1-D-\varepsilon h(\varepsilon))^3}, \tag{25}$$

and

$$S_\Theta = \frac{\mathrm{d}^2 \Theta}{\mathrm{d}U^2} = \frac{(k+1)^2\varepsilon(\varepsilon h'(\varepsilon) + 2h(\varepsilon))(k+1-D)}{2(k+1-D-\varepsilon h(\varepsilon))^3}, \tag{26}$$

where

$$h'(\varepsilon) = \frac{\mathrm{d}h(\varepsilon)}{\mathrm{d}\varepsilon}. \tag{27}$$

Under the MF approximation, S_ε is equivalent to $S_{H_1^*}$ which was defined for the experiments before.

From Eqs. (24), (25), and (26), it can be seen that all sensitivities are dependent on the distribution function $h(\varepsilon_c)$. Next, the effect of distribution function, such as single-peak distribution function and multi-peak distribution function, on sensitivity will be discussed.

(a) Single-peak distribution function, for example, Weibull distribution function (WEIBULL, 1951)

$$h^*(\sigma_c^*) = \frac{m}{\eta^*}\left(\frac{\sigma_c^*}{\eta^*}\right)^{m-1} e^{-\left(\frac{\sigma_c^*}{\eta^*}\right)^m}, \tag{28}$$

or its normalized form

$$h(\sigma_c) = h(\varepsilon_c) = m\varepsilon_c^{m-1}e^{-\varepsilon_c^m}, \tag{29}$$

where m is the shape factor and η^* is the position factor of Weibull distribution. For Weibull distribution, the three sensitivities, S_D (Eq. (24)), S_ε (Eq. (25)), and S_Θ (Eq. (26)) can be expressed by

$$S_D = \frac{(k+1)^2 m\varepsilon^{m-2}e^{-\varepsilon^m}\left[(m-1-m\varepsilon^m)\left(k+e^{-\varepsilon^m}\right)+2m\varepsilon^m e^{-\varepsilon^m}\right]}{\left(k+(1-m\varepsilon^m)e^{-\varepsilon^m}\right)^3}, \tag{30}$$

$$S_\varepsilon = \frac{(k+1)^2 m\varepsilon^{m-1}e^{-\varepsilon^m}(m+1-m\varepsilon^m)}{\left(k+(1-m\varepsilon^m)e^{-\varepsilon^m}\right)^3}, \tag{31}$$

and

$$S_\Theta = \frac{(k+1)^2 m\varepsilon^m e^{-\varepsilon^m}(m+1-m\varepsilon^m)\left(k+e^{-\varepsilon^m}\right)}{2\left(k+(1-m\varepsilon^m)e^{-\varepsilon^m}\right)^3}. \tag{32}$$

(b) Multi-peak distribution function

$$\tilde{h}(\varepsilon_c) = \sum_{i=1}^{N} a_i \frac{m_i}{\eta_i}\left(\frac{\varepsilon_c - b_i}{\eta_i}\right)^{m_i-1} e^{-\left(\frac{\varepsilon_c - b_i}{\eta_i}\right)^{m_i}} \theta(\varepsilon_c - b_i), \tag{33}$$

where N, m_i, a_i, b_i, and η_i ($i = 1, 2, \ldots, N$) are undetermined coefficients and

$$\theta(x) = \begin{cases} 1 & x \geq 0 \\ 0 & x < 0 \end{cases}. \tag{34}$$

Suppose $m_i = m$, and b_i, a_i, and η_i be determined as follows. Give a definite interval $\Delta > 0$ and let

$$b_i = (i-1)\Delta, \quad i = 1, 2, \ldots N \text{ and } b_{N+1} = \infty. \tag{35}$$

In accordance with Eqs. (29) and (33),

$$a_i = \int_{b_i}^{b_{i+1}} h(\varepsilon_c)\mathrm{d}\varepsilon_c = e^{-b_i^m} - e^{-b_{i+1}^m}, \tag{36}$$

and

$$\eta_i = \left(\int_{b_i}^{b_{i+1}} \varepsilon_c h(\varepsilon_c)\mathrm{d}\varepsilon_c \Big/ a_i - b_i\right) \Big/ \Gamma\left(1 + \frac{1}{m}\right). \tag{37}$$

In this situation, it can be easily seen that the two distribution functions satisfy the unitary condition Eq. (22) and have the same mean value

$$\int_0^\infty \varepsilon_c h(\varepsilon_c)\mathrm{d}\varepsilon_c = \int_0^\infty \varepsilon_c \tilde{h}(\varepsilon_c)\mathrm{d}\varepsilon_c. \tag{38}$$

Clearly, there is $\tilde{h}(\varepsilon_c) = h(\varepsilon_c)$ as $N=1$ and $\Delta \to \infty$, i.e. $h(\varepsilon_c)$ is a special sample of $\tilde{h}(\varepsilon_c)$. Figure 6(a) shows the distribution function of mesoscopic strength, single-peak distribution function $h(\varepsilon_c)$ with $m = 2$ (bulk solid line) and multi-peak distribution function $\tilde{h}(\varepsilon_c)$ with $m = 2$, $N = 100$, and $\Delta = 0.04$ (solid line). Figure 6(b) shows the curves of normalized nominal stress σ versus dimensionless boundary displacement U with $k = 0.3$, single-peak distribution function (■ and corresponding line) and multi-peak distribution function (○ and corresponding line). Clearly, many small events appear prior to final catastrophe with multi-peak distribution function, while there is only final catastrophe with single-peak distribution function. Thus, the damage fraction D, nominal strain ε and the released energy Θ are continuous and smooth functions of boundary displacement prior to final catastrophe (■ and corresponding line in Figures 7(a), 8(a), and 9(a)) for the case of single-peak distribution function, but discontinuous for multi-peak distribution function (○ and corresponding line in Figs. 7(a), 8(a), and 9(a)).

The variations of sensitivities, S_D, S_ε, and S_Θ, versus dimensionless boundary displacement U are shown in Figures. 7(b), 8(b), and 9(b). It is found that S_D, S_ε, and S_Θ for the two different distribution functions display critical sensitivity, that is to say, they increase sharply prior to catastrophic rupture. It can also be seen that, in the case of single-peak distribution function $h(\varepsilon_c)$, the sensitivity is monotonic and smooth (■ and corresponding line in Figs. 7(b), 8(b), and 9(b)), but the sensitivity for multi-peak distribution function displays fluctuations (○ and corresponding line in Figs. 7(b), 8(b), and 9(b)). Since the distribution function of real materials may present complicated structure and cannot be modeled by a continuous single-peak function, the distribution function with multiple peaks might be more appropriate to

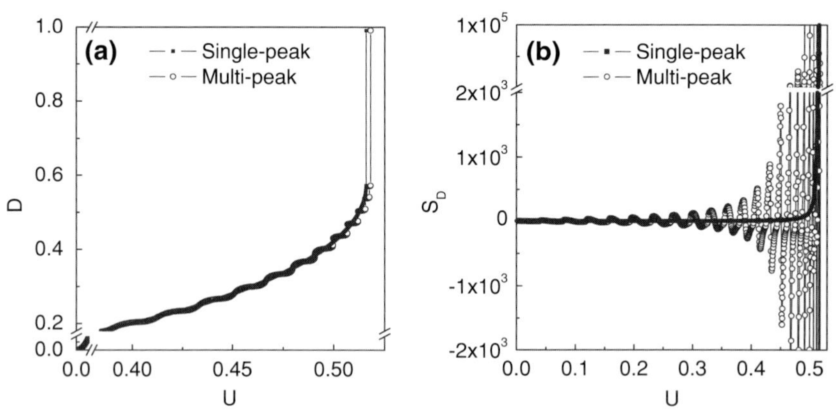

Figure 7
(a) Damage fraction D versus boundary displacement U. (b) Sensitivity S_D calculated from damage. In (a) and (b), single-peak function (■ and corresponding line) and multi-peak function (○ and corresponding line).

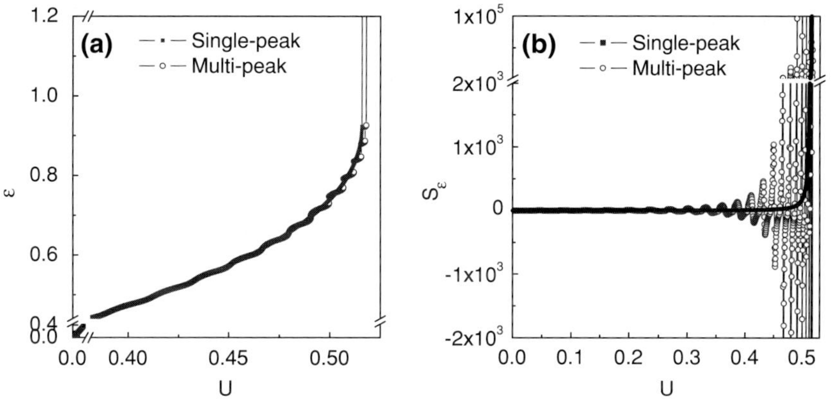

Figure 8
(a) Nominal strain ε versus U. (b) Sensitivity S_ε calculated from nominal strain. In (a) and (b), single-peak function (■ and corresponding line) and multi-peak function (○ and corresponding line).

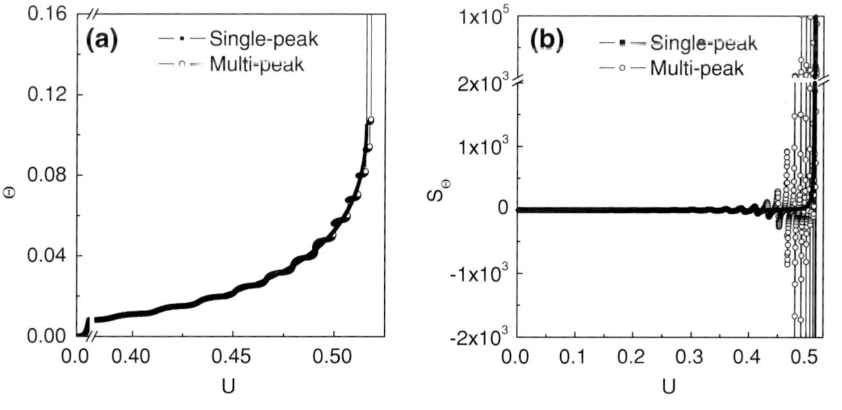

Figure 9
(a) Accumulated released energy Θ versus U. (b) Sensitivity S_Θ calculated from released energy. In (a) and (b), single-peak function (■ and corresponding line) and multi-peak function (○ and corresponding line).

represent real materials, and the corresponding results are more similar to that observed in experiments. Thus, the multi-peak structure of the distribution function might be the reason for the fluctuations shown in the critical sensitivity.

4. *Conclusions*

The prediction of catastrophic rupture, like forecasting of large earthquakes or prediction of materials failure, is a problem of great scientific and societal concern.

However, it is also a very difficult problem due to its richness of dynamical complexity. A possible approach to catastrophe prediction is to search universal features of catastrophe. Based on analytical and numerical studies, it is found that critical sensitivity might be a common precursory feature of catastrophe in heterogeneous brittle media, and to monitor the sensitivity of a system may give helpful clues to rupture prediction.

Experimental studies on critical sensitivity are presented in this paper. It is found that: (1) critical sensitivity is a common precursor prior to catastrophic rupture; (2) the observed fluctuations in sensitivity can be attributed to the multi-peak structure of the distribution function of mesoscopic strain threshold. These results provide experimental evidence for critical sensitivity.

The series of sensitivity presents sample specificity, i.e., they are different from sample to sample for similar samples under the same loading condition. In particular, the maximum values of sensitivity prior to catastrophe show strong diversity for these samples. It is inappropriate to predict catastrophe based on a common threshold of the sensitivity. We think that an effective method of catastrophe prediction might be based on the universal behaviors of sensitivity as the system is approaching its catastrophic transition point. For example, a theoretical analysis shows that, in the case of fast loading, the sensitivity displays a peak prior to catastrophic rupture (ZHANG *et al.*, 2004). The catastrophe prediction in the case of fast loading might be based on the appearance of the peak in the sensitivity curve. The experimental study of loading effect on critical sensitivity will be discussed in another paper.

Acknowledgements

This work is supported by the National Natural Science Foundation of China (Grant No. 10172084, 10232050, 10372012 and 10472118) and the Major State Research Project "Nonlinear Science" G200007735. The authors would like to thank Haisheng Yang for assistance with the experimentation and Dr Can Yin for editing this paper for grammar.

REFERENCES

BOWMAN, D.D., OUILLON, G., SAMMIS, C.G., SORNETTE, A., and SORNETTE, D. (1998), *An observation test of the critical earthquake concept*, J. Geophys. Res. *103*, 359–372.

JAYATILAKA, A.S., *Fracture of Engineering Brittle Materials*. (Applied Sciences Publishers Ltd., London, 1979).

JAUMÉ, S.C. and SYKES L.R. (1999), *Evolving towards a critical point: A review of accelerating seismic moment/energy release prior to large and great earthquakes*, Pure Appl. Geophys. *155*, 279–305.

MA, S.P., XU, X.H., and ZHAO, Y.H. (2004), *The Geo-DSCM system and its application to the deformation measurement of rock materials*, Int. J. Rock Mech. Mining Sci. *41*, 411–412.

PETER, W.H. and RANDSON, W.F. (1981), *Digital imaging techniques in Experimental stress analysis*, Opt. Eng. *21*, 427–431.

RUNDLE, J.B. (2000a), *Precursory seismic activation and critical-point phenomena*, Pure Appl. Geophys. *157*, 2165–2182.

RUNDLE, J.B. and KLEIN, W. (2000b), *Linear pattern dynamics in nonlinear threshold systems*, Phys. Rev. E *61*, 2418–2431.

STEIN, R.S. (1999), *The role of stress transfer in earthquake occurrence*, Nature *402*, 605–609.

WEIBULL, W. (1951), *A statistical distribution function of wide applicability*, J. Appl. Mech. ASME *18*, 293–297.

XIA, M.F., KE, F.J., and BAI, Y.L. (2000), *Evolution induced catastrophe in a nonlinear dynamical model of materials failure*, Nonlinear Dynamics *22*, 205–224.

XIA, M.F., WEI, Y.J., KE F.J., and BAI, Y.L. (2002), *Critical sensitivity and transscale fluctuations in catastrophe rupture*, Pure Appl. Geophys. *159*, 2491–2509.

XU, X.H., MA, S.P., XIA, M.F., KE, F.J., and BAI, Y.L. (2004), *Damage evaluation and damage localization of rock*, Theor. Appl. Fract. Mech. *42*, 131–138.

YIN, X.C., WANG, Y.C., PENG, K.Y., BAI, Y.L., WANG, H.T., and YIN, X.F. (2000), *Development of a new approach to earthquake prediction: Load/Unload Response Ratio (LURR) Theory*, Pure Appl. Geophys. *157*, 2365–2383.

ZHANG, X.H., XU, X.H., XIA, M.F., and BAI, Y.L. (2004), *Critical sensitivity in driven nonlinear threshold systems*, Pure Appl. Geophys. *161*, 1931–1944.

(Received November 12, 2004, revised October 6, 2005, accepted November 7, 2005)

Pure appl. geophys. 163 (2006) 1769–1785
0033–4553/06/091769–17
DOI 10.1007/s00024-006-0096-0

Pure and Applied Geophysics

Implementation of Particle-scale Rotation in the 3-D Lattice Solid Model

YUCANG WANG,[1,2] STEFFEN ABE,[1,2] SHANE LATHAM,[1,2] and PETER MORA[1,2]

Abstract—In this study, 3-D Lattice Solid Model (LSMearth or LSM) was extended by introducing particle-scale rotation. In the new model, for each 3-D particle, we introduce six degrees of freedom: Three for translational motion, and three for orientation. Six kinds of relative motions are permitted between two neighboring particles, and six interactions are transferred, i.e., radial, two shearing forces, twisting and two bending torques. By using quaternion algebra, relative rotation between two particles is decomposed into two sequence-independent rotations such that all interactions due to the relative motions between interactive rigid bodies can be uniquely decided. After incorporating this mechanism and introducing bond breaking under torsion and bending into the LSM, several tests on 2-D and 3-D rock failure under uni-axial compression are carried out. Compared with the simulations without the single particle rotational mechanism, the new simulation results match more closely experimental results of rock fracture and hence, are encouraging. Since more parameters are introduced, an approach for choosing the new parameters is presented.

Key words: 3-D Particle Rotation, decomposition of rotation, quaternion, the Lattice Solid Model, Distinct Element Method.

Introduction

Particle-based models have been extensively used in recent years to model a diverse range of processes from rock failure (HUZZARD *et al.*, 2000, 2002; CHANG *et al.*, 2002), particle flow (LANGSTON *et al.*, 2004), mining (PROCHAZKA, 2004), to tectonic processes (HU *et al.*, 2004). These models typically deal with a system which can be considered as an assemblage of discrete particles interacting with one another. The algorithm to extrapolate the positions of each particle in the assemblage involves application of equations of motion for these particles followed by application of a force-displacement law at contact. Although different kinds of models have been

[1] QUAKES, Earth Systems Science Computational Centre, The University of Queensland, St. Lucia, Brisbane, QLD 4072, Australia.
E-mail: wangyc@esscc.uq.edu.au or yucang-wang@hotmail.com; steffen@esscc.uq.edu.au
[2] Australian Computational Earth Systems Simulator (ACcESS), MNRF, The University of Queensland, St. Lucia, Brisbane, QLD 4072, Australia.

used, they are more or less similar to the Discrete Element Model first proposed by CUNDALL *et al.* (1979). Among these models, some are discrete models for granular media in which particles are not bonded, which means that they cannot transmit tensile forces and only interact with one another when they come to contact, just as the original one used by CUNDALL *et al.* (1979). These kinds of models are suitable to model motions of powder (SHENG *et al.*, 2004) and behaviors of particulate materials (SITHARAM, 2000).

Another kind of model permits the neighboring particles to be linked together by elastic bonds or molecular bonds or springs (e.g., MORA *et al.*, 1993; CHANG *et al.*, 2002; HUZZARD *et al.*, 2000, 2002), which can transfer elastic forces, both attractive or repulsive, when the bond is pulled or pushed. Bonds break if the distance between the particle pair exceeds certain threshold breaking distance. After bond breakage, only pure repulsive elastic forces and frictional force exist between particles. Therefore these models are capable of modeling processes involved in the failure of intact material and frictional processes.

The Lattice Solid Model (LSM) is among the second category and is similar to the Discrete Element Model (CUNDALL *et al.*, 1979) and Molecular Dynamics (MD, ALLEN *et al.*, 1987) but involves a different computational approach (PLACE and MORA, 1999). It was developed to simulate the nonlinear dynamics of earthquakes (MORA and PLACE, 1993, 1994, 2000). It has been applied to the study of physical processes such as rock fracture (MORA and PLACE, 1993; PLACE *et al.*, 2002), stick-slip friction behavior and friction (MORA *et al.*, 1994; PLACE *et al.*, 1999), granular dynamic (MORA *et al.*, 1998, 1999) and heat-flow paradox (MORA *et al.*, 1998, 1999), localization phenomena (PLACE *et al.*, 2000) and Load-Unload Response Ratio theory (MORA *et al.*, 2002a, WANG *et al.*, 2004) and Critical Point system (MORA *et al.*, 2002b).

Past work in 2-D has demonstrated that rotation of grains consisting of particle groupings (MORA and PLACE, 1998, 1999) and rotation of round particles (MORGON, 2001) modify the micromechanics of fault gouge layers and result in lower effective friction (MORA *et al.*, 1999, 2000; MORGON, 2001). However, 3-D simulations need to be performed to continue these studies. At this stage, the paralleled LSM code (ABE *et al.,* 2004) has been extended to 3-D and computations of gouge layer dynamics and compression tests are starting to become feasible (LATHAM *et al.*, 2006).

However, the current 3-D LSM model is still too simple since it includes only radial interactions between two linked particles and involves only translational motions of particles. Theoretically, six degrees of freedoms are needed to completely describe a single 3-D particle, and therefore six kinds of independent relative movements are also required between two interacting particles. Sometimes it becomes necessary to consider not only the stretching of bonds, but also bending and twisting of bonds. These effects must be treated properly within the classical frame, but were thought to be quite unrealistic more than one decade ago (ALLEN *et al.*, 1987). The challenge mainly comes from the difficulty of calculating the interactions due to relative motions between linked rigid bodies, especially when each of the two bonded rigid bodies has an arbitrary rotation. The solution to this problem requires decomposition of relative rotations between two rigid bodies in such a way that the torques and forces caused by such relative rotations can be uniquely determined.

To our knowledge, very few 3-D particle models involve full rigidity between bonded particles. Some models take into account the radial interaction (MORA, 1993, 1994; TOOMEY, 2000), some models include radial and shearing interaction (SCOTT, 1996) but without rotation mechanism for single particles. However, it is necessary to point out that without rotational degrees of freedom at particle scale, half the shearing mechanism is lost since rotations of both particles also contribute to shearing interaction besides relative translational motions. It is found both from experiments and numerical simulations that rolling resistance at contacts has a significant influence on behavior of granular media (ODA *et al.*, 1982; ODA *et al.*, 1998; IWASHITA *et al.*, 1998; TORDESILLAS *et al.*, 2002; KUHN *et al.*, 2004). By now, PFC3D is one of few 3-D particle models we know which includes full rigidity (POTYONDY *et al.*, 2004). In their algorithm, incremental method is used to calculate interaction caused by relative rotations and tangential motion between bonded particles. However, we point out that when dealing with finite rotations, incremental method is not as stable and accurate as total method used in our model (WANG *et al.*, 2006).

In this study, we extend the LSM by introducing full rigidity and particles with full degrees of freedom. We first give the solution to the decomposition of relative rotations between two bodies or systems, then present an algorithm for solving the rotational equations. The detailed theory and algorithm can be found in another paper (WANG *et al.*, 2006). Last we incorporate this technique into LSM and present some preliminary simulations. We also show that the model is improved and results in more realistic fracturing behavior.

The New Model

In the new model, for each particle, we use three parameters x, y, z to represent the position of center of the mass, and three other parameters φ, θ, ψ (such as Euler angles) to represent rotations around the center. In case of small deformation, the relationship between interactions (forces or torques) and relative displacements between two bonded particles could be written in the linear form as

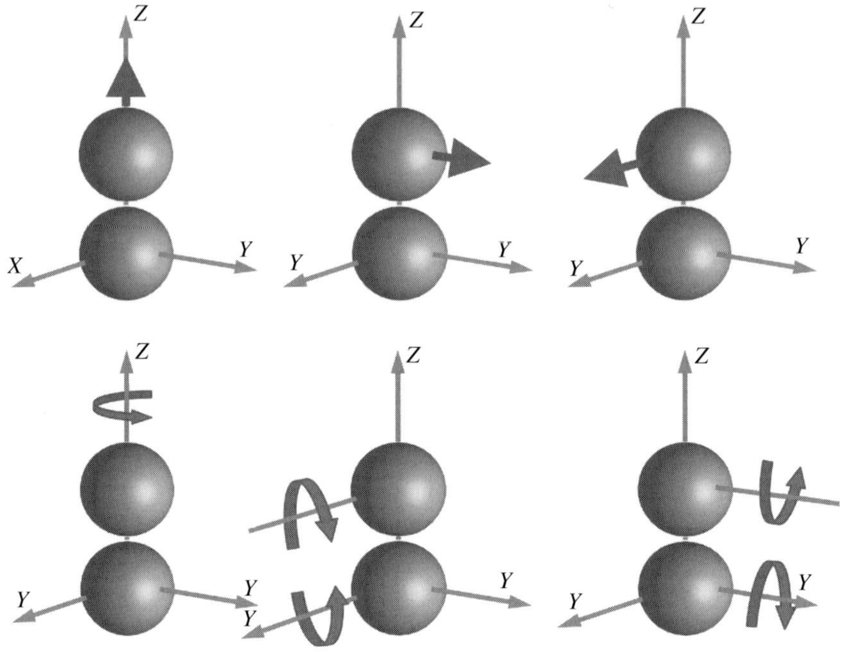

Figure 1
Six kinds of interactions between bonded particles: Pulling or pressing in radial direction (left, above), shearing in two tangential directions (middle and right, above), twisting (left, bottom) and bending around two axes (middle and right, bottom).

$$\begin{aligned} F_r &= K_r \, \Delta \, r \\ F_{s1} &= K_{s1} \, \Delta \, s_1 \\ F_{s2} &= K_{s2} \, \Delta \, s_2 \\ M_t &= K_t \, \Delta \, \alpha_t \\ M_{b1} &= K_{b\,1} \, \Delta \, \alpha_{b1} \\ M_{b2} &= K_{b2} \, \Delta \, \alpha_{b2}, \end{aligned} \tag{1}$$

where F, M, K, $\Delta\, r$ (and $\Delta\, s$) and $\Delta\, \alpha$ represent forces, torques, rigidity constant, relative translational and angular displacement, subscript r, s, t and b stand for radial, shearing, twisting and bending components.

In the isotropic case, $K_s = K_{s1} = K_{s2}$ and $K_b = K_{b1} = K_{b2}$. Therefore only four rigidity parameters are required.

Equations and Representation of Orientation

Particle motion can be decomposed into two completely independent parts, translational motion of the center of mass and rotation about the center. The former is governed by the Newtonian equations and integrated using a conventional

Molecular Dynamic scheme (MORA *et al.*, 1993, 1994). The latter is governed by the equations

$$\dot{\omega}_i^b = \frac{\tau_i^b}{I_{ii}} \quad (i = 1, 2, 3), \tag{2}$$

where ω_i^b, τ_i^b and I_{ii} are angular velocities, torques and principle moment of inertia measured in the body-fixed frame.

There are several different ways to represent orientation, such as Euler angles and rotation matrix and vector, and quaternion. Due to the existence of singularities and numerical instability caused by Euler angles near certain values, the unit quaternion $q = q_0, q_1, q_2, q_3) = q_0 + q_1 i + q_2 j + q_3 k$ is often used to describe the particle orientation (EVANS, 1977; EVANS and MURAD, 1977) where

$$\begin{aligned} i^2 = j^2 = k^2 = ijk = -1 \\ ij = -ji = k \\ jk = -kj = i \\ ki = -ik = j \end{aligned} \tag{3}$$

and

$$\sum_{i=0}^{3} q_i^2 = 1. \tag{4}$$

The physical meaning of a quaternion is that it represents a one-step rotation around the vector $q_1\hat{i} + q_2\hat{j} + q_3\hat{k}$ with a rotation angle of $2\arccos(q_0)$ (KUIPERS, 1998). It is a very convenient tool due to some of its interesting features, such as the fact that sequences of rotations can be conveniently represented as quaternion product (KUIPERS, 1998).

Quaternion for each particle satisfy the following equations of motion (EVANS, 1977; EVANS and MURAD, 1977)

$$\dot{q} = \frac{1}{2} Q(q)\,\omega \tag{5}$$

where

$$\dot{q} = \begin{pmatrix} \dot{q}_0 \\ \dot{q}_1 \\ \dot{q}_2 \\ \dot{q}_3 \end{pmatrix}, \quad Q(q) = \begin{pmatrix} q_0 & -q_1 & -q_2 & -q_3 \\ q_1 & q_0 & -q_3 & q_2 \\ q_2 & q_3 & q_0 & -q_1 \\ q_3 & -q_2 & q_1 & q_0 \end{pmatrix}, \quad \omega = \begin{pmatrix} 0 \\ \omega_x^b \\ \omega_y^b \\ \omega_z^b \end{pmatrix}.$$

Equations (2) and (5) govern the rotation of a particle and can be integrated numerically in the next section.

Algorithms of Solving Rotational Equations

We integrate equations (2) and (5) a using Fincham's leap-frog algorithm step by step (FINCHAM, 1992), which is simple and proved to be accurate enough. The approach obtains $q(t+dt)$ from $q(t)$ using

$$q(t+dt) = q(t) + dt\,\dot{q}(t) + \frac{dt^2}{2}\ddot{q}(t) + O(dt^3) = q(t) + dt\,\dot{q}(t+dt/2) + O(dt^3). \quad (6)$$

Hence, quaternion derivative at mid-step $\dot{q}(t+dt/2)$ is required. Equation (5) indicates that $q(t+dt/2)$ and $\omega(t+dt/2)$ are required, the former can be easily calculated using

$$q(t+dt/2) = q(t) + \dot{q}(t)\,dt/2, \quad (7)$$

where $\dot{q}(t)$ is again gained by (5), prior to which $\omega(t)$ can be calculated using

$$\omega(t) = \omega(t-dt/2) + I^{-1}\tau^b(t)dt/2 \quad (8)$$

and the later, $\omega(t+dt/2)$, can be gained using

$$\omega(t+dt/2) = \omega(t-dt/2) + I^{-1}dt\,\tau^b(t). \quad (9)$$

I is the inertia tensor expressed in the body-fixed frame. Hence, in this algorithm, only $\omega(t-dt/2)$, $q(t)$ need to be stored, the other quantities, such as $q(t+dt/2)$, $\omega(t)$, $\dot{q}(t)$, $\dot{q}(t+dt/2)$ are treated as temporary and auxiliary values.

To avoid the buildup errors, it is a common practice to renormalize quaternions at frequent intervals (usually done every step). The entire algorithm is as follows

step 1: calculate torque $\tau^b(t)$ at time t.
step 2: using the stored $\omega(t-dt/2)$, update $\omega(t)$ using (8)
step 3: gain $\dot{q}(t)$ using (5)
step 4: using the stored $\omega(t-dt/2)$,calculate $\omega(t+dt/2)$ using (9)
step 5: compute $q(t+dt/2)$ using (7)
step 6: evaluate $\dot{q}(t+dt/2)$ using (5)
step 7: calculate $q(t+dt)$ using (6)
step 8: renormalize the quaternion $q(t+dt)$.

In recent years, some more advanced and complicated algorithms have been proposed which focus on conserving the symplectic structure, time reversible properties and energy (OMELYAN, 1998a, b; DULLWEBER *et al.*, 1997; KOL *et al.*, 1997).

Calculation of Interactions due to Relative Motions

One important problem remained is that, given the position and orientation for each particle, how can one calculate the interactions between neighboring bodies due to relative motion between bonded particles? This involves decomposition of the relative motions in such a way that interactions can be uniquely calculated. Since it is easy to decompose the relative translation (WANG *et al.*, 2004), here we mainly focus on decomposition of relative rotation.

Suppose $X_2'Y_2'Z_2'$ is an auxiliary local frame of particle 2. With its Z_2'-axis pointing to particle 1, $X_1Y_1Z_1$ is the body-fixed frame of particle 1 (only Z_1-axis is drawn in Fig. 2). We consider decomposition of the rotation from $X_2'Y_2'Z_2'$ to $X_1Y_1Z_1$. As a first intuition, one might factorize the relative rotation between two particles into three rotations around three axis, such as XYZ-type successive rotations, which starts from a rotation of θ about the X_2' axis, followed by a rotation of φ about the new Y_2' axis, and then finally a rotation of ψ about the new Z_2' axis. However, further study shows that these kinds of successive rotations are sequence-dependent, which means that a sequence YXZ provides orientation angles different from an sequence of XYZ rotations. Therefore, θ, φ and ψ decided by this method are not mutually independent. For this reason, they are not suitable for deciding the torques and moments between the two bonded particles.

Using quaternion algebra, we proved that (WANG *et al.*, 2006) an arbitrary rotation between two rigid bodies or two coordinate systems cannot be decomposed into three mutually independent rotations around three orthogonal axes, but hopefully it can be decomposed into two rotations, one pure axial rotation of angle ψ around its Z_2' axis, and one rotation of Z_2' axis over θ on certain plane controlled by another parameter φ Figure 2 describes such decomposition. The two rotations, corresponding to the relative twisting and bending between two bodies in our model, are sequence-independent. Such a two-step rotation is controlled by three independent parameters (ψ, θ and φ just mentioned in this paragraph), which can be decided as follows

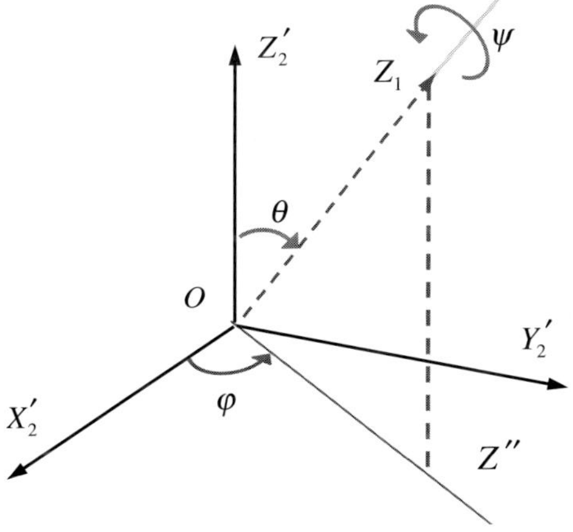

Figure 2
An arbitrary rotation between two rigid bodies or two coordinate systems can be decomposed into two-step rotations, one pure axial rotation of angle ψ around its Z axis, and one rotation of Z axis over θ on certain plane controlled by another parameter φ. O is the centre of particle 2. OZ_2' points to particle 1. $X_1Y_1Z_1$ defines the orientation of particle 1 (only Z_1-axis is drawn here).

$$\cos\frac{\psi}{2}=\frac{r_0}{\sqrt{r_0^2+r_3^2}}\qquad \sin\frac{\psi}{2}=\frac{r_3}{\sqrt{r_0^2+r_3^2}}$$
$$\cos\theta=r_0^2-r_1^2-r_2^2+r_3^2 \tag{10}$$
$$\cos\varphi=\frac{r_1r_3+r_0r_2}{\sqrt{(r_0^2+r_3^2)(r_1^2+r_2^2)}}\qquad \sin\varphi=\frac{r_2r_3-r_0r_1}{\sqrt{(r_0^2+r_3^2)(r_1^2+r_2^2)}}$$

where $r=r_0+r_1i+r_2j+r_3k$ is the quaternion representing the relative rotation between two particles, that is, a rotation from $X_2'Y_2'Z_2'$ to $X_1Y_1Z_1$.

It is noted that, rotation represented by angle θ, or bending, cannot be further decomposed into two sequence-independent bending rotations as we would expect to do in Eq. (1). However, in the isotropic case, this seems not to be a problem, since once θ and φ are gained one can still easily calculate the torque caused by this bending without further decomposition. Using such two-step decomposition, all interactions due to the relative translational and rotational motions between interactive rigid bodies can be uniquely determined (WANG *et al.*, 2006).

Incorporation into LSM and Parameter Calibration

In our new model, a bond is permitted to break under purely extension if the force exceeds certain threshold F_{r0}, but it does not break under compression. Similarly, a bond breaks under pure shear load when shear force reaches F_{s0}, or under pure twisting load if twisting torque exceeds M_{t0}, or under pure bending load if bending torque exceeds value M_{b0}. When all the interactions exist at the same time, the following criterion is used to judge whether or not a bond is going to break

$$\frac{F_r}{F_{r0}}+\frac{|F_s|}{F_{s0}}+\frac{|M_t|}{M_{t0}}+\frac{|M_b|}{M_{b0}}\geq 1, \tag{11}$$

where $\overrightarrow{F_s}=\overrightarrow{F_{s1}}+\overrightarrow{F_{s2}}$, $\overrightarrow{M_b}=\overrightarrow{M_{b1}}+\overrightarrow{M_{b2}}$. We set F_r positive under extension and negative under compression, therefore the effects of normal force on breakage of the bond has been taken into account.

Our input parameters include the particle mass m, radius R, rigidity parameters K_r, K_s, K_b and K_t and fracture parameters F_{r0}, F_{s0}, M_{t0} and M_{b0}. Since we introduce more parameters, the question arises: How should one choose these parameters? It is obvious that they cannot be chosen arbitrarily, otherwise some unphysical results will be produced. At this stage, it is not clear how these parameters should be chosen aside from detailed calibration of the numerical model against laboratory experiments. However, we can still have an approximate estimation as follows.

In the 2-D case, when equal-sized particles are arranged as tight and regular triangular pattern, analytical relation between microscopic rigidity parameters and macroscopic elasticity constant can be gained (MORA *et al.*, 2000; ABE *et al.*, 2000).

In the current model, there are three rigidity parameters K_r, K_s and K_b, three fracture parameters F_{r0}, F_{s0} and M_{b0}. According to some studies (WANG *et al.*, 2000; BATHURST *et al.*, 1988), K_r should be related to macroscopic Young's modulus E and Poisson ratio v and $k_1 = K_s/K_r$ controls Poisson ratio v

$$K_r = \frac{\sqrt{3}E}{3(1-v)}$$
$$k_1 = \frac{K_s}{K_r} = \frac{1-3v}{1+v} \tag{12}$$
$$K_b = \frac{\sqrt{3}\,ER^2}{18}.$$

However, there is no analytical result that provides insight into how to choose fracture parameters, but it is easy to imagine that F_{r0} can be chosen according to tensile strain threshold. CHANG *et al.* (2002) suggested that another ratio $k_2 = F_{s0}/F_{r0}$ may effect the fracture mode, they found that larger k_2 generate brittle failure in uni-axial compression tests. According to rock experiments (ATTEWELL *et al.*, 1976), compressive strength is generally ten times of tensile strength, and shear strength (cohesion) is nearly one to two times of tensile strength. If these relation still exist at particle bond scale, k_2 should be chosen around 1–2. However, a wider range of k_2 values will be used in the next section to test its influences.

It is possible to relate M_{b0} to F_{r0} in case of pure bending. Consider a 2-D beam (SOKOLNIKOFF, 1956) subjected to pure bending load M_b at two ends (Fig. 3), the maximum tensile stress is $\sigma_{\max} = 6M_b/A^2$. In case of two equal-sized particles, $L = 4R$ and $A = 2R$, let $\sigma_{\max} = F_{r0}/2R$, we have

$$M_{b0} = F_{r0}\,R/3. \tag{13}$$

In the 3-D case, it is very difficult to have analytical results between microscopic rigidity parameters and macroscopic elasticity constant. The reason is that geometrically in 3-D there is no regular tight pack for equal-sized spheres similar to the triangular pattern in 2-D case which yields macroscopic isotropic elasticity, in other words, dense packings of equal spheres are intrinsically disordered. LIU *et al.* (2003) derived the following formulas to determine normal and shear spring stiffness from closest-packed 3-D arrangement

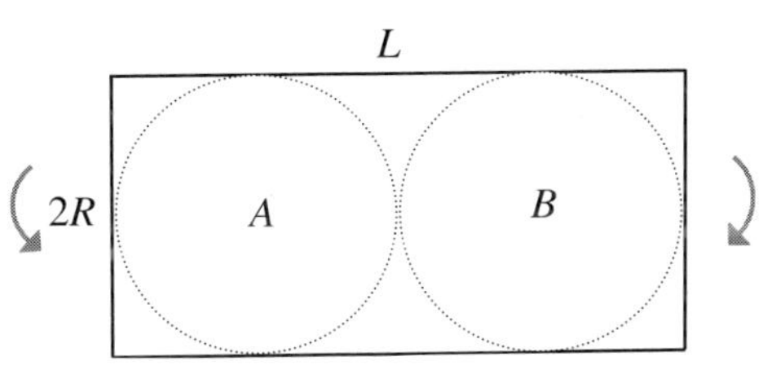

Figure 3
Pure bending of 2-D beam, which is used to have an approximated estimation of the parameters.

$$K_r = \frac{\sqrt{2}ER}{2(1-2v)}$$
$$k_1 = \frac{K_s}{K_r} = \frac{1-5v}{1+v}. \tag{14}$$

K_t can be estimated according to the elastic solution of torsion of a circular shaft. Consider elastic deformation of a cylindrical bar of circular cross section of radius R and length $L = 4R$ that is being twisted by a moment M_t at the right end and is fixed at the left end (SOKOLNIKOFF, 1956). The rotation along the z axis is $\theta = \alpha z$ where α is a constant. $M_t = \pi G \alpha R^4/2$, where G is shear modulus. Rotation angles at point A and B are $\theta_a = \alpha R$ and $\theta_b = 3\alpha R$, then $\Delta\theta = \theta_b - \theta_a = 2\alpha R$, so $M_t = (\pi G R^3/4)\,\Delta\theta$, we have

$$K_t = \pi G R^3/4 = \pi ER^3/8(1+v). \tag{15}$$

The maximum shear stress is $\tau_{\max} = 2M_t/\pi R^3$, let $\tau_{\max} = F_{s0}/\pi R^2$, we have

$$M_{t0} = F_{s0} R/2. \tag{16}$$

K_b and M_{b0} can be estimated using similar method

$$K_b = \pi ER^3/2, \tag{17}$$

$$M_{b0} = F_{r0} R/4. \tag{18}$$

Note that Eqs. (13)–(18) are not strict derivations for these parameters and further study is required regarding how to theoretically choose these parameters. However these formulas are still used as an approximate estimation in this paper. In practice, theoretical relationships would be used as a guide to choose the parameters combined with calibration of the numerical model against laboratory observations if one wishes to use the model to study the detailed behavior of a specific rock.

Preliminary Simulation Results

After incorporating the mechanism and algorithm above into the 3-D paralleled code of the LSM, we carried out several tests of 2-D and 3-D rock failure under compression.

Our first example aims to reproduce the classical experimental results obtained by BRACE (1960) which involves a brittle rock with an oblique pre-existed crack subjected to uni-axial compression. Parameters in this simulation are: $m = 1.0$, $R = 1.0$, $K_r = 10.0$, $K_s = 3.0$, $K_b = 1.0$, $F_{r0} = 0.005$, $F_{s0} = 0.025$, $M_{b0} = 0.002$, time step $dt = 0.01$ and loading velocity at top and bottom $v = 0.125$, number of particle is 2348, the dimension of the sample is 78×108, pre-existed fault size is 30, and orientation is 30 degree with the loading direction. Figure 4 shows our 2-D

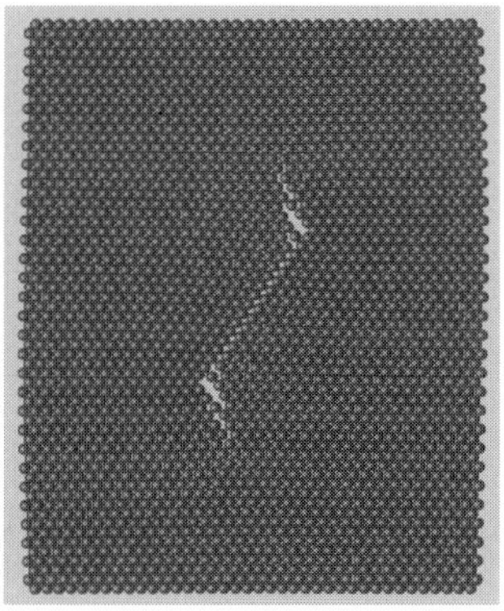

Figure 4
Simulated extension of cracks in a brittle rock with an oblique pre-existed crack subjected to uni-axial compression. The growth of cracks from the crack tips is similar to those obtained in laboratory experiments.

simulation result, which suggests that the form of stable tensile crack growth away from the initial crack tips is very similar to those obtained in the laboratory. In contrast, it is difficult to reproduce this experimental result without the single particle rotational mechanism and shearing interaction between bonded particles. This fact may imply that at least in certain circumstances, particle-scale rotation, shearing and bending interactions cannot be ignored, unless particles represent units on the atomic or molecular scale, in which case it is reasonable to assume that only radial interaction can be transmitted between particles.

Another example models brittle failure of a rock sample subjected to uni-axial compression. Parameters in this simulation are: $m = 1.0$, $R = 1.0$, $K_r = 10.0$, $K_s = 3.0$, $K_b = 1.0$, $F_{r0} = 0.005$, $F_{s0} = F_{r0}\,k_2$, $M_{b0} = 0.002$, time step $dt = 0.005$ and loading velocity at top and bottom $v = 0.025$, number of particle is 1976, the

Figure 5
Different failure modes corresponding to different k_2, left: $k_2 = 1.0$, middle: $k_2 = 1.5$, right $k_2 = 2.0$.

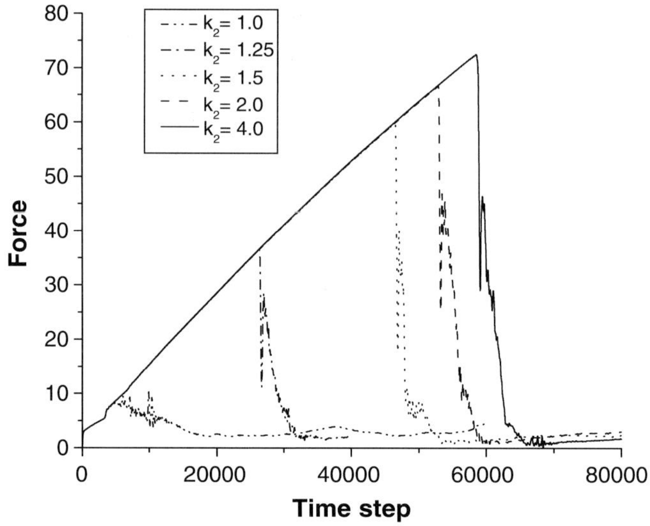

Figure 6
Force versus time step curves for different k_2, parameters are the same as Figure 5.

dimension of the sample is 60 × 118. Different $k_2 = F_{s0}/F_{r0}$ values are tested in this experiment. Figures 5 and 6 show failure moods and stress-strain curves for different k_2. It is found that when $k_2 < 1.25$, we mainly get ductile failure, with low macroscopic strength (less than 10) and narrow X-shape shear bands (Fig. 5, left). However when $k_2 > 1.25$ abrupt failure moods are produced, with significant high macroscopic strength (40 to 75), which is similar with Chang's results (CHANG *et al.*, 2002). Figure 6 shows with the increase of ratio k_2, macroscopic strength increase, and the shear bands seem to be wider. In fact, our simulations reproduced more

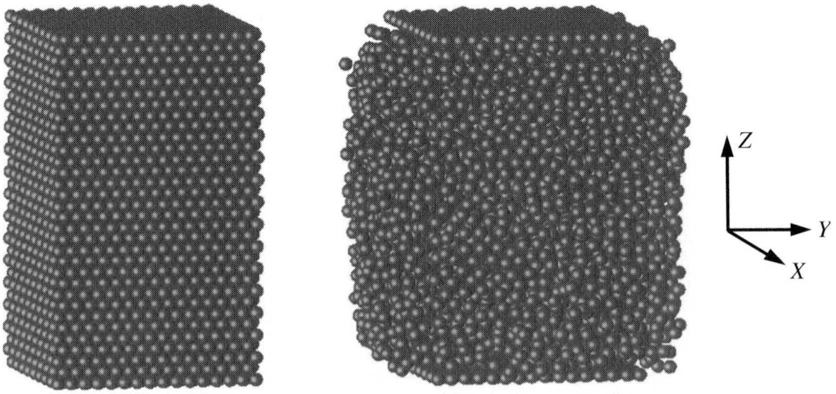

Figure 7
3-D simulation under uni-axial compression, left: time step = 1, initial state; right: time step = 225000, just after failure.

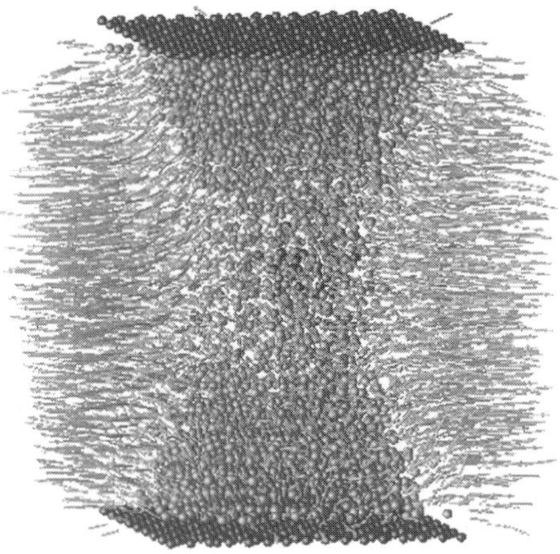

Figure 8
Displacement vector field at time step = 225000. The discontinuity of displacement field shows that an X-shape fracture appeared.

realistic brittle behavior, since forces applied to the two ends drop to zero or values very close to zero in short time, with two relative intact triangular areas on the top and bottom and two fractured areas on two sides (Fig. 5, middle and right).

Figure 7 shows another test of 3-D failure simulation. Parameters in this simulation are: $m = 1.0$, $R = 1.0$, $K_r = 80000$, $K_s = 48000$, $K_b = 20000$, $K_t = 20000$, $F_{r0} = 80$, $F_{s0} = 240$, $M_{t0} = 30$ and $M_{b0} = 30$, time step $dt = 0.0001$ and loading velocity at top and

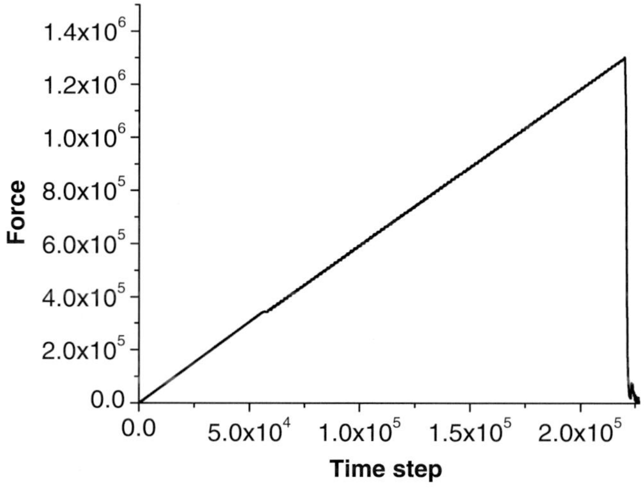

Figure 9
Force versus time step curve for Figure 7. Abrupt brittle fracture mode is seen.

bottom $v = 0.000002$, number of particle is 8948, the dimension of the sample is $32 \times 32 \times 52$. Since it is not easy to visualize the fracture patterns, displacement vector are plotted in Figure 8, which clearly demonstrate that after peak stress particles move inward and outward, and the discontinuity of displacement field means that an X-shape fracture is produced. This result is quite similar to the laboratory tests. Figure 9 is force versus time step, from which abrupt brittle fracture is observed.

Discussions and Conclusions

We incorporated particle-scale rotation and full interactions between bonded particles into the Lattice Solid Model. The preliminary simulation tests indicate that compared with the simulations using the original central force model, the new simulation results for fracture are closer to those observed in the laboratory, and are encouraging. The results also show that introduction of full rigidity and full degree of freedom permit us to model not only global elastic properties, but also local elasticity, especially stress field near the crack tip, which is important for crack extension. Meanwhile, it provides new mechanisms for bonds to break, for example, bonds can still break even when two bonded particles are being compressed if shearing forces are large enough, which is expected to be useful when modeling fracture within the crust is subjected to large confining pressure.

However, since the new approach involves more parameters, it introduces extra complexities and computational costs. We have presented an approximate approach to select the new parameters. From (12), (16) and (17), one can see that when particle radius R becomes very small, the contribution of bending and twisting to the macroscopic elasticity can be ignored. However, the effects of bending and twisting must be taken into account if the particle sizes are large enough, in which case the model and algorithm presented in this paper should be used. It is also shown in (12) and (14) that shear stiffness is of the same order as the normal stiffness, and should always be taken into account. In fact shear rigidity plays an important role not only in macroscopic elasticity in that the ratio of shear stiffness and normal stiffness controls Poisson ratio, but also in fracture behaviors in that the ratio of shear strength and normal strength significantly effects macroscopic strength and fracture modes. Ignoring shear rigidity, or $K_s = 0$, corresponds to a special case of certain Poisson ratio (1/3 in 2-D case and 1/5 in 3-D case according to Eqs. (12) and (14)).

Further study is still required to determine how these parameters should be chosen properly and how do they effect the macroscopic fracture behavior. In this paper, only equal-sized particles are used, in the future, random-sized particles should be used to avoid limitations caused by regular arrangement (PLACE *et al.*, 2001). Nondimensionlized simulations are going to be considered to compare with the experimental results. More large-scale simulations are required to further confirm the results.

Acknowledgments

Funding support is gratefully acknowledged by the Australian Computational Earth Systems Simulator Major National Research Facility, The University of Queensland and SGI. The ACcESS MNRF is funded by the Australian Commonwealth Government and participating institutions (Univ. of Queensland, Monash U, Melbourne U., VPAC, RMIT) and the Victorian State Government. Computations were made using the ACcESS MNRF supercomputer – a 208 processor 1.1 TFlops SGI Altix 3700 – which was funded by the Queensland State Government Smart State Research Facility Fund and SGI.

References

Abe, S., Mora, P., and Place, D. (2000), *Extension of the Lattice Solid Model to incorporate temperature related effects*, Pure Appl. Geophys. *157*, 1867–1887.

Abe, S., Place, D., and Mora, P. (2004), *A parallel implication of the Lattice Solid Model for the simulation of rock mechanics and earthquake dynamics*, Pure Appl. Geophys. *161*, 2265–2277.

Allen, M. P. and Tildesley, D. J. *Computer Simulation of Lliquids* (Oxford Science Press, Oxford, 1987).

Attewell, P. B. and Farmer, I. W. *Principles of Engineering Geology* (Chapman and Hall, London 1976).

Bathurst, R. J. and Rothenburg, L. (1988), *Micromechanical aspects of isotropic granular assemblies with linear contact interactions*, J. Appl. Mech. *55*, 17–23.

Brace, W. F. (1960), *An extension of the Griffith theory of fracture to rocks*, J. Geophys. Rev. *65*, 3477–3480.

Chang, S. H., Yun, K. J., and Lee, C. I. (2002), *Modeling of fracture and damage in rock by the bonded-particle model*, Geosystem Eng. *5*(4), 113–120.

Cundall, P. A. and Stack, O. D. L. (1979), *A discrete numerical model for granular assemblies*, Geotechnique *29*, 47–65.

Dullweber, A., Leimkuhler, B., and McLachlan, R. (1997), *Symplectic splitting methods for rigid-body molecular dynamic*, J. Chem. Phys. *107*(15), 5840–5851.

Evans, D. J. (1977), *On the representation of orientation space, Molecular Physics*, *34*, 317–325.

Evans, D. J. and Murad, S. (1977), *Singularity free algorithm for molecular dynamic simulation of rigid polyatomice*, Molecular physics *34*, 327–331.

Fincham, D. (1992), *Leapfrog rotational algorithm*, Molec. Simul. *8*, 1165.

Huzzard, J. F. and Young, R. P. (2000), *Simulation acoustic emissions in bonded-particle models of rock*, Inter. J. Rock Mech. Mining Sci. *37*, 867–872.

Huzzard, J. F. and Young, R. P. (2002), *3-D numerical modeling of Acoustic emissions*, Proc. 5th International Workshop on the Application of Geophysics in Rock Engineering, Toronto, Canada.

Hu, J. C. and Angelier, J. (2004), *Stress permutations: Three-dimensional distinct element analysis accounts for a common phenomenon in brittle tectonics*, J.G.R. *109* (B9).

Iwashita, K. and Oda, M. (1998), *Rolling resistance at contacts in simulation of shear-band development by DEM*, J. Eng. Mech. *124*, 285–292.

Kol, A., Laird, B. B., and Leimkuhler, B. J. (1997), *A symplectic method for rigid-body molecular simulation*, J. Chem. Phys. *107*(7), 2580–2588.

KUHN, M. R. and BAGI, K. (2004), *Contact rolling and deformation in granular media*, Int. J. Solid Struc. *41*, 5793–5820.

KUIPERS, J. B. *Quaternion and Rotation Sequences* (Princeton University Press, Princeton, New Jersey. 1998).

LATHAM, S., ABE, S., and MORA, P. (2002), *Parallel 3D simulation of a fault gouge using the Lattice Solid Model.*

LANGSTON, P. A., AL-AWAMLEH, M. A., FRAIGE, F. Y., and ASMAR, B. N. (2004), *Distinct element modeling of non-spherical frictionless particle flow*, Chem. Eng. Sci. *59*(2), 425–435.

LIU, K. X. and GAO, L. T. (2003), *The application of discrete element method in solving three-dimensional impact dynamics problems*, Acta Mech. Sol. Sin. *16*, 256–261.

MORA, P. and PLACE, D. (1993), *A lattice solid model for the nonlinear dynamics of earthquakes*, Int. J. Mod. Phys. *C4*, 1059–1074.

MORA, P. and PLACE, D. (1994), Simulation *of the frictional stick-slip instability*, Pure Appl. Geophys. *143*, 61–87.

MORA, P., WANG, Y. C., YIN, C., and PLACE, D. (2002a), *simulation of the load-unload response ratio and critical sensitivity in the Lattice Solid Model,* Pure Appl. Geophys, *159*, 2525–2536.

MORA, P. and PLACE, D. (2002b), *Stress correlation function evolution in lattice solid elasto-dynamic model of shear and fracture zones and earthquake prediction* Pure. Appl Geophys. *159*, 2413–2427.

MORA, P., PLACE, D., ABE, S., and JAUMÉ, S. *Lattice solid simulation of the physics of earthquakes: the model, results and directions.* In *GeoComplexity and the Physics of Earthquakes* (Geophysical Monograph series; no. 120), (eds. Rundle, J.B., Turcotte, D.L. and Klein, W.,) pp. 105–125 (American Geophys. Union, Washington, DC 2000).

MORA, P. and PLACE, D. (1999), *The weakness of earthquake faults*, Geophys. Res. Lett. *26*, 123–126.

MORA, P. and PLACE, D. (1998), *Numerical simulation of earthquake faults with gouge: Towards a comprehensive explanation for the heat flow paradox*, J. Geophys. Res. *103* 21, 067–21, 089.

MORGAN, J. K. (2001), *Distinct element simulations of granular shear zones: Micromechanics of localization and frictional behavior.* In 2nd ACES Workshop Proceedings (eds.) M. Matsu'ura, K. Nakajima, and P. Mora, APEC Cooperation for Earthquake Simulation, 83–90.

ODA, M., KONISHI, J., and NEMAT-NASSER, S. (1982), *Experimental micromechanical evaluation of the strength of granular materials: Effects of particle rolling*, Mech. Mater. *1*, 269–283.

ODA, M. and KAZAMA, H. (1998), *Microstructure of shear bands and its relation to the mechanisms of dilatancy and failure of dense granular soils*, Geotechnique *48*(4), 465–481.

OMELYAN, I. P. (1998a), *Algorithm for numerical integration of the rigid-body equations of motion*, Phys. Rev. E *58*(1), 1169–1172.

OMELYAN, I. P. (1998b), *On the numerical integration of motion for rigid polyatomics: The modified quaternion approach*, Comp. Phys. *12*(1), 97–103.

PLACE, D. and MORA, P. (1999), *A lattice solid model to simulate the physics of rocks and earthquakes: Incorporation of friction*, J. Comp. Phys. *150*, 1–41.

PLACE, D. and MORA, P. (2000), *Numerical simulation of localization in a fault zone*, Pure Appl. Geophys. *157*, 1821–1845.

PLACE, D. and MORA, P. (2001), *A random lattice solid model for simulation of fault zone dynamics and fracture process. In Bifurcation and Localization Theory for Soil and Rock'99* (eds. Muhlhaus, H.B., Dyskin, A.V., and Pasternak, E.) (AA Balkema, Rotterdam/ Brookfield, 2001).

PLACE, D., LOMBARD, F., MORA, P., and ABE, S. (2002), *Simulation of the micro-physics of rocks using LSMearth*, Pure Appl. Geophys. *159*, 1933–1950.

PROCHAZKA, P. P. (2004), *Application of discrete element methods to fracture mechanics of rock bursts*, Eng. Fracture Mech. *71*, 601–618.

POTYONDY, D. and CUNDALL, P. (2004), *A bonded-particle model for rock*, Internat. J.Rock Mech. Mining Sci. *41*, 1329–1364.

SCOTT, D. R. (1996), *Seismicity and stress rotation in a granular model of the brittle crust*, Nature *381*, 592–595.

SHENG, Y., LAWRENCE, C. J., and BRISCOE, B. J. (2004), *Thornton, numerical studies of uniaxial powder compaction process by 3D DEM*, Eng. Comp. *21*, 304–317.

SITHARAM, T. G. (2000), *Numerical simulation of particulate materials using discrete element modeling*, Current Science *78*(7), 876–886.

SOKOLNIKOFF, I. S., *Mathematical Theory of Elasticity*, 2nd ed. (New York, McGraw-Hill, 1956).

TOOMEY, A. and BEAN, C. J. (2000), *Numerical simulation of seismic waves using a discrete particle scheme*, Geophys. J. Inter. *141*, 595–604.

TORDESILLAS, A. and WALSH, D.C. S. (2002), *Incorporating rolling resistance and contact anisotropy in micromechanical models of granular media*, Powder Technology *124*, 106–111.

WANG, Y. C., YIN, X. C., KE, F. J., XIA, M. F., and PENG, K. Y. (2000), *Numerical simulation of rock failure and earthquake process on mesoscopic scale*, Pure Appl. Geophys. *157*, 1905–1928.

WANG, Y. C., MORA, P., YIN, C., and PLACE, D. (2004), *Statistical tests of Load-Unload Response Ratio signals by Lattice Solid Model: Implication to tidal triggering and earthquake prediction*, Pure Appl. Geophys. *161*, 1829–1839.

WANG, Y. C. and MORA, P. (2006), *A complete scheme for modeling 3-D interactive rigid body motions with six degrees of freedom and full rigidity*, to be submitted.

(Received November 11, 2004, revised August 18, 2005, accepted September 20, 2005)
Published Online First: August 31, 2006

Pure appl. geophys. 163 (2006) 1787–1801
0033–4553/06/091787–15
DOI 10.1007/s00024-006-0097-z

Pure and Applied Geophysics

Numerical and Experimental Study on Progressive Failure of Marble

Robina H. C. Wong,[1] M. R. Jiao,[1,2] and K. T. Chau[1]

Abstract—This study presents a framework for numerical simulations based upon micromechanical parameters in modeling progressive failures of heterogeneous rock specimens under compression. In our numerical simulations, a Weibull distribution of the strength and elastic properties of the finite elements is assumed, and the associated Weibull parameters are estimated in terms of microstructural properties, such as crack size distribution and grain size, through microscopic observations of microcracks. The main uncertainty in this procedure lies on the fact that various ways can be used to formulate a "microcrack size distribution" in relating to the Weibull parameters. As one possible choice, the present study uses the number of counted cracks per unit scanned volume per grain size to formulate the crack distributions. Finally, as a tool, the Rock Failure Process Analysis code (RFPA^{2D}) is adopted for simulating the progressive failure and microseismicity of heterogeneous rocks by using an elastic-damage finite-element approach. To verify our framework, compression tests on marble specimens are conducted, and the measured acoustic emissions (AE) are compared with those predicted by the numerical simulations. The mode of failure, compressive strength and AE pattern of our simulations basically agree with experimental observations.

Key words: Heterogeneous rocks, Weibull parameters, crack distributions, acoustic emissions, microseismicity, numerical simulation.

Introduction

Brittle fracture has been intensively investigated in the areas of rock mechanics and rock engineering. A number of studies have been done to investigate the fracturing process of rock using an optical microscope (Mowar *et al.*, 1996; Homand *et al.*, 2000), or scanning electron microscope (SEM) (Wu *et al.*, 2000; Heidelbach *et al.*, 2000). These kinds of studies can provide the knowledge on the progressive failure of rock. However, the microscopic observations are limited to the surface of the specimen. To capture the failure process within a rock specimen and the induced microseismicities during failure process, acoustic emission (AE) detection (Young and Collins, 2001; Dobson *et al.*, 2004) or an X-ray computerized tomography (CT) scanning system have been used, (Kawakata *et al.*, 1999;

[1] Civil and Structural Engineering Department, The Hong Kong Polytechnic University, Hung Hom, Hong Kong, China. E-mail: cerwong@polyu.edu.hk; cektchau@polyu.edu.hk
[2] Liaoning Seismological Bureau, Shenyang, 110031, China. E-mail: mrjiao@yahoo.com

FENG *et al.*, 2004). Incorporation of AE or CT measurements with surface observations leads to a better understanding of the progressive failure of rocks under compression.

Recently, a number of numerical models were developed to capture the growth, interaction and coalescence of the pre-existing microcracks. Successful numerical models have been proposed to simulate rock failure study by considering the heterogeneity of materials using the Weibull distribution (TANG *et al.*, 1998; FANG and HARRISON, 2002; YUAN and HARRISON, 2004). In the Weibull distribution model for heterogeneity, both elastic moduli and strength parameters can be characterized by a set of Weibull parameters as (m and σ_o), where m and σ_o are the heterogeneous index and the average mechanical property (either modulus or strength) at the element level. Physically, a smaller m implies a more heterogeneous distribution of properties, and vice versa. One main limitation of such an approach is that the values of both m and σ_o are estimated purely by trial and error procedure in previous studies (LIN *et al.*, 2000; LIU *et al.*, 2004; TANG *et al.*, 2001) since there has been no theoretical basis for estimating these Weibull parameters. Recently, WONG *et al.* (2004, 2006) and JIAO *et al.* (2004) proposed a theoretical framework to determine m and σ_o quantitatively in terms of these microstructural attributes, such as grain size and microcrack statistics. Preliminary trials have been conducted by WONG *et al.* (2004) and JIAO *et al.* (2004) in relating the Weibull parameters to Yuen Long marbles found in Hong Kong. However, we subsequently realize that there are various ways of interpreting the microcrack statistics distribution in getting m and σ_o (WONG *et al.*, 2006).

The purpose of this paper is to propose another way of interpreting the Weibull parameters from the microcrack data. In particular, the main difference of the present microcrack data analysis from the previous one (WONG *et al.*, 2004, 2006; JIAO *et al.*, 2004) lies on the fact that the crack distribution is expressed in terms of crack number per total scanned area per unit thickness per grain size, whereas in our preliminary analysis crack number per unit volume and per unit length are assumed (see Fig. 1 in JIAO *et al.*, 2004). The rationale for the present change is that the number of crack count should be sensitive to the scanned volume as well as the grain size (note that, as to be shown later, the crack size distribution has to be in the unit of crack number per length to the power four).

For the present study, a particular type of marble from mainland China was used as an example. The average gain size of this Chinese marble is 0.52 mm, compared to 0.36 mm for Yuen Long marble used in our earlier study. For our laboratory study, acoustic emissions (AE) during the cracking process are monitored and recorded. A commercial code called "Rock Failure Process Analysis" ($RFPA^{2D}$) (TANG *et al.*, 1998) is used for numerical simulations in the present study, however we expect that similar results will be produced even if another similar numerical code is used. The main issue here is how to interpret the Weibull parameters from raw microcrack data, instead of the use of a particular finite element method (FEM). The $RFPA^{2D}$ is

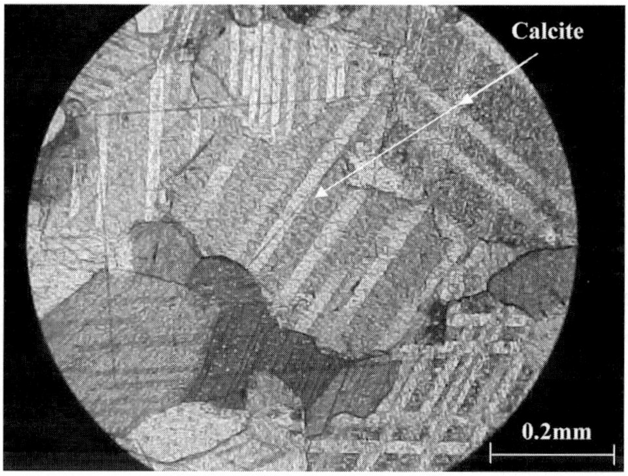

Figure 1
Marble specimen under polarizing microscope.

selected here because microseismicity (or acoustic emissions) are generated during the simulations, and they can be used directly to compare with our experiments on this Chinese marble.

Micromechanical-based Weibull Parameters

The determination of the Weibull parameters (m and σ_o) in terms of the micromechanical properties was first derived by Professor Teng-fong Wong of the State University of New York as presented in reference (WONG *et al.*, 2004, 2006). Thus, in the present section, only the essence is summarized.

Because of the grain-scale heterogeneity, the failure strength of polycrystalline rock can vary for the same type and size of rock specimen. To analyze the statistical variation of the bulk failure strength in heterogeneity material, WEIBULL (1951) adopted the statistics of the cumulative probability function (WEIBULL, 1951) to characterize the failure strength σ:

$$P(\sigma) = 1 - \exp\left[-\left(\frac{\sigma}{\sigma_o}\right)^m\right] \tag{1}$$

in terms of the Weibull parameters m and σ_o. Physically, this distribution given in (1) reflects the inevitable heterogeneity of solids (i.e., no solid is absolutely homogeneous). For rocks, the heterogeneity is caused by the probabilistic aspects of the microstructures, and particularly of the microcrack size distribution. It is well known that a brittle solid containing a longer crack is easier to load to failure than a solid containing a shorter crack LAWN (1993). Thus the length statistics of the long cracks can be described by an extreme value distribution of the Cauchy type: $g(a) = [q^*/a]^Z$, where a is half crack length, z is a microcrack size distribution index or called the fractal index, q^* is an external parameter depending on the types of rocks. To interpret this distribution, we assume that the cumulative probability function for the cracks in the elemental volume V_0 to have size longer than a_1 is given by $V_o \int_{a_1}^{\infty} g(a)\,da$ (ARGON, 1974; MCCLINTOCK and ARGON, 1966). For a solid of volume V that contains many elemental volumes V_0, if it is assumed that the failure in each elemental volume represents an independent event, then the cumulative probability for failure in V to arise from cracks with lengths longer than a_1 is as follows (ARGON, 1974; MCCLINTOCK and ARGON, 1966):

$$P = 1 - \exp\left[V \int_{a_1}^{\infty} g(a)da\right] = 1 - \exp\left[-\frac{(q*)^z}{(z-1)}\frac{V}{(a_1)^{z-1}}\right]$$
$$= 1 - \exp\left[-\frac{V}{V_o}\left(\frac{a'}{a_1}\right)^{z-1}\right] \tag{2}$$

where the parameter a' is defined by the following relation

$$(a')^{z-1}/V_o = (q*)^z/(z-1). \tag{3}$$

Again, V_o is the representative element volume. Considering that the crack size distribution can be related to fracture probability by applying fracture mechanics concepts, the probability that the stress intensity factor $K = Y\sigma\sqrt{a}$ ($Y = \sqrt{\pi}$ if we assume the crack is isolated and under uniform far-field stress) is equal to the critical value K_{IC} at stress less than σ is given by:

$$P(\sigma) = 1 - \exp\left[-\frac{V}{V_o}\left(\frac{\sqrt{a'}\,\sigma\,Y}{K_{Ic}}\right)^{2z-2}\right]. \tag{4}$$

By equating equations (1) and (4), the Weibull parameter m can be related to the crack size distribution as $m = 2z - 2$. The Weibull parameter σ_o that represents the mean strength for the volume V of an element in the finite-element grid can be evaluated as

$$\sigma_o = (V/V_o)^{-1/m} K_{Ic}/(\sqrt{a'}\,Y). \tag{5}$$

Therefore, the Weibull parameters (m, σ_o) can now be determined from micromechanical parameters. The actual evaluation of these parameters from microcrack data will be discussed next.

Experimental Studies

Physical Properties Testing of Marble and AE Measurement

The rock specimens used in this study are white marble purchased from mainland China. The mechanical and physical properties of this Chinese marble that have been tested include uniaxial compressive strength σ_c, tensile strength σ_t, Young's modulus E, Poission's ratio υ and density ρ. These tests were conducted in accordance with the testing standards of ASTM (1995a–d). Fractures toughness K_{IC} is determined in accordance with the standard testing method of ISRM (International Society for Rock Mechanics) using Chervon Bend (V-notch) specimen (ISRM, 1988). The results of these tests are listed in Table 1.

Microcrack Study for Marble

The micromechanical basis of Weibull parameters (m, σ_0) [$m = 2z - 2$ and $\sigma_o = (V/V_o)^{-1/m} K_{IC}/\sqrt{a}'Y$)] is determined by using the microcrack statistics. To investigate the statistics of microcracking in marbles, an optical microscope with fine resolutions has been used for microcrack counting. The magnification of the optical microscope used in this study is 100x. A thin section of rock specimen with a size of 45 mm $\times$ 25 mm $\times$ 3 mm is prepared according to the standard method of ISRM (Brown, 1981). The final thickness of the thin section is grounded to 0.03 mm. Ten areas (each area of 7.8 mm^2) are randomly selected on the thin section for measuring the distribution of crack length. Typically each scanned area under the optical microscopy covers a minimum of six grains (Fig. 1) and therefore both intragranular and transgranular cracks can be observed. The crack statistics is done based on a sample of more than 60 grains. The Chinese marble specimen consists of uniformly distributed recrystallized calcites. The average grain sizes (D) and crack length $2c$ for the marble are listed in Table 2. Most of the pre-existing microcracks are intragranular, but there are also some transgranular cracks.

The sample measurement is covered over a total area A = 10 $\times$ 7.8 mm^2, and the half crack length data are mainly grouped into a length interval of 0.025 mm. The microcrack data are fitted to a power law to determine the parameters z and $(q^*)^z$. It

Table 1

Physical properties of marble

σ_c (MPa)	σ_t (MPa)	E (GPa)	υ	K_{IC} (MPa m$^{1/2}$)	ρ (kg/m^3)
65.74	5.06	47.74	0.289	0.865	2713

Table 2

Microcrack statistics and micromechanical parameters

Grain size D (mm)	Average crack length $2C$ (mm)	z [Fig. 1b]	m (m = 2z – 2)	$(q^*)^z$ (mm^{z-4}) [Fig. 1b]	V_o ($V_o = \pi \times D^2/4 \times t$) (mm^3)	Characteristic crack length a' (3) (mm)	K_{Ic} of calcite (Potyondy *et al.*, 1996) (MPa m$^{1/2}$)	σ_0 (5) (MPa)
0.52	0.123	1.878	1.76	0.0017	0.229	0.00015	0.2	290.2

is noted from eq. (3) $[(a')^{z-1}/V_o = (q*)^z/(z-1)]$ that the unit for g(a) $[g(a) = (q^*/a)^z]$ is mm^{-4}. Therefore, it can be inferred that the values of $(q^*)^z$ is in unit of $mm^{(z-4)}$. As mentioned in the Introduction, there is a flexibility of choosing the dimension for yielding the resulting unit of mm^{-4}. In an attempt to improve our earlier studies, the crack length distribution g(a) is evaluated as the number of microcracks (N) observed from a length interval over the total volume (measured area times uniform thickness $A \times t$; A $= 10 \times 7.8$ mm^2; $t = 1$) and the average grain size D [i.e., $g(a) = N(a)/(10 \times 7.8 \times 1 \times D)$]. This choice is based on the rationale that the number of counted cracks should be scaled by the actual scanned area, that the microcracks are mainly two-dimensional in the thin section, and that the number of microcracks is controlled by grain size. However, we should emphasize that the current combination is only one possible choice in setting up $g(a)$, and it is by no means unique. The curve fitting and Weibull parameters inferred from microcrack statistics is shown in Figure 2 and listed in Table 2.

The z parameter is determined by the power-law function of $g(a) = 0.0017x^{-1.8776}$ (see Fig. 2). Therefore, z is equal to 1.8776 and thus $(q^*)^z = 0.0017$ mm^{z-4}, respectively. Consequently, m ($m = 2z - 2$) value is then calculated as 1.76. The characteristic crack length a' is calculated by (3) as $(a')^{z-1}/V_o = (q^*)^z/(z-1)$. In a two-dimensional model, the ratio elemental volume is equal to $V_o = \pi \times D^2/4 \times t$. The characteristic crack length a' is calculated of 0.00015 mm. The parameter of σ_o is determined by $\sigma_o = (V/V_o)^{-1/m} K_{IC}/\sqrt{a'}Y$. The parameter Y is a geometric constant and is chosen as $\sqrt{\pi}$ by neglecting the crack-crack interactions. Since the average crack length is smaller than the average grain size of calcite, fracture toughness K_{IC} of calcite mineral of 0.2 MPa $m^{1/2}$ is selected (Atkinson, 1987). It is further assumed that the size of the finite-element grid is the same as the size of the grain ($V/V_o = 1$).

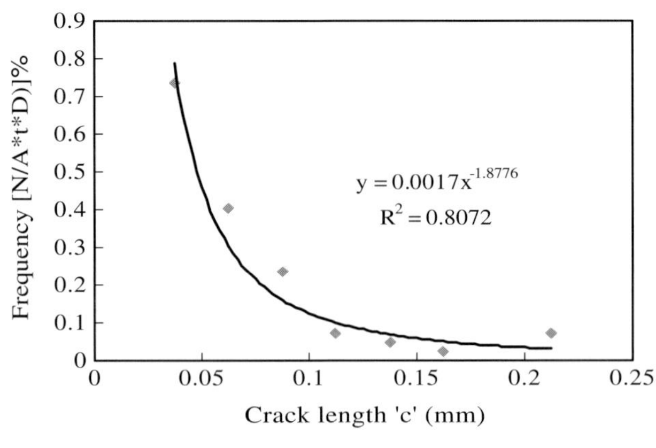

Figure 2
Frequency of crack vs. half size of crack length. The frequency of crack is defined as a number of microcracks (N) over total measured volume ($A \times t = 10 \times 7.8 \times 1$) and the average grain size (D).

With the known parameters, the average elemental strength σ_0 is calculated of 290.2 MPa from (5).

AE Measurement

The uniaxial compression test was conducted using a servo-control MTS compression machine with the loading rate of 0.0005 mm/s at the Rock Mechanics Laboratory of Hong Kong University. A rectangular specimen (50 mm × 25 mm × 110 mm) was compressed to failure with the associated acoustic emissions recorded by four sensors (5 mm diameter) with a frequency of 480 kHz. These sensors were mounted onto the rock surface by epoxy resin. The detected AE wave signals were pre-amplified and then fed into transient recorders at a sampling rate of 1 MHz. The noise threshold is set at 40 dB. The AE signals are recorded by using the MISTRAS-2001 AE recording software and the data are transferred to a computer. Figure 3(a) shows the stress-strain curve of the tested marble versus the AE events plotted on Fig. 3(b). The uniaxial compressive strength of the Chinese marble is 58.51 MPa (Fig. 3(a)) with a mixed shear-tensile mode of final failure, as shown in Figure 3(c). The maximum number of AE events at failure is 534.

A total of 1289 AE events were recorded by the sensors, but most of the released energy was released during a few events near the peak strength. According to MOGI (1962), the seismic pattern of the Chinese marble is a "main-shock type". In particular, MOGI (1962) studied fracture phenomena for various brittle materials with different degrees of heterogeneity. He found that the main-shock type seismic pattern occurred in relatively homogeneous rocks. Our finding agrees with his general conclusion. To compare our experimental finding with our micromechanical-based interpretation of Weibull parameters of heterogeneity, a finite-element code RFPA has been used to simulate the uniaxial compressive strength, the failure patterns and the associated AE events. The results of these numerical simulations will be summarized in the next section.

Numerical Simulation

RFPA^{2D} Model

The code RFPA^{2D} (Rock Failure Process Analysis) (TANG *et al.*, 1998) is based on an elastic finite-element model incorporated with a damage criterion to judge the failure of individual elements, and it has been successfully used to simulate the deformation and failure process of brittle solids (LIN *et al.*, 2000; LIU *et al.*, 2004; TANG *et al.*, 2001). A brief outline of RFPA^{2D} is presented here.

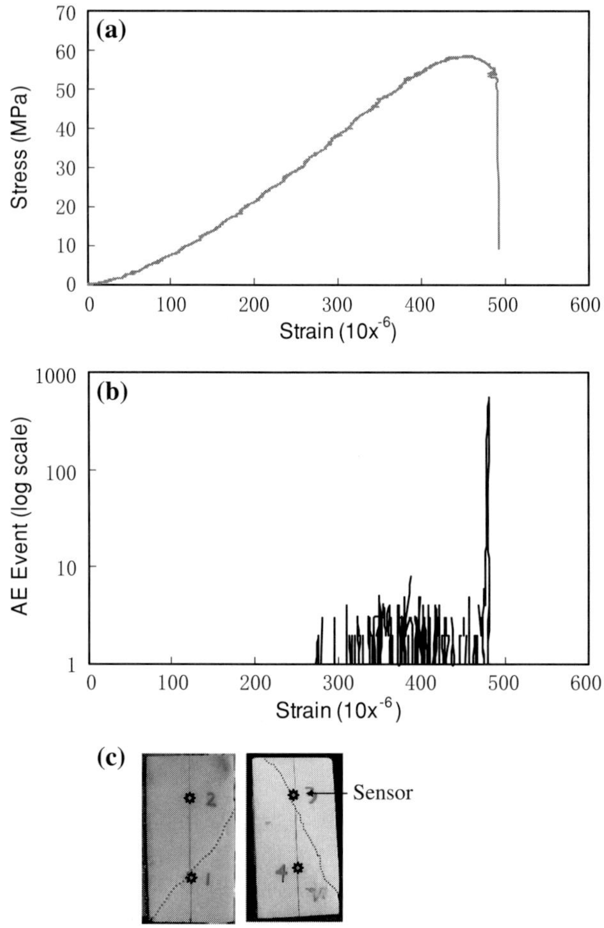

Figure 3
Experimental study on marble specimen with AE measurement. (a) Stress strain curve of marble under uniaxial compression, (b) AE events vs strain of specimen under compression, (c) failure mode of the specimen.

RFPA^{2D} is a progressive "elastic-damage model." Using elastic stress analysis, stress and deformation in each element of a sample is calculated for every loading step. When the stress in some elements within the specimen reaches a critical strength, these elements will fail either in shear or in tension. A Coulomb criterion with a tension cut-off is used as a failure criterion to determine whether each element fails in shear or in tension. When each element fails, an AE event is reported. To overcome the numerical instabilities, the failed elements are given with a small, but non-zero, elastic modulus. Thus, in reality the failed elements were neither removed from the mesh nor from the stress analysis. The elastic modulus of element reduces to $(1-D_e)E_o$ from E_o after failure, where $D_e = 1 - \sigma/(E_o\varepsilon)$ depending on the current reduced stress level (normally taken as 1/10 of the peak strength) and strain at the element. The Poisson's ratio v for all the elements is assumed to be constant throughout the deformation. When an element fails, this element will not resist pressure because stress will redistribute around this damaged element instantaneously. Within some elements at other positions, where stress increases due to stress redistribution, the local stress may exceed the critical value and hence further failure will result. The iterative process of checking failure is repeated until no more failure at other elements occurs. Then, the next increment of loading can be applied and the same iterative procedure can be used to trace the failed elements in the next loading step. Once the global load-bearing capacity of the specimen drops, the strength can be obtained. In this model, owing to a constant Poisson's ratio that is assumed, no anisotropy failure occurs in a element level. But for global damage, this model has anisotropy failure. The friction angle is not changed after failure of a residual friction angle is assumed in the numerical model.

The two dimensional FEM mesh for a specimen of 50 mm × 100 mm (width × height) is a 92 × 184 mesh (or 16,928 elements); that is, element size equal to the grain size of the Chinese marble. The calibrated input parameters for the numerical simulations are listed in Table 3. The value of σ_0 and m is the result of microcracks statistics obtained from Table 2. The value of E, υ and σ_c/σ_t is the experimental result obtained from Table 1. The specimen is assumed to undergo plane strain deformation. An external displacement is applied at a constant rate of 0.0005 mm/step in the axial direction.

Seismic Events

The RPFA^{2D} code can simulate the damage of initiation and propagation causing seismic energy release during the unstable failure of brittle rock (TANG and KAISER, 1998). It is assumed that the radiated energy is equal to the energy stored in the element before failure is triggered. Due to the heterogeneity of rock properties, seismic quiescence may occasionally occur with the nucleation zone. The cumulative seismic damage calculated based on the seismic event rate form the simulation can be used as a damage parameter to describe the damage evolution.

The energy e_f released by the failure of each element can be calculated by the peak strength of the failed element.

$$e_f = \frac{1}{2E_f}\sigma_{cf}^2 . v_f \tag{6}$$

Table 3

Material parameters for numerical simulation

Element number	σ_0 (MPa)	m	E (GPa)	υ	σ_c/σ_t	ϕ (°)
(50/0.54 × 100/0.54) = 92 × 184 = 16,928	290	1.76	47.74	0.29	10	30

where E_f is the elastic modulus for the element, σ_{cf} is the peak strength of the failed element and v_f is the volume of the individual failed element. The cumulative seismic energy can then be obtained by

$$e = \sum e_f = \sum \frac{1}{2E_f} \sigma_{cf}^2 . v_f = \frac{v_f}{2} \sum \frac{\sigma_{cf}^2}{E_f}. \tag{7}$$

The magnitude, m_f of an individual microseismic event or the magnitude M_f of a seismic event cluster can then be obtained by the following equations:

$$m_f = \log(e_f) + C, \tag{8a}$$

and

$$M_f = \log\left(\sum_{i=1}^{n} e_{fi}\right) + C, \tag{8b}$$

where C is a constant and i is the number of events in an event cluster. By using the calculated results form Equations (6) to (8), a frequency-magnitude relation can be obtained for each step and for both microseismic events and seismic events of cluster.

The results of the numerical simulations and the comparison with the experiment is presented in the next section.

Discussions

Using the micromechanical parameters, numerical simulations have been conducted and the results are given in Figure 4. More specifically, stress-strain curve is given in Figure 4(a) and the AE events associated with elemental failures are plotted in Figure 4(b). To illustrate the process of progressive failure, both the failed elements and the associated AE events were shown in Figure 5. The simulated uniaxial compressive strength is 48.83 MPa which is about 16% lower than that of the experiment (58.51 MPa). Furthermore in Figure 4(a), it is observed that there is still residual stress of 10–15 MPa after peak stress, although such residual stress is not observed in the experiment which is under compression stress without confining pressure (Fig. 3(a)). The residual stress existing in the modelling specimen is due to the elements which are confined by the adjacent elements. This confinement can be treated as a kind of confining pressure to the elements within the specimen. Thus, residual strength may exist. In addition, it is observed that the stiffness increases at the beginning of the loading (Fig. 3(a)), while a softening regime appears before the brittle failure. However, these two features are not captured by the numerical model. Actually, the marble specimen contains pores and cracks. The increased stiffness at the beginning of the load may be due to the closing of the microcracks and pores. The stress-strain curve from plastic deformation changes to elastic deformation.

However for the modeling specimen, the heterogeneity of rock is generated by using the Weibull distribution. Strong elements and weaker elements are generated according to the m value (heterogeneity index). No real cracked element is generated in the modeling specimen. Thus, no cracks or pores close the process in the modeling specimen at the beginning of the applied loading. For the softening regime appearing before the brittle failure of the experiment, as it is known before the brittle failure some of the cracking may start jointing together. Thus, stress would not increase and the softening regime would appear. The size of the softening regime depends on the rate of crack growth and coalescence process. However, for the modeling specimen, the softening regime depends on the number of failure elements. It was observed that a small amount of elements fails before the brittle failure, and stress drops slightly. After two steps displacement, a large number of elements fail and form a rapid drop of stress-strain curve. The modeling result has a softening regime but appears only within three steps (Fig. 4(a)). Although there are discrepancies in the strength and

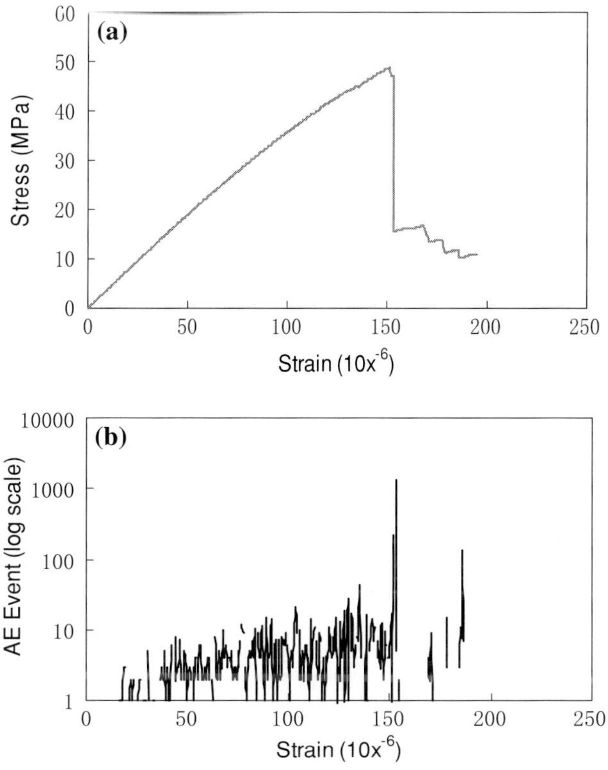

Figure 4

Numerical simulation using the micromechanical Weibull parameters. (a) Stress strain curve, (b) AE counts.

stress-strain curve, the general and failure patterns for both experiment and simulation are similar (compare Figs. 3(a) and 4(a)). In particular, the stress-strain curves drop rapidly after the peak for both experiment and simulation, and this characterizes the behavior of a brittle rock.

For AE pattern, both experiments (Fig. 3b) and numerical simulation (Fig. 4b) show a main-shock type seismic pattern. In the experiment of Fig. 3(b), the maximum AE event appears before the brittle failure. It is believed some of the AE signal data loss occurs during the failure process, because there are too many AE signals induced during the failure process. Thus, some of data are overlapped and cause the data loss. For the numerical model, the AE signal can be captured before and after the failure.

The failure process of the specimen is shown in Figure 5. The first row shows the stress distribution of the specimen and the second row shows the AE location under compression. The black color in the specimens of the first row represents the failed element. The white and red colors in the specimens of the second row represent the failure nature of each element under compression and tension, respectively. The radius of each circle is proportional to the stress drop of the element. It is noted that both tensile and compressive failure are induced under compressive loading. The ultimate failure is mainly in the tensile mode but with macroscopic shear coalescence. The failure pattern of numerical simulation agrees well with the experiment (Fig. 3(c)).

It is moreover to confirm that there are various ways to formulate a microcrack size distribution in relating to the Weibull parameters to simulate the progressive

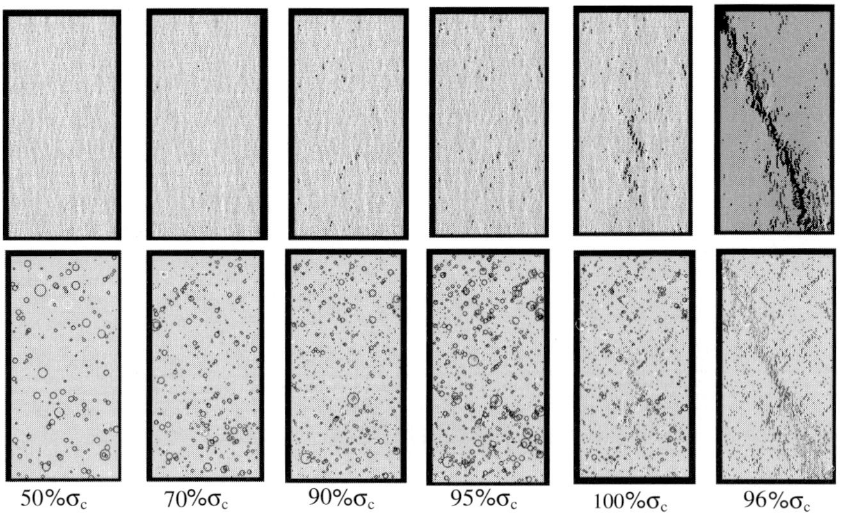

Figure 5

Failure process under compression (numerical simulation). The σ_c represents the peak strength. The first row shows stress distribution. The second row shows the AE location where the white and red color represents the failure element under compression and tension, respectively.

failures of heterogeneous rock specimens under compression. It is worthwhile for us to use the micromechanical parameters to apply to the other numerical models, which were developed to capture the progressive failure of rock. For example, the discrete element method (DEM, (CUNDALL and STRACK, 1979). This model can produce many features of rock such as fracturing, acoustic emission, damage accumulation, etc. (POTYONDY *et al.*, 1996). The strength property determined by using the statistic of crack distribution can be used as microproperties of strength parameter for the particles.

Conclusions

This study presents a realistic numerical simulation using micromechanical parameters. The micromechanical parameters (or Weibull parameters) related to grain size and crack statistics are input into the RFPA^{2D} (Rock Failure Process Analysis code) to generate the heterogeneity of natural rocks and to study the failure process of rock. A compression test was conducted on a specimen of marble found in mainland China, and the acoustic emissions (AE) associated with cracking are recorded. Based on the numerical simulation and experimental studies, although there are discrepancies in the strength and stress-strain curve, the general and failure patterns for both experiment and simulation are similar in which a rapid drop of stress is observed after the peak stress. The AE pattern in both experiment and simulation are the main-shock type. The observed and predicted failure modes are similar. The failure process is mainly tensile but with macroscopic shear coalescence. This paper further confirms that the micromechanical parameters approach for numerical simulation is a more realistic method to simulate the failure process of rock.

The present paper proposes to plot the crack statistics using the crack count per scanned volume per grain size to yield the required Weibull parameters. It is found that the crack statistics will change with the scanned area and the length interval. Thus, it is of value to carry out a detailed examination to interpret the crack data. In addition, more carefully designed experiments will also be conducted on various kinds of rocks such as granite and sandstone with various grain sizes. It is of interest to explore the extent to which the influences on compressive failure behaviour are similar to the observations in this study. Furthermore, it is beneficial for us to use the micromechanical parameters to apply to the other numerical models which were developed to capture the progressive failure of rock.

Acknowledgments

The study was supported by Hong Kong Polytechnic University to RHC Wong (A-PD49 and A-PE31). We are grateful to Professor Teng-fong Wong, the Professor of State University of New York at Stony Brook, for formulating the relations between microcracks and Weibull parameters and on how to interpret microcrack data, and to Prof. L. G. Tham of HKU for allowing us to use the facilities at their rock mechanics laboratory.

References

Argon, A.S. (1974), *Statistical aspects of fracture*, Composite Mater. *5*, 153–189.

ASTM (1995a), *Standard test method for unconfined compressive strength of intact rock core specimens*. Test Designation D2938-95 Annual book of ASYM standards, American Society for Testing and Materials, Philadelphia (*14*) 02.

ASTM (1995b), *Standard test method for elastic moduli of undrained intact rock core specimens in biaxial compression*. Test Designation D3148-95 Annual book of ASYM standards, American Society for Testing and Materials, Philadelphia. (*14*) 02.

ASTM (1995c), *Standard test method for splitting tensile strength of intact rock core specimens*. Test Designation D3967-95a Annual book of ASYM standards, American Society for Testing and Materials, Philadelphia (*14*) 02.

ASTM (1995d), *Standard test method for triaxial compressive strength of undrained rock core specimens without pore pressure measurements*. Test Designation D2664-95 Annual book of ASYM standards, American Society for Testing and Materials, Philadelphia (*14*) 02.

Atkinson, B.K. *Fracture Mechanics of Rock,* Academic Press Geology Series (Academic Press Inc., London Ltd., (1987)).

Brown, E.T. (1981), *Rock Characterization Testing and Monitoring, ISRM Suggested Methods*: *Suggested Method for Petrographic Description of Rocks,* International Society for Rock Mechanic, pp. 73–77.

Cundall, P.A. and Strack, O.D.L. (1979), *A discrete numerical model for granular assemblies,* Geotechnique *29*, 47–65.

Dobson, D.P., Philip, G., Meredith, S., and Boon, A. (2004), *Detection and analysis of microseismicity in multi-anvil experiments,* Phys. Earth Planet. Interiors 143–144, 337–346.

Fang, Z. and Harrison, J.P. (2002), *Development of a local degradation approach to the modeling of brittle fracture in heterogeneous rocks*, Int. J. Rock Mech. Min. Sci. *39*, 443–457.

Feng, X.T., Chen, S.L., and Zhou, H. (2004), *Real-time computerized tomography* (*CT*) experiments on sandstone damage evolution during triaxial compression with chemical corrosion, Int. J. Rock Mech. Min. Sci. *41*, 181–192.

Heidelbach, F., Kunze, K., and Wenk, H.R. (2000), *Texture analysis of a recrystallized quartzite using electron diffraction in the scanning electron microscope,* J. Struct. Geology *22*(1), 91–104.

Homand, F.D., Hoxha, T., Belem, M., Pons, N., and Hoteit, N. (2000), *Geometric analysis of damaged microcracking in granites*, Mech. Mater. *32*, 361–376.

ISRM (1988), *Suggested Method for determining fracture toughness of rock* (*F. Ouchterlony, Working Group Coordinator*). Int. J. of Rock Mech. Min. Sci. Geomech. Abstr. *25*, 71–96.

Jiao, M.R., Wong, R.H.C., Wong, T.F., Chau, K.T., and Tang, C.A. (2004), *Numerical simulation of the influence of grain size on the progressive development of brittle failure in Yuen Long marbles*, Key Eng. Mater. 261–263, 1511–1516.

Kawakata, H., Cho, A., Kiyama, T., Yanagidani, T., Kusunose, K., and Shimada, M. (1999), *Three-dimensional observations of faulting process in Westerly granite under uniaxial and triaxial conditions by X-ray CT scan*, Tectonophysics *313*, 293–305.

Lawn, B. *Fracture of Brittle Solids*, Second Edition (Cambridge University Press, Cambridge, (1993)) 378 pp.

Lin, P., Wong, R.H.C., Chau, K.T., and Tang, C.A. (2000), *Multi-crack coalescence in rock-like material under uniaxial and biaxial loading*, Key Engin. Mater. 183–187, 809–814.

Liu, H.Y., Kou, S.Q., Lindqvist, P.A., and Tang, C.A. (2004), *Numerical studies on the failure process and associated microseismicity in rock under triaxial compression*, Tectonophysics *384*, 149–174.

MOGI, K. (1962), *Study of elastic shocks caused by the fracture of heterogeneous materials and their relation to earthquake phenomena,* Bull. Earthq. Res. Inst., University Tokyo *40*, 125–173.

MCCLINTOCK, F.A. and ARGON, A.S. *Mechanical Behavior of Materials*. Addison-Wesley, Reading, MA., (1966) 504–508. pp.

MOWAR, S.M., ZAMAN, D., STEARNS, W., and ROEGIERS, J.-C. (1996), *Micro-mechanisms of pore collapse in limestone*, J. Petroleum Sci. Eng. *15*, 221–235.

POTYONDY, D.O., CUNDALL, P.A., and LEE, C. (1996), *Modeling rock using bonded assemblies of circular particles*, Proc. North Am. Rock Mech. Symp. *2*, 1937–1944.

TANG, C.A. WANG, W.T., FU, Y.F., and XU, X.H. (1998), *A new approach to numerical method of modeling geological processes and rock engineering problems-continuum to discontinuum and linearity to nonlineartiy*, Eng. Geology *49*(3–4), 207–214.

TANG, C.A. and KAISER, P.K. (1998), *Numerical simulation of cumulative damage and seismic energy release during brittle rock failure - Part I: Fundamentals*, Int. J. Rock Mech. Min. Sci. *35*(2) 113–121.

TANG, C.A., LIN, P. WONG, R.H.C., and CHAU, K.T. (2001), *Analysis of crack coalescence in rock-like materials containing three flaws-Part II: Numerical approach*, Int. J. Rock Mech. Min. Sci. *38*, 925–939.

WEIBULL, W. (1951), *A statistical distribution function of wide applicability*, J. Appl. Mech., *18*, 293–297.

WONG, T-F., WONG, R.H.C., JIAO, M.R., CHAU, K.T., and TANG, C.A., (2004), *Micro-mechanics and rock failure process analysis*, Key Eng. Mater. 261–263, 39–44.

WONG, T-F., WONG, R.H.C., CHAU, K.T., and TANG, C.A. (2005), *Microcrack statistics, Weibull distribution and micromechanical modeling of compressive failure in rock*, J. Mech. Mater. *38*, 664–681.

WU, X.Y., BAUD, P., and WONG, T.F. (2000), *Micromechanics of compressive failure and spatial evolution of anisotropic damage in Darley Dale sandstone*, Int. J. Rock Mech. Min. Sci. *37*, 143–160.

YOUNG, R.P. and COLLINS, D. S. (2001), *Seismic studies of rock fracture at the Underground Research Laboratory, Canada*, Int. J. Rock Mech. Min. Sci. *38*(6), 787–799.

YUAN, S.C. and HARRISON, J.P. (2004), *Numerical modelling of progressive damage and associated fluid flow using a hydro-mechanical local degradation approach,* Int. J. Rock Mech. Min. Sci. *41*(3), paper 2A 01.

(Received December 28, 2004, revised July 8, 2005, accepted August 29, 2005)
Published Online First: September 2, 2006

Pure appl. geophys. 163 (2006) 1803–1817
0033–4553/06/091803–15
DOI 10.1007/s00024-006-0098-y

Pure and Applied Geophysics

3-D Simulation of Tectonic Loading at Convergent Plate Boundary Zones: Internal Stress Fields in Northeast Japan

Chihiro Hashimoto and Mitsuhiro Matsu'ura

Abstract—The evidence of east-west compression in northeast Japan has been reported by many investigators on the basis of geodetic, geologic and geomorphic data, but its origin still remains far from understood. In the present study we have proposed a mechanical model of tectonic loading at convergent plate boundary zones, and demonstrated its validity through the numerical simulation of internal stress fields in northeast Japan with realistic 3-D geometry of plate interfaces. At convergent plate boundary zones, in general, a part of plate convergence is consumed by steady slip along plate interfaces, and the remaining part by inelastic deformation (seismic faulting, aseismic faulting, and active folding) of overriding plates. Such a plate boundary process to be called "partial collision" can be quantitatively described by introducing a collision rate defined as $c = 1 -$ *steady slip rate at plate interfaces/plate convergence rate*. By this definition, we can simply represent the mechanical process of partial collision, which includes total subduction ($c = 0$) and total collision ($c = 1$) as two extreme cases, in terms of steady slip rates at plate interfaces. On the basis of elastic dislocation theory, first, we numerically computed the internal stress fields in northeast Japan produced by the total subduction of the Pacific plate beneath the North American plate, however the computed stress pattern was opposite in sense to observations. Then, we computed the internal stress fields by taking $c = 0.1$ on average, and succeeded in reproducing the observed east-west compression in northeast Japan. This indicates that the concept of partial collision is essential to understand the mechanism of intraplate tectonic loading.

Key words: Plate convergence, partial collision, collision rate, tectonic loading, elastic dislocation theory.

1. Introduction

Plate convergence brings about diverse crustal activities such as earthquakes, volcanism, and mountain building. The essential cause of these crustal activities is in physicochemical interaction at plate boundaries. In the present study we focus on the long-term stress build-up due to mechanical interaction at plate boundaries. In and around Japan, where the four plates of Pacific, North American, Philippine Sea and Eurasian interact with each other, large earthquakes occur repeatedly not only at

Department of Earth and Planetary Science, The University of Tokyo, 7-3-1 Hongo, Bunkyo-ku, Tokyo 113-0033, Japan. E-mail: hashi@eps.s.u-tokyo.ac.jp; matsuura@eps.s.u-tokyo.ac.jp plate interfaces but also at intraplate major active faults. As for interplate earthquake generation, physics-based 3-D computer simulation models have been developed (e.g., HASHIMOTO and MATSU'URA, 2000, 2002; FUKUYAMA *et al.*, 2002; MATSU'URA, 2004). In these models earthquake generation is rationally represented as a stress accumulation-release process at strongly coupled regions on a plate interface, driven by relative plate motion. On the other hand, the physical process of intraplate earthquake generation, especially the mechanism of intraplate tectonic loading, still remains far from understood. In modelling the process of tectonic loading at convergent plate boundary zones, 3-D geometry of plate interfaces is essential, because mechanical interaction at plate interfaces strongly depend on their geometry. Recently, HASHIMOTO *et al.* (2004) constructed a realistic 3-D structure model of plate interfaces in and around Japan by applying an inversion technique to ISC hypocenter data. With this structure model, given relative velocities among the plates, we can now numerically simulate internal stress fields in and around Japan.

The northeast Japan arc lies to the west of the Japan trench, where the Pacific plate descends beneath the North American plate at a rate of about 8 cm/yr. As shown in Figure 1, the tectonic structure of northeast Japan is characterized by the thrust faults and folds with their strikes parallel to North-South direction (e.g., RESEARCH GROUP FOR ACTIVE FAULTS OF JAPAN, 1991; SATO, 1994; TAJIKARA, 2003). The N-S trending thrust faults and folds in northeast Japan suggest the long-term duration of East-West compression there. Actually, many investigators have reported the E-W compression in northeast Japan on the basis of recent GPS data (e.g., KATO *et al.*, 1998; SAGIYA *et al.*, 2000), triangulation and trilateration data for the last 100 years (e.g., Geographical Survey Institute, 1996), and historic intraplate earthquake data and Quaternary active fault data (e.g., WESNOUSKY *et al.*, 1982; NOHARA, 2000). The estimated compressive strain rates are of the order of 10^{-7}/yr from geodetic observations and 10^{-8}/yr from geologic and geomorphic observations. Here, it would be reasonable that the short-term strain rates are larger than the long-term strain rates, since the effects of interseismic strain accumulation related to interplate earthquake cycles cannot be cancelled over the geodetic time-scale ($< 10^2$ yr).

In the present study, we propose a mechanical model of tectonic loading at convergent plate boundary zones, and examine the validity of the proposed model through the computation of internal stress fields in northeast Japan. In Section 2, we show a basic idea for representing mechanical interaction at convergent plate boundaries, and develop a 3-D simulation model for tectonic loading in and around Japan. In modelling we consider the three different types of mechanical interaction; that is, (1) frictional resistance to steady-state plate subduction, (2) reaction to steady plate bending, and (3) collision between adjacent plates. In Sections 3 and 4, we evaluate the effects of the first and the second types of mechanical interaction, respectively. In Section 5, we introduce the concept of partial collision, and define a collision rate to treat the second and the third types of mechanical interaction in

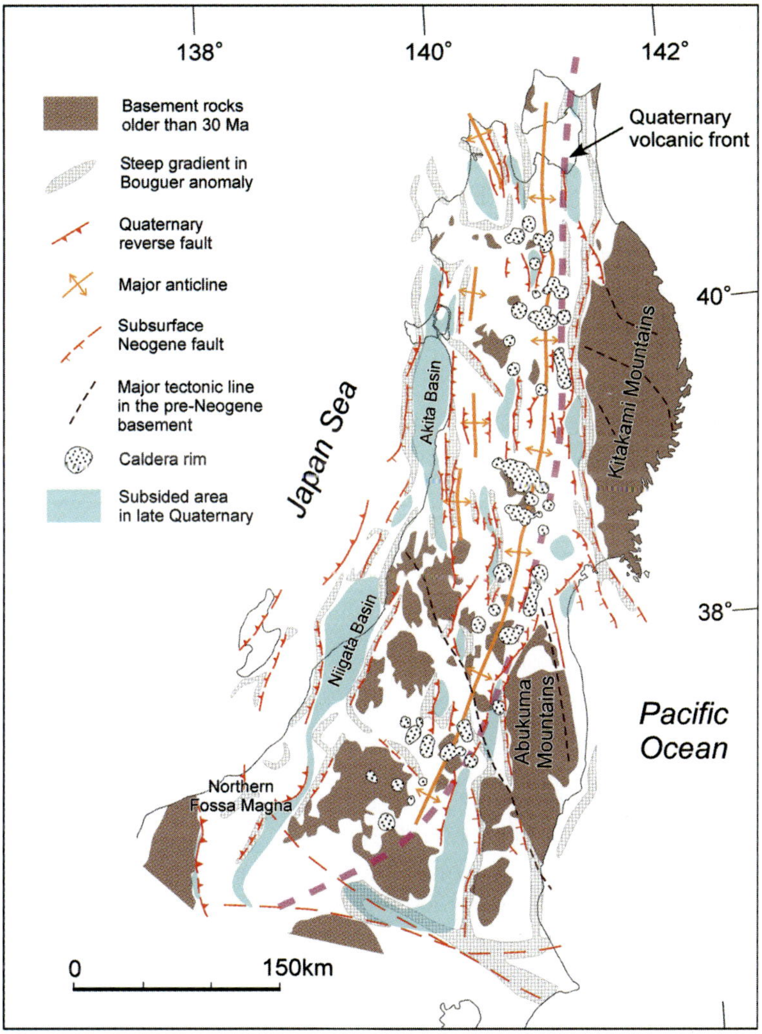

Figure 1
The geologic structure of northeast Japan (Tajikara, 2003, modified from Sato, 1994).

a unified way. Then, we numerically simulate the stress field due to partial collision in northeast Japan. In Section 6, we discuss the mechanism of tectonic loading due to partial collision at convergent plate boundary zones.

2. 3-D Modelling of Long-term Tectonic Loading at Convergent Plate Boundary Zones

At subduction zones oceanic plates descend beneath the overriding plates at constant rates on a long-term average. The occurrence of interplate earthquakes can be regarded as its perturbation. Mechanical interaction at a plate interface can be rationally represented by the increase of tangential displacement discontinuity across the interface (MATSU'URA and SATO, 1989). The displacement discontinuity (fault slip) is mathematically equivalent to the force system of a double couple without moment (MARUYAMA, 1963; BURRIGE and KNOPOFF, 1964), which has no net force and no net torque. Such a property must be satisfied for any force system acting on plate interfaces, which is produced by a dynamic process within the earth. In this Section, on the basis of such a general idea, we develop a 3-D simulation model for long-term tectonic loading at convergent plate boundary zones.

2.1 Mathematical Expressions of Internal Stress Fields due to Steady Plate Subduction

We model the lithosphere-asthenosphere system by an elastic surface layer overlying a Maxwellian viscoelastic half-space. The constitutive equation of the elastic surface layer is given by

$$\sigma_{ij} = \lambda^{(1)} \varepsilon_{kk} \delta_{ij} + 2\mu^{(1)} \varepsilon_{ij} \tag{1}$$

and that of the underlying viscoelastic half-space by

$$\dot{\sigma}_{ij} + \frac{\mu^{(2)}}{\eta} \left(\sigma_{ij} - \frac{1}{3} \sigma_{kk} \delta_{ij} \right) = \lambda^{(2)} \dot{\varepsilon}_{kk} \delta_{ij} + 2\mu^{(2)} \dot{\varepsilon}_{ij}. \tag{2}$$

Here, σ_{ij}, ε_{ij}, and δ_{ij} are the stress tensor, the strain tensor, and the unit diagonal tensor, respectively. The dot indicates differentiation with respect to time. $\lambda^{(i)}$ and $\mu^{(i)}$ $(i = 1, 2)$ denote the Lamé elastic constants of each medium, and η is the viscosity of the underlying half-space. The standard values of structural parameters used for computation are listed in Table 1.

We introduce a 3-D curved interface Σ that divides the elastic lithosphere into two plates. Denoting the *ij*-component of internal stress at a point $\boldsymbol{x}$ and time t due to a unit step slip at a point $\boldsymbol{\xi}$ and time τ by $T_{ij}(\boldsymbol{x}, t; \boldsymbol{\xi}, \tau)$, we can calculate the viscoelastic internal stress field due to arbitrary fault slip w on the plate interface by using the technique of hereditary integral as

Table 1

Structural parameters used in numerical simulations

	h [km]	ρ [kg/m^3]	λ [GPa]	μ [GPa]	η [Pa s]
Lithosphere	40	3000	40	40	∞
Asthenosphere	∞	3400	90	60	5×10^{18}

h: thickness, ρ: density, λ and μ: Lamé elastic constants, η: viscosity

$$\sigma_{ij}(\boldsymbol{x},t) = \int_{-\infty}^{t}\int_{\Sigma} T_{ij}(\boldsymbol{x},t-\tau;\xi,0)\,\frac{\partial w(\xi,\tau)}{\partial\tau}\,d\xi d\tau. \tag{3}$$

The concrete expressions of the slip-response functions T_{ij} for the gravitating elastic-viscoelastic layered structure model are given in MATSU'URA *et al.* (1981) and FUKAHATA and MATSU'URA (2005).

Now we decompose the fault slip w into the steady slip at a plate convergence rate v_{pl} and its perturbation Δw as

$$w(\xi,\tau) = v_{pl}(\xi)\,\tau + \Delta w(\xi,\tau). \tag{4}$$

Substituting this expression into Eq. (3), we obtain

$$\sigma_{ij}(\boldsymbol{x},t) = \int_{0}^{t}\int_{\Sigma} T_{ij}(\boldsymbol{x},t-\tau;\xi,0)v_{pl}(\xi)d\xi d\tau + \int_{0}^{t}\int_{\Sigma} T_{ij}(\boldsymbol{x},t-\tau;\xi,0)\frac{\partial\Delta w(\xi,\tau)}{\partial\tau}d\xi d\tau. \tag{5}$$

Here, we supposed that the slip motion started at $t = 0$ and the system has been in a state of deviatric stress free before that. The first and the second terms on the right-hand side of Eq. (5) represent the contributions from the steady slip and its perturbation, respectively. If we focus on the long-term stress increase rates, we can neglect the second perturbation term, since it does not bring about secular stress changes. Furthermore, assuming a perfectly elastic lithosphere, we may regard T_{ij} to be constant in time for sufficiently large t. Thus, the stress increase rates at a point $\boldsymbol{x}$ due to steady plate subduction are given by

$$\dot{\sigma}_{ij}(\boldsymbol{x}) = \int_{\Sigma} T_{ij}(\boldsymbol{x},\infty;\xi,0)\,v_{pl}(\xi)\,d\xi. \tag{6}$$

Here, the dot indicates differentiation with respect to time t. It should be noted that the slip response functions T_{ij} become zero at $t = \infty$, if the source is located in the viscoelastic half-space (MATSU'URA and SATO, 1989). Hence, in the case of steady plate subduction, we may consider only the effects of slip motion at the interface between the elastic plates.

2.2 A Standard Model for 3-D Plate Interface Geometry in and Around Japan

Given the geometry of Σ and the steady slip rates v_{pl} over the whole plate interface, we can compute the increase rates of internal stress due to steady plate subduction by using Eq. (6). For the geometry of Σ, recently, HASHIMOTO *et al.* (2004) has precisely modelled the 3-D geometry of plate interfaces in and around Japan on the basis of the topography of ocean floors and the hypocenter distribution of earthquakes (Fig. 2). The modelled region extends from 125 °E to 155 °E in

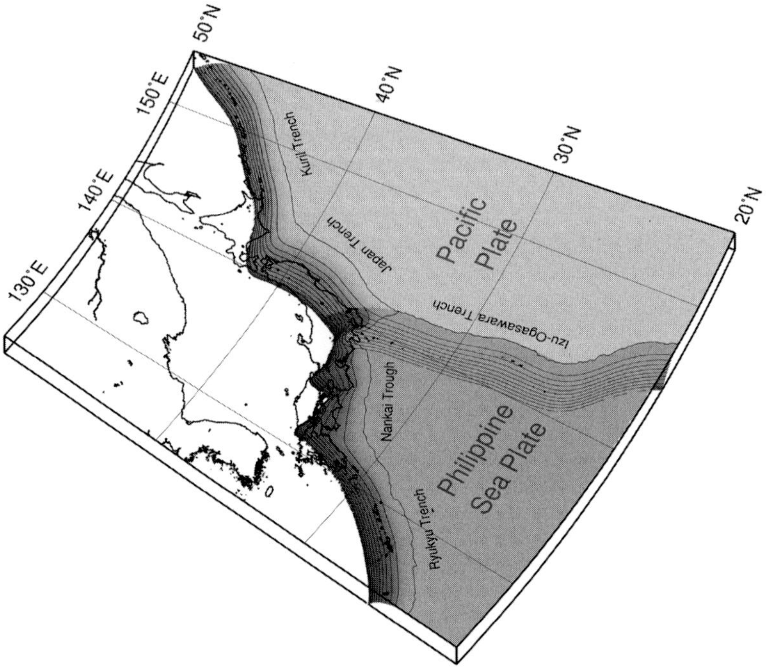

Figure 2
A standard model for 3-D geometry of plate interfaces in and around Japan. The intervals of the iso-depth contours are 10 km. The geometry of plate interfaces (upper boundaries of the descending Pacific and Philippine Sea plates) is determined from ISC hypocenter data by using the inversion technique. Each plate interface is represented by the superposition of bi-cubic B-splines with 8 km equally spaced local supports.

longitude and from 20 °N to 50 °N in latitude (about 3000 km × 3000 km). In and around Japan, the Pacific (PA) plate is descending beneath the North American (NA) and the Philippine Sea (PH) plates, and the Philippine Sea plates are descending beneath the North American and the Eurasian (EU) plates. Since the boundary between the North American and the Eurasian plates is not clear, we discard this plate boundary in our modelling. The 3-D geometry of plate interfaces is represented by the superposition of bi-cubic B-splines with 8 km equally spaced local supports. To determine the coefficients of superposition, first, we applied an inversion technique developed by Yabuki and Matsu'ura (1992) to ISC (International Seismological Centre) hypocenter location data. After that, we slightly modified the plate interface geometry by using JMA (Japan Meteorological Agency) unified hypocenter data.

In order to determine the direction of steady slip vectors $\boldsymbol{v}_{pl}$ on the given four plate interfaces, Σ_1 (PA-NA), Σ_2 (PA-PH), Σ_3 (PH-NA) and Σ_4 (PH-EU), first, we calculate the relative plate motion vectors $\boldsymbol{v}$ at every plate boundary from NUVEL-1A (DeMets *et al.*, 1994) as shown in Figure 3. Then, we project the calculated

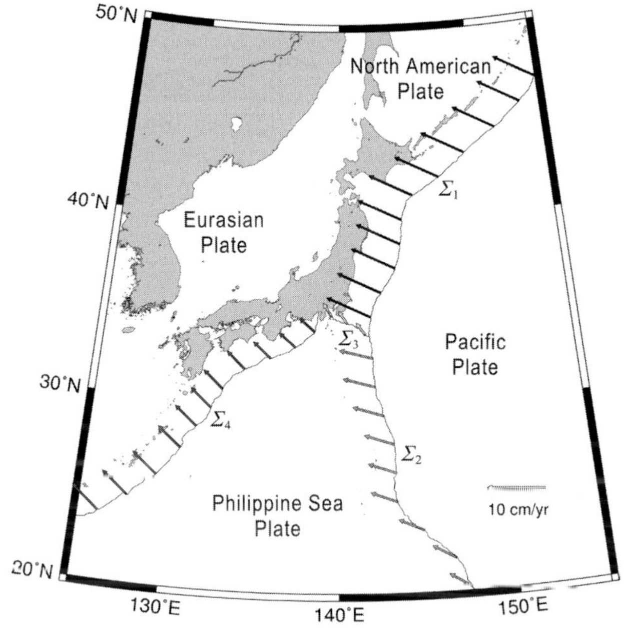

Figure 3

Plate boundaries and relative plate motion vectors in and around Japan. The elastic lithosphere is divided into four plates; the Pacific (PA), the North American (NA), the Philippine Sea (PH), and the Eurasian (EU) plates. These four plates interact with each other at four plate interfaces; Σ_1 (PA-NA), Σ_2 (PA-PH), Σ_3 (PH-NA), and Σ_4 (PH-EU). The relative plate motion vectors (thick arrows) are calculated from the global plate motion model NUVEL-1A (DeMets *et al.*, 1994). The convergence rate between the Pacific and the North American plates is about 8 cm/yr in northeast Japan, and that between the Philippine Sea and the Eurasian plates is about 4 cm/yr in southwest Japan.

relative plate motion vectors $\boldsymbol{v}$ onto the plate interfaces (Fig. 4). As to the magnitude v_{pl} of the steady slip vector, we take the absolute value $|\boldsymbol{v}|$ of the relative plate motion vector to satisfy the condition of mass conservation for the descending slab. Here, given the normal vector of the plate interface and the relative plate motion vector, we can describe the steady slip vector $\boldsymbol{v}_{pl}$ on the plate interface with the magnitude v_{pl} and the slip direction ϕ. In our mathematical expressions, the slip direction ϕ is included in the slip response function T_{ij}. The detailed expressions of steady slip vectors are given in Hashimoto *et al.* (2004).

3. The Stress Field Formed by Steady-state Frictional Sliding

In general, steady plate subduction is subjected to a certain frictional resistance corresponding to the absolute shear strength of plate interfaces. This type of mechanical interaction forms a steady-state stress field around the plate interface. In this Section, first, we show an idea for modelling the steady-state frictional

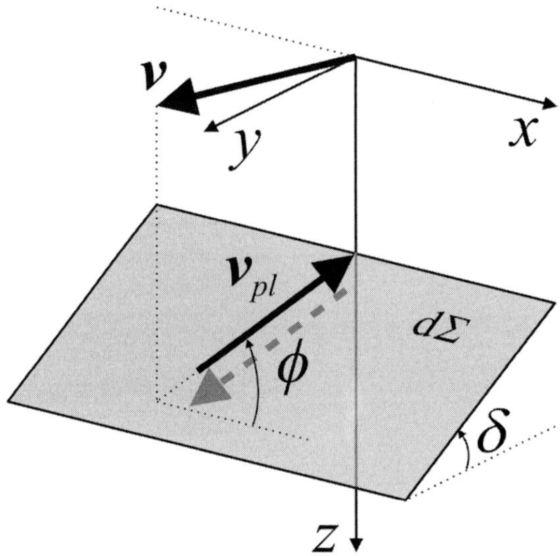

Figure 4

Determination of the steady slip vector $\boldsymbol{v}_{pl}$ on an element of plate interface Σ in the Cartesian coordinate system (x, y, z). The rectangle $d\Sigma$ represents an element of the plate interface. The dip-angle of $d\Sigma$ is denoted by δ, and the slip-angle ϕ is measured counterclockwise on $d\Sigma$. The direction of the steady slip vector $\boldsymbol{v}_{pl}$ is taken to be parallel to the direction of the projected relative plate motion vector $\boldsymbol{v}$ on Σ. The magnitude v_{pl} of the steady slip vector is taken to be equal to $|\boldsymbol{v}|$.

sliding at the plate interface. Then, on the basis of the idea, we numerically compute the steady-state internal stress field formed by frictional sliding in northeast Japan.

We assume the initial state of elastic plates to be deviatric stress free. Given the steady slip vectors and a frictional strength distribution at a plate interface, we can obtain a steady-state slip deficit distribution that satisfies the strength boundary conditions over the plate interface through numerical computation. After the steady-state plate subduction has been realized, the slip deficit distribution does not change in time, unless the absolute strength of plate interfaces changes. This means that the frictional resistance to steady-state plate subduction is represented by the steady-state slip deficit distribution at the plate interface. Denoting the steady-state slip deficit w_f at the plate interface, we can compute the internal stress field as

$$\sigma_{ij}^{(f)}(\boldsymbol{x}) = \int_{\Sigma} T_{ij}(\boldsymbol{x}, \infty; \boldsymbol{\xi}, 0)\, w_f(\boldsymbol{\xi})\, d\boldsymbol{\xi}. \tag{7}$$

Here, $T_{ij}(\boldsymbol{x}, \infty; \boldsymbol{\xi}, 0)$ are constant in time, and so the stress field formed by the steady-state frictional resistance to plate subduction does not change in time.

In the following computation, for simplicity, we assumed the steady-state slip deficit distribution shown in Figure 5b. The slip deficit at the interface between the

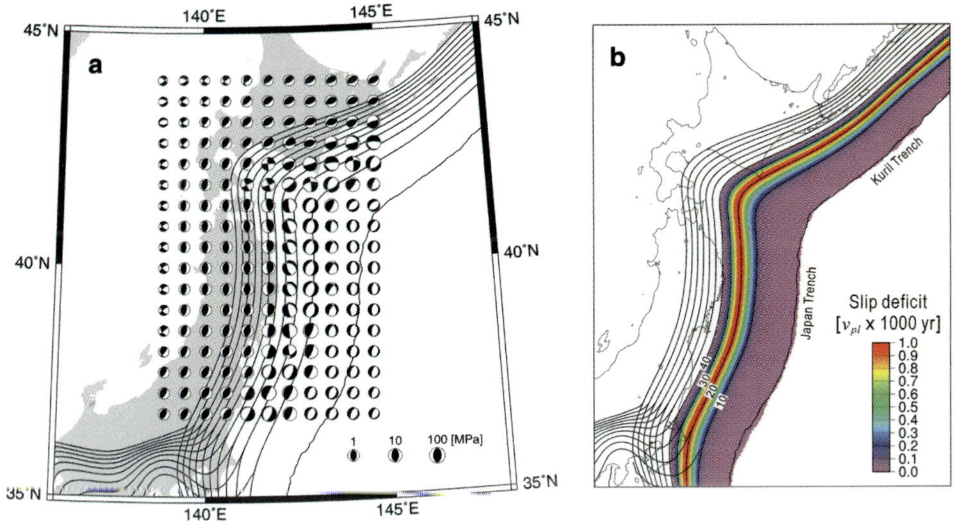

Figure 5
The internal stress field formed by frictional resistance to steady-state slip at the interface between the Pacific and the North American plates in northeast Japan. The geometry of plate interfaces is represented by the iso-depth contours with 10 km-intervals. (a) The pattern of the stress field at the depth of 10 km, which is represented with the upper focal hemispheres (white: compression, black: extension). The size of focal spheres is taken to be in proportion to the log of maximum shear stress. (b) The assumed slip deficit distribution on the plate interface. The maximum value of slip deficit is taken to be $v_{pl} \times 1000$ yr.

Pacific and the North American plates varies with depth; the slip deficit takes its maximum at the depth of 20 km and tapers off at both the deeper and the shallower ends of the plate interface. The maximum value of the slip deficit is assumed to be $v_{pl} \times 1000$ yr, which corresponds to the absolute strength level of about 100 MPa at the plate interface.

Figure 5a shows the computed internal stress fields at the depth of 10 km in northeast Japan. The steady-state frictional sliding at the interfaces between the Pacific and the North American plates forms the steady-state E-W compressive stress field in northeast Japan. The steady-state stress field formed in north Japan is NW-SE compression. The orientation of the maximum compressive principal stress is nearly perpendicular to the strike of trench axes. The stress field formed by steady-state frictional sliding well explains the observed E-W compression in northeast Japan, but it does not increase with time.

4. The Stress Field due to Steady Plate Subduction without Friction

In this Section we consider the stress field caused by steady plate subduction without friction. If the plate interface has a curvature in the direction of fault slip, steady slip along the plate interface produces an internal force (dislocation source) that causes crustal deformation and stress build-up (Hashimoto *et al.*, 2004). Therefore, we may regard the internal force due to steady plate subduction (steady plate bending) as the second type of mechanical interaction at convergent plate boundaries. As shown in Section 2, given relative velocity vectors at the plate boundaries, we can compute the increase rates of internal stress due to steady plate subduction in and around Japan by using Eq. (6).

It is widely accepted that the Izu peninsula has been colliding with the Japan arc at the base of the Izu peninsula since about 3 million years ago. The collision between the Izu peninsula and the Japan arc obstructs the steady subduction of the Philippine Sea plate beneath the North American plate. In order to represent such a state, following Matsu'ura (1998), we assume total collision (0% steady slip) at the base of the Izu Peninsula and total subduction (100% steady slip) over the whole plate interfaces except there.

Figure 6 shows the computed stress accumulation patterns at the depth of 10 km in and around Japan. The steady plate subduction produces the increase of internal stress at a constant rate. The computed stress accumulation pattern is accordant with the stress field estimated from active faults in the south Kanto-Tokai region (Matsuda, 1977), but not in other regions. For example, in northeast Japan, the computed stress field is E-W extension, while the observed stress field is E-W compression. Such a clear discrepancy between the observed and the computed stress fields suggests the existence of another mechanism of tectonic loading in northeast Japan.

5. The Stress Field Produced by Partial Collision

In Sections 3 and 4 we evaluated the effects of the two different types of mechanical interaction at plate interfaces; that is, frictional resistance to steady-state plate subduction and reaction to steady plate bending. The first type of interaction produces the E-W compressive stress in northeast Japan, but it does not increase with time. The second type of interaction produces the secular increase of internal stress, but its sense is opposite to the observed E-W compression.Then, we consider the third type of mechanical interaction, collision between a descending oceanic plate and the overriding plate. In this section, first, we introduce the concept of partial collision and define a collision rate. Then, applying this concept to the case of northeast Japan, we examine the effects of partial collision on the internal stress field.

At convergent plate boundaries, in general, a part of plate convergence is consumed by steady slip along plate interfaces, but the remaining part by inelastic deformation (seismic faulting, aseismic faulting, and active folding) of overriding plates. At the collision boundary extending along the Himalayas, for example, about 40% of the total plate convergence (50 mm/yr) is consumed by steady subduction of

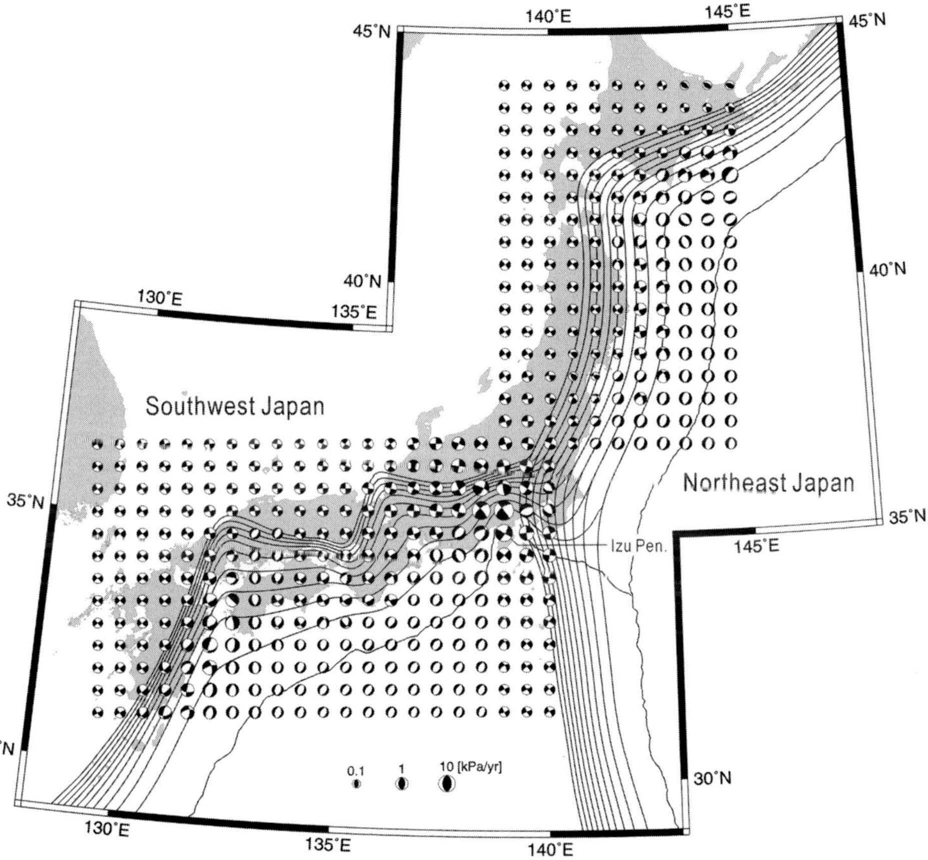

Figure 6
The increase rates of internal stress in and around Japan produced by steady plate subduction without friction. The stress pattern at the depth of 10 km is represented with the upper focal hemispheres (white: compression, black: extension). The size of focal spheres is taken to be in proportion to the log of maximum shear stress. The geometry of plate interfaces is represented by the iso-depth contours with 10 km-intervals.

the Indian plate beneath the Eurasian plate, and the rest of about 60% is consumed by the internal deformation of the Eurasian plate (e.g., TAKADA and MATSU'URA, 2004). Such a plate boundary process to be called "partial collision" can be quantitatively described by introducing a collision rate defined as

$$c = 1 - \textit{steady slip rate at plate interfaces/plate convergence rate} \tag{8}$$

According to this definition, total subduction ($c = 0$) and total collision ($c = 1$) can be regarded as the two extreme cases of partial collision, which is represented by the increase of slip deficit at plate interfaces.Then, we rewrite the fault slip w in Eq. (4) as

$$w(\xi, \tau) = [1 - c(\xi)]\ v_{pl}(\xi)\,\tau + \Delta w(\xi, \tau) \tag{9}$$

Substituting Eq. (9) into Eq. (3), we obtain the increase rate of internal stress as

$$\dot{\sigma}_{ij}(\boldsymbol{x}) = \int_{\Sigma} [1 - c(\boldsymbol{\xi})]\ T_{ij}(\boldsymbol{x}, \infty; \boldsymbol{\xi}, 0)\ v_{pl}(\boldsymbol{\xi})\, d\boldsymbol{\xi}. \tag{10}$$

Thus, given the spatial distribution of collision rates at plate interfaces, we can compute the increase rate of internal stress due to partial collision by using Eq. (10).

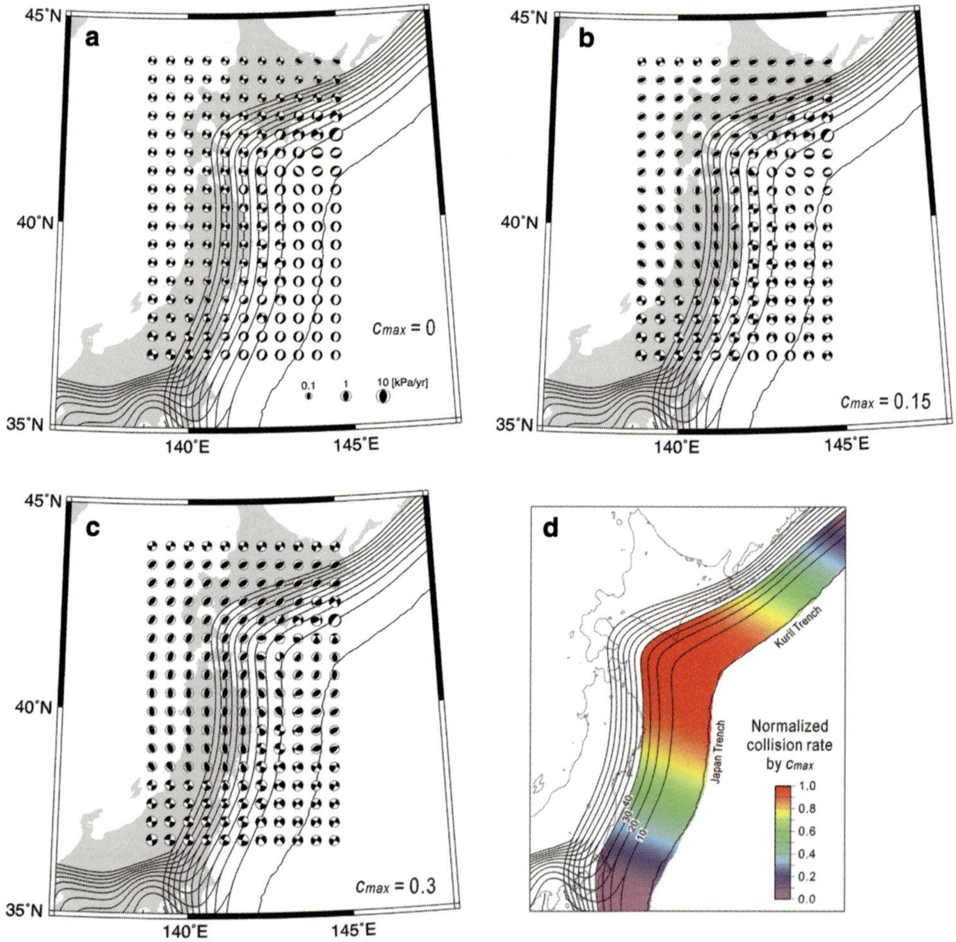

Figure 7

The increase rates of internal stress produced by partial collision in northeast Japan. The geometry of plate interfaces is represented by the iso-depth contours with 10 km-intervals. (a) The increase rate of internal stress at the depth of 10 km, due to total plate subduction ($c = 0$). (b) The increase rate of internal stress at the depth of 10 km, due to partial collision with the maximum collision rate $c_{\max}$ of 0.15. (c) The increase rate of internal stress at the depth of 10 km, due to partial collision with $c_{\max} = 0.3$. (d) The relative collision rate distribution used for computation. In any case of a, b, and c, the stress pattern is represented with the upper focal hemispheres (white: compression, black: extension). The size of focal spheres is taken to be in proportion to the log of maximum shear stress.

Now we take the collision rate distribution on the interface between the Pacific and the North American plates in northeast Japan as shown in Fig. 7d; that is, the collision rate takes its maximum at the junction of the Japan trench and the Kuril trench and tapers off at both ends of the modelled region. In the following numerical computation we examine the three cases with the different maximum collision rates (c_{max}) of 0.0, 0.15, and 0.3. The increase rates of internal stress computed for these three cases are shown in Figures 7a, b, and c, respectively.

Figure 7a shows the stress accumulation pattern at the depth of 10 km in the case of total subduction ($c = 0$). As demonstrated in Section 4, the steady plate subduction cannot reproduce the observed E-W compression. In the cases of partial collision with the maximum collision rates of 0.15 and 0.3 (Fig. 7b and c), on the other hand, we obtained the computed stress pattern in accordance with observations; that is, E-W compression in northeast Japan and NW-SE compression in north Japan. The orientation of the maximum compressive principal stress is nearly perpendicular to the strike of trench axes. The computed increase rates of shear stress are 0.4×10^3 Pa/yr in the case of $c_{max} = 0.15$ and 0.8×10^3 Pa/yr in the case of $c_{max} = 0.3$. Here, it should be noted that the maximum collision rates (c_{max}) of 0.15 and 0.3 correspond to the average collision rates of 0.1 and 0.2, respectively. Then, from these computations, we may conclude that at least the 10% partial collision on average is needed to reproduce the observed E-W compressive stress field in northeast Japan.

6. Discussion and Conclusions

In the present study we proposed a mechanical model of tectonic loading at convergent plate boundary zones. We examined the effects of three different types of mechanical interaction (frictional resistance, plate bending, and collision) at plate interfaces on intraplate tectonic loading. The first type of interaction can be rationally represented by the steady-state slip deficit distribution at the plate interface. In the case of northeast Japan, this type of interaction forms the steady-state E-W compressive stress field, but it does not change in time unless the absolute strength of plate interfaces changes. The second and third types of interaction can be treated in a unified way by introducing the collision rate defined by $c = 1 -$ *steady slip rate at plate interfaces/plate convergence rate*. According to this definition, we can regard total subduction ($c = 0$) and total collision ($c = 1$) as the two extreme cases of partial collision. In the case of total subduction ($c = 0$), the computed stress field is E-W extension in northeast Japan, which is opposite in sense to observations. In the case of partial collision with the average collision rate of 0.1, on the other hand, we succeeded in reproducing the observed E-W compressive stress field in northeast Japan. This means that at least 10% of the total plate convergence between the Pacific and the North American plates is consumed by the internal deformation of the North American plate in northeast Japan. Thus, the concept of partial collision is essential to understand the mechanism of intraplate tectonic loading. On long-term average the internal stress level must be held constant by the balance of secular stress accumulation due to mechanical interaction at plate interfaces and secular stress release due to inelastic crustal deformation including seismic and/or aseismic slip on active faults.

Partial collision produces a trench-normal compressive stress field along the plate boundary. The increase rate of the compressive stress field is roughly proportional to the plate convergence rate and the collision rate, and the trench-normal extent of the compressive stress zone to the along-plate-boundary scale of the partial collision zone. Actually, the trench-normal extent of the compressive stress field produced by the 500 km-long partial collision in northeast Japan is much wider than that by the 50 km-long partial collision at the base of the Izu Peninsula. According to Takada and Matsu'ura (2004), the 2500 km-long partial collision along Himalayas produces the compressive stress field extending over the Tibetan plateau. The quantity obtained by multiplying the collision rate by the plate convergence rate gives the rate of crustal shortening. Therefore, we can examine the validity of the concept of partial collision through the estimation of long-term crustal shortening. In the case of northeast Japan, the crustal shortening rate estimated from the geologic/geomorphic data is about 5–8 mm/yr (Wesnousky *et al.*, 1982; Tajikara, 2003). This estimate is consistent with the 10% partial collision in northeast Japan, where the convergence rate between the Pacific and the North American plates has been estimated to be 80 mm/yr by DeMets *et al.* (1994).

Acknowledgments

We thank Masayoshi Tajikara for permission to use the figure in Tajikara (2003). Computation of viscoelastic slip-response functions was performed on the Earth Simulator at the Earth Simulator Center, Japan Agency for Marine-Earth Science and Technology (JAMSTEC).

References

Burridge, R. and Knopoff, L. (1964), *Body force equivalents for seismic dislocations*. Bull. Seismol. Soc. Am. *54*, 1875–1888.

DeMets, C., Gordon, R. G., Argus, D. F., and Stein, S. (1994), *Effect of recent revisions to the geomagnetic reversal time scale on estimates of current plate motions*, Geophys. Res. Lett. *21*, 2191–2194.

Geographical Survey Institute (1996), Horizontal crustal deformation in the Japanese islands, Rep. Coord. Comm. Earthquake Predict. *55*, 658–665, (in Japanese).

Fukahata, Y. and Matsu'ura, M. (2005), *General expressions of internal deformation fields due to a dislocation source in a multilayered elastic half-space*, Geophys. J. Int. *161*, 507–521.

Fukuyama, E., Hashimoto, C., and Matsu'ura, M. (2002), *Simulation of the transition of earthquake rupture from quasi-static growth to dynamic propagation*, Pure Appl. Geophys. *159*, 2057–2066.

HASHIMOTO, C. and MATSU'URA, M. (2000), *3-D physical modelling of stress accumulation and release processes at transcurrent plate boundaries*, Pure Appl. Geophys. *157*, 2125–2147.

HASHIMOTO, C. and MATSU'URA, M. (2002), *3-D simulation of earthquake generation cycles and evolution of fault constitutive properties*, Pure Appl. Geophys. *159*, 2175–2199.

HASHIMOTO, C., FUKUI, K., and MATSU'URA, M. (2004), *3-D Modelling of plate interfaces and numerical simulation of long-term crustal deformation in and around Japan*, Pure Appl. Geophys. *161*, 2053–2067.

KATO, T., EL-FIKY, G. S., OWARE, E. N., and MIYAZAKI, S. (1998), *Crustal strains in the Japanese islands as deduced from dense GPS array*, Geophys. Res. Lett. *25*, 3445–3448.

MARUYAMA, T. (1963), *On the force equivalents of dynamical elastic dislocations with reference to the earthquake mechanism*, Bull. Earthq. Res. Inst., Tokyo Univ. *41*, 467–486.

MATSUDA, T. (1977), *Estimation of future destructive earthquakes from active faults on land in Japan*, J. Phys. Earth *25*, Suppl., S251–S269.

MATSU'URA, M. (1998), *Modelling and forecast of earthquake generation processes*, Jour. Seis. Soc. Japan (Zisin) *50*, 213–227 (in Japanese with English abstract).

MATSU'URA, M. (2004), *Reproducing core-mantle dynamics and predicting crustal activities through advanced computing*, J. Earth Simulator *1*, 67–74.

MATSU'URA, M. and SATO, T. (1989), *A dislocation model for the earthquake cycle at convergent plate boundaries*, Geophys. J. Int. *96*, 23–32.

MATSU'URA, M., TANIMOTO, T., and IWASAKI, T. (1981), *Quasi-static displacements due to faulting in a layered half-space with an intervenient viscoelastic layer*, J. Phys. Earth. *29*, 23–54.

NOHARA, T., KORIYA, Y., and IMAIZUMI, T. (2000), *An estimation of the crustal strain rate using the active fault GIS data*, Active Fault Res. *19*, 23–32 (in Japanese with English abstract).

RESEARCH GROUP FOR ACTIVE FAULTS OF JAPAN (1991), Active faults in and around Japan, sheet maps and inventories (revised edition), Univ. of Tokyo Press, Tokyo.

SAGIYA, T., MIYAZAKI, S., and TADA, T. (2000), *Continuous GPS array and present-day crustal deformations of Japan*, Pure Appl. Geophys. *157*, 2303–2322.

SATO, H. (1994), *The relationship between late Cenozoic tectonic events and stress field and basin development in northeast Japan*, J. Geophys. Res. *99*, 22261–22274.

TAJIKARA, M. (2003), *Vertical crustal movements of the northeast Japan arc in late Quaternary time*, Dr. thesis, Univ. of Tokyo.

TAKADA Y., and MATSU'URA, M. (2004), *A unified interpretation of vertical movement in Himalaya and horizontal deformation in Tibet on the basis of elastic and viscoelastic dislocation theory*, Tectonophysics *383*, 105–131.

YABUKI, T. and MATSU'URA, M. (1992), *Geodetic data inversion using a Bayesian information criterion for spatial distribution of fault slip*, Geophys. J. Int. *109*, 363–375.

WESNOUSKY, S. G., SCHOLTZ, C. H., and SHIMAZAKI, K. (1982), *Deformation of an island arc: rates of moment release and crustal shortening in intraplate Japan determined from seismicity and Quaternary fault data*, J. Geophys. Res. *87*, 6829–6852.

(Received November 29, 2004, revised August 26, 2005, accepted September 6, 2005)
Published Online First: September 2, 2006

Pure appl. geophys. 163 (2006) 1819–1846
0033–4553/06/091819–28
DOI 10.1007/s00024-006-0099-x

Pure and Applied Geophysics

Virtual California: Fault Model, Frictional Parameters, Applications

P. B. Rundle,[1] J. B. Rundle,[1] K. F. Tiampo,[2] A. Donnellan,[3] and D. L. Turcotte[4]

Abstract—Virtual California is a topologically realistic simulation of the interacting earthquake faults in California. Inputs to the model arise from field data, and typically include realistic fault system topologies, realistic long-term slip rates, and realistic frictional parameters. Outputs from the simulations include synthetic earthquake sequences and space-time patterns together with associated surface deformation and strain patterns that are similar to those seen in nature. Here we describe details of the data assimilation procedure we use to construct the fault model and to assign frictional properties. In addition, by analyzing the statistical physics of the simulations, we can show that that the frictional failure physics, which includes a simple representation of a dynamic stress intensity factor, leads to self-organization of the statistical dynamics, and produces empirical statistical distributions (probability density functions: PDFs) that characterize the activity. One type of distribution that can be constructed from empirical measurements of simulation data are PDFs for recurrence intervals on selected faults. *Inputs* to simulation dynamics are based on the use of *time-averaged* event-frequency data, and *outputs* include PDFs representing measurements of *dynamical variability* arising from fault interactions and space-time correlations. As a first step for productively using model-based methods for earthquake forecasting, we propose that simulations be used to generate the PDFs for recurrence intervals instead of the usual practice of basing the PDFs on standard forms (Gaussian, Log-Normal, Pareto, Brownian Passage Time, and so forth). Subsequent development of simulation-based methods should include model enhancement, data assimilation and data mining methods, and analysis techniques based on statistical physics.

Key words: Earthquakes, simulations, fault Systems, forecasting, statistics.

1. Introduction—Virtual California

Virtual California is a model representing the structure and dynamics of the vertical, strike-slip earthquake fault system in California (Rundle, 1988; see Rundle *et al.*, 2004 for a recent description). It is a type of model called a *backslip model* (see the Appendix for details), so-called because the loading in the model arises

[1] Center for Computational Science and Engineering and Departments of Physics and Engineering, University of California at Davis, Davis, CA, U.S.A.
E-mail: prundle@cse.ucdavis.edu; jbrundle@ucdavis.edu
[2] Department of Earth Sciences, University of Western Ontario, London, ON, Canada.
E-mail: ktiampo@uwo.ca
[3] Earth and Space Science Division, Jet Propulsion Laboratory, Pasadena, CA, U.S.A.
E-mail: andrea.donnellan@jpl.nasa.gov
[4] Department of Geology, University of California at Davis, Davis, CA, U.S.A.
E-mail: turcotte@geology.ucdavis.edu

from "negative slip" or "backslip" applied to each fault segment at its geologically observed long-term rate of offset, $V(\mathbf{x})$. We note that in this context, a "fault segment" is regarded simply as a "degree of freedom", rather than a spatially coherent entity with geological meaning. All of the rectangular fault segments are embedded in an elastic half space, and they interact with each other by means of quasistatic elastic interactions, whose stress Greem's functions are computed by means of a Boundary Element Method.

Frictional coefficients are assigned to each fault segment, along with other frictional parameters, by means of a *data assimilation* technique described in this paper. When the model is used to produce a simulation, the result is a history of slip on the fault segments in response to the driving forces. The interactions between the fault segments serve to organize the system so that, instead of a sequence of single segments breaking individually, multiple segments break simultaneously, producing large earthquakes. The dynamical evolution in the model is also produced by means of a stochastic, cellular automaton method, in which a random overshoot or undershoot component is added at the time of sliding of a segment. Once an earthquake history is computed, the surface deformation can be computed as well by using the appropriate kinematic Green's functions, which at the moment are also computed by means of Boundary Element Methods.

As an aside, we note that, since the Green's function code is an entirely separate code from the actual time-stepping, dynamical evolution simulation code, the stress and kinematic Green's functions could also be computed, for example, by a Finite Element Method. For that reason, this approach to earthquake simulation is quite general and in principle, any level of geometric or topological complexity is possible in the structure of the fault system. In addition, this generality will allow us to include any general type of faulting, including thrust faults, normal faults, and dipping strike-slip faults, in future models.

The usefulness of models such as Virtual California is that: 1) They allow the user to investigate all of the multitude of time and space scales in the problem that are inaccessible to direct human observation; 2) they allow the basic physics of the system to be investigated through the use of a "numerical laboratory" approach to earthquake fault system science; 3) they allow the user to have access not only to the observable surface deformation, but in addition to all the other physical variables that are generally impossible to observe, such as internal stresses and strains anywhere within the medium; and 4) they allow the user to develop a number of forecasting and prediction methodologies, such as statistical methods, ensemble methods, and model steering methods, based on data assimilation and parameter estimation and refinement techniques.

A similar type of simulation, the *Standard Physical Earth Model* (*SPEM*) has been developed and used by WARD (2000, 1996) and WARD AND GOES (1993). *SPEM* has two important differences from *Virtual California*. The first is that whereas *Virtual California* utilizes the full equations of three-dimensional elasticity, *SPEM* is a plane-strain model. The second difference is in the form of the friction laws. The equations describing the laboratory-based friction model for *Virtual California* are detailed in the appendix here. In contrast, *SPEM* uses two scales: an inner and an outer scale, which control the conditions for rupture initiation and healing. Despite these important differences, it can be shown that many of the statistical results produced by the two simulations are very similar (RUNDLE *et al.*, 2006). This indicates that the self-organization of the system induced by the elastic interactions is probably the dominant feature of these types of simulations, and that details of the local frictional physics are probably less important in the system-level physics.

In the remainder of this paper, we describe details of the construction of the Virtual California fault and friction models at a greater level of detail than has been previously published. We describe details of the data assimilation method by which the frictional parameters are set for each fault segment. In an appendix, we also describe details of the physics of the Virtual California simulation, in which the friction and other parameters are introduced.

What is described here is the "Standard Model" for Virtual California. Minor variations of this model have been used in recent work (RUNDLE *et al.*, 2004) to improve computational performance and efficiency. Moreover, since several versions of the model have appeared in the literature over recent years (e.g., RUNDLE *et al.*, 2001, 2004), part of this paper is a chronology of successive versions of the model, and how they were constructed.

2. *Compilation of the Fault Geometry*

A fault segment in the Virtual California model is geometrically represented by a two-dimensional rectangular object embedded in an elastic half space. Only strike-slip faults with the potential to be involved in magnitude $M \geq 6$ earthquakes are included, which allows us to reasonably assume that each fault segment has a vertical dip and horizontal rake. Furthermore, although we collect depth measurements for each segment (in order to prepare for the inclusion of depth-dependent dynamics), for now we assume that all faults run from 0 to 20 km deep. Thus, the two endpoints of the surface trace, measured in latitude and longitude[1], define a fault segment.

Our earliest model (RUNDLE *et al.*, 2001), used in simulations through 1999, is based mostly on the data in Table 1 of DENG and SYKES (1997). As such, it is limited to southern California and a small area of northern Mexico, from roughly 36.2° to 31.8° N. The Mojave block and offshore regions are noticeably incomplete. Our model includes only strike-slip faults, so just the segments labeled RL (right-lateral strike-slip) and LL (left-lateral strike-slip) are used. In the DENG and SYKES (1997)

[1] North latitudes and west longitudes are given positive values.

Table 1

Segments and geologic rates of offset for the modified version of VC 2001 used in Figures 1–7

Fault or Fault System Name	Segment Nos.		Chart Distance (km)		Average Slip Rate (mm/yr)
	Begin	End	Begin	End	
Bartlett Springs	0	7	0.0	84.7	6
Calaveras	8	22	84.7	238.9	15 (8- > 17) 6 (18- > 22)
Collayomi	23	25	238.9	266.8	.6
Concord-Green Valley	26	31	266.8	322.2	6
Death Valley	32	55	322.2	569.6	5 (32- > 49) 4 (50- > 55)
Garberville-Briceland	56	59	569.6	609.2	9
Greenville	60	66	609.2	682.2	2
Hayward	67	77	682.2	793.3	9 (67- > 74) 3 (75- > 77)
Hunter Mtn.-Saline Val.	78	84	793.3	861.3	2.5
Hunting Creek-Berryessa	85	90	861.3	920.3	6
Lake Mountain	91	93	920.3	953.7	6
Maacama	94	111	953.7	1133.3	9
Monterey Bay-Tularcitos	112	119	1133.3	1213.6	.5
Ortigalita	120	126	1213.6	1280.1	1
Owens Valley	127	138	1280.1	1401.6	1.5
Palo Colorado-Sur	139	146	1401.6	1479.8	3
Panamint Valley	147	156	1479.8	1584.5	2.5
Quien Sabe	157	158	1584.5	1607.6	1
Rinconada	159	177	1607.6	1796.9	1
Rodgers Creek	178	183	1796.9	1858.9	9
Round Valley	184	189	1858.9	1914.3	6
San Gregorio	190	198	1914.3	2003.3	5
Sargent	199	203	2003.3	2056.0	3
West Napa	204	206	2056.0	2085.9	1
White Mountains	207	216	2085.9	2186.5	1
San Andreas North	217	263	2186.5	2653.6	24 (217- > 248) 17 (249- > 263)
San Andreas Creeping	264	273	2653.6	2751.3	34
San Andreas South	274	335	2751.3	3330.7	34 (274- > 298) 30 (299- > 312) 24 (313- > 321) 25 (322- > 335)
San Jacinto	336	364	3330.7	3622.1	12 (336- > 352) 14 (353- > 364)
Elsinore	365	388	3622.1	3857.5	3 (365- > 368) 5 (369- > 384) 4 (385- > 388)
Imperial Valley	389	406	3857.5	4020.0	30
Laguna Salada	407	416	4020.0	4118.5	4

Table 1

(Contd.)

Garlock	417	440	4118.5	4353.0	−5 (417- > 426)−7 (427- > 440)
Palos Verdes	441	447	4353.0	4428.6	3
Santa Cruz Island	448	452	4428.6	4481.9	-3
Brawley	453	457	4481.9	4533.8	25
Santa Monica	458	468	4533.8	4653.3	-3
Cleghorn	469	470	4653.3	4676.4	-3
Tunnel Ridge	471	472	4676.4	4695.6	-1.3
Helendale	473	481	4695.6	4781.7	.8
Lenwood-Lockhart	482	499	4781.7	4955.2	.8
Pipes Canyon	500	501	4955.2	4970.8	.7
Gravel Hills-Harper	502	509	4970.8	5051.2	.9
Blackwater	510	516	5051.2	5113.0	2
Camp Rock-Emerson	517	527	5113.0	5227.2	1 (517- > 524) .6 (525- > 527)
Homestead Valley	528	530	5227.2	5254.4	.6
Johnson Valley	531	536	5254.4	5320.4	.6
Calico-Hidalgo	537	549	5320.4	5455.5	1 (537) 1.7 (538) 2.6 (539- > 545) .6 (546- > 549)
Pisgah-Bullion	550	562	5455.5	5571.2	1
Mesquite Lake	563	564	5571.2	5592.2	1
Pinto Mountain	565	573	5592.2	5676.0	-1
Morongo Valley	574	574	5676.0	5690.6	-.5
Burnt Mountain	575	576	5690.6	5707.6	.6
Eureka Peak	577	578	5707.6	5725.8	.6
Hollywood-Raymond	579	582	5725.8	5763.7	-1 (579- > 580) -.5 (581- > 582)
Inglewood-Rose Cyn	583	604	5763.7	5979.2	1 (583- > 590) 1.5 (591- > 604)
Coronado Bank	605	623	5979.2	6179.5	3
San Gabriel	624	637	6179.5	6310.8	3 (624- > 628) 2 (630- > 633) 1 (634- > 637)
Big Pine	638	644	6310.8	6379.5	-4
White Wolf	645	649	6379.5	6427.6	-5

data tabulation, the surface traces are defined by a midpoint (latitude/longitude), a length (km), and a strike angle (degrees clockwise from North), so a simple algorithm is used to find the segment endpoints.

Assume a spherical Earth with radius 6371 km. Define midpoint longitude θ, midpoint latitude ϕ, strike angle ω, and length L. Take $\omega' = 90° - \omega$. The endpoint latitudes are

$$\phi \pm \frac{L}{2} \sin \omega' \frac{180°}{\pi (6371\text{km})}. \tag{1}$$

The endpoint longitudes are

$$\theta \pm \frac{L}{2}\cos\omega' \frac{180^\circ}{\pi(6371\text{km})\cos\frac{\phi+\phi_C}{2}}, \tag{2}$$

where ϕ_C is the latitude of the corresponding endpoint. The $\cos\phi + \phi_C/2$ term is a compromise to avoid integration.

DENG and SYKES (1997) do not contain the Landers fault, consequently three Landers segments were added, based on WALD and HEATON (1994). Measurements are taken directly from their map of three fault segments labeled Figure 1.

In 2000, the model geometry was again revised to include northern California faults with the added segments ranging from approximately 35.5° to 41.3° N. The great majority of fault segments are taken from a table of fault parameters used in (http://eqhazmaps.usgs.gov/html/faults.html) USGS, 1996 Seismic Hazard maps, compiled by BARNHARD and HANSON (1996). Faults included are those in California that are, for the most part, north of latitude 35.5°N. Any faults without a dip angle

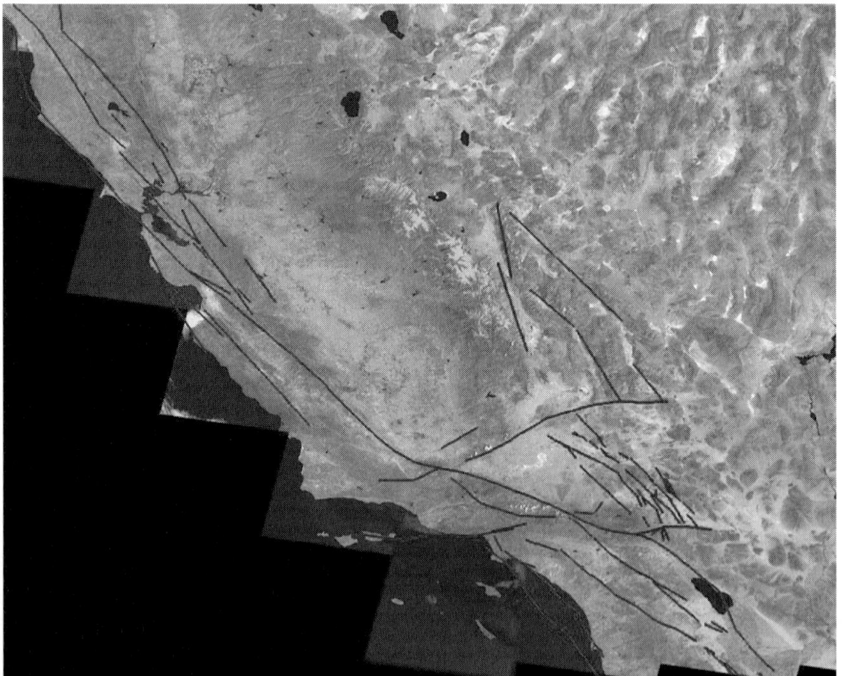

Figure 1
VC 2001 superposed on map of California. (Map of California courtesy of About.com.)

of 90° are excluded, based on the assumption that they are not strike-slip[2]. The endpoints of the faults were entered as they appeared.

Data for the northern San Andreas fault segments were assembled from several sources. First, the northern endpoint of the DENG and SYKES (1997) segment SACreeping1 is connected to the southern endpoint of the BARNHARD and HANSON (1996) segment San Andreas Fault, 1906. Both endpoints of San Andreas Fault, 1906 are used, however, because of this model segment's enormous span, intermediate endpoints are added to better reflect the natural trace. The southernmost such endpoints belong to the BARNHARD and HANSON (1996) segments 'San Andreas Fault, Santa Cruz Mtn.' and San Andreas Fault, Peninsula Segment. Four additional intermediate endpoints are added between 39.1° and 40.1°N' based on measurements taken from the map in Figure 2A of WGCEP (1999). Finally, north of the San Andreas Fault, 1906 segment, there is an intermediate endpoint near the Mendocino Triple Junction measured from Figure 1 of WARD (2000), and a final endpoint is taken from the westernmost endpoint of BARNHARD and HANSON (1996) Mendocino Fault Zone.

The final major revision to the Virtual California fault model came in 2001. The focus was the full realization of southern California, compensating for the omissions of DENG and SYKES (1997). The principal changes are that the Mojave block is fleshed out and offshore faults are added. The majority of the fault trace information for this revision is taken directly from the clickable fault maps on the Southern California Earthquake Data Center website (http://www.data.scec.org/faults/fault-map.html). The digitized maps were enlarged and endpoints were specified so as to reasonably approximate the fault trace shapes. These endpoints were measured and, where possible, corrected against the source material for BARNHARD and HANSON (1996), Appendix A of USGS Open-File Report 96-706 by PETERSEN *et al.*, (1996). Some smaller faults, as well as continuations of offshore faults beyond the scope of the SCEDC maps, are taken directly from PETERSEN *et al.* (1996). Non-strike-slip faults are excluded, based on information on the SCEDC website and the first column of PETERSEN *et al.* (1996). In addition, some strike-slip faults are excluded if they are inactive during modern times, so that they have not ruptured during the Holocene epoch. Length and geometrical significance are also considered, especially if a fault seems likely to transfer stress from one fault subsystem to another. The reconstruction of the Mojave block also necessitated the removal of the WALD and HEATON (1994) Landers faults and the DENG and SYKES (1997) Pisgah fault.

Once the fault segment endpoints in a version of the Virtual California model are compiled, they are converted from latitude and longitude to distances (in km) north and east of an arbitrary reference point—31°N, 121°W. The resulting plot is an equal-area sinusoidal projection based on a central meridian of 121°W. As relative

[2] The table does not explicitly specify fault type. It should also be noted that the table lists no dip angles between 90° and 75°, making 90° the logical cutoff point.

distances in such a plot grow more distorted farther from the central meridian, future models should move the central meridian eastward to minimize distortion. Currently, linear distortion does not exceed 2% for any portion of the model, and segment strike angles are distorted by no more than 10°, and generally much less.

As the Virtual California simulation code is designed to work with fault segments of approximately 10 km in length, the length of a $M \approx 6$ earthquake, the final step is to split the segments into appropriate pieces. The following algorithm is used. The length L of a given fault with endpoints ($[x_1, y_l]$, $[x_2, y_2]$) is computed by $L = \sqrt{(x_2 - x_1)^2 + (y_2 - y_1)^2}$. L is then divided by 10 and rounded to the nearest integer. This new value N represents the number of smaller segments into which the initial fault will be split. The endpoints of the ith smaller segment, $1 \le i \le N$, are then

$$\left(\left[x_1 + \frac{i-1}{N}(x_2 - x_1) y_1 + \frac{i-1}{N}(y_2 - y_1)\right], \left[x_1 + \frac{i}{N}(x_2 - x_1) y_1 + \frac{i}{N}(y_2 - y_1)\right]\right).$$

In the event that the fault-specific depth parameters are used in future models, the appropriate sources are hereby stated: DENG and SYKES (1997) faults, BARNHARD and HANSON (1996) faults, and PETERSEN *et al.* (1996) faults are all associated with depth values in the tables from which the endpoints are taken. The WALD and HEATON (1994) Landers faults are assigned depths of 15 km based upon that paper's Figure 5 and comments from the text. Depth values for faults taken from the SCEDC clickable fault maps are based on corresponding or similar neighboring faults in PETERSEN *et al.* (1996) tables. For depth values along the northern San Andreas, the respective sources (traveling from south end to north end) are: DENG and SYKES (1997), SACreeping1; BARNHARD and HANSON (1996), San Andreas Fault, Santa Cruz Mtn.; BARNHARD and HANSON (1996), San Andreas Fault, Peninsula Segment; WGCEP (1999), Table 2, SAF—North Coast South; WGCEP (1999), Table 2, SAF–North Coast North; BARNHARD and HANSON (1996), Mendocino Fault Zone.

3. Activity and Friction Parameters

The dynamics of the Virtual California model are largely governed by the frictional parameters associated with individual faults. Each model segment is

Table 2

Mean (λ), standard deviation (θ), and Coefficient of Variation (CoV) for inter-event time interval statistics measured from the simulation data in Figures 4 and 5

Fault (Magnitude Range)	λ (years)	θ (years)	CoV
Northern SAF ($M \ge 5.8$)	47	28	.59
Southern SAF ($M \ge 5.8$)	57	36	.63
Northern SAF ($M \ge 7.5$)	181	100	.56
Southern SAF ($M \ge 7.5$)	221	116	.52

associated with three values: slip rate, aseismic fraction, and mean recurrence interval. The fault slip rate, in cm/year, is the long-term mean rate and includes both aseismic and coseismic slip. Right-lateral slip is considered positive and left-lateral slip negative. As before, we use only vertical strike-slip faults, allowing us to assume horizontal rake in all cases. The aseismic fraction is the fraction of the fault's total slip that is attributable to creep, and can theoretically take on any value from 0 to 1. We set a minimum aseismic fraction of 0.1 to enhance the verisimilitude of our results (the reasons are complex; see RUNDLE *et al.* (2001) for details). The mean recurrence interval is the mean time, in years, between large seismic events on the fault. We bound the mean recurrence interval between 1 and 5000 years for reasons of model functionality. Presently, we are interested only in the most active fault segments, therefore we do not consider segments that slip on average more infrequently than once every 5000 years. While slip rates, including aseismic and coseismic components, are relatively easy to find in literature, mean recurrence intervals are much harder to quantify; no fault is free to rupture in isolation. It was necessary to develop a special method to compute tractable recurrence intervals, which we discuss below.

With the exception of the northern San Andreas and Mojave block systems, assignment of total slip rate (combined aseismic and coseismic) values to faults is straightforward. For any fault in the model, we note that we enforce a minimum long-term slip rate value of 0.1 mm/yr. Slip rates for DENG and SYKES (1997) faults are taken from Table 1 of DENG and SYKES (1997); slip rates for the WALD and HEATON (1994) Landers faults are assumed to be 0.3 mm/yr; and slip rates for BARNHARD and HANSON (1996) faults are taken from BARNHARD and HANSON (1996).

For the northern San Andreas, the respective sources for slip rate values (traveling from south end to north end) are: DENG and SYKES (1997), SACreeping1; WGCEP (1999), Table 2, SAF – Santa Cruz Mtns; BARNHARD and HANSON (1996), San Andreas Fault, Peninsula Segment; WGCEP (1999), Table 2, SAF – North Coast South; WGCEP (1999), Table 2, SAF – North Coast North; BARNHARD and HANSON (1996), Mendocino Fault Zone. For the Mojave block segments based on the SCEDC clickable fault maps, slip rate values are taken from corresponding entries in the accompanying Alphabetical Fault Index whenever possible. For uncertain or missing slip values, PETERSEN *et al.* (1996) and Table 1 of PETERSEN and WESNOUSKY (1994) were consulted, and the final slip rate value is something of a consensus.

The aseismic slip factors α for fault segments (RUNDLE *et al.*, 2001) are less stringently based on the field data. Early in the construction of the model, it became clear that this parameter has a disproportionately significant impact on the simulated seismicity patterns. RUNDLE *et al.* (2001) found that the effect of this parameter, which is modeled after laboratory results of TULLIS (1996) and KARNER and MARONE (2000), generally acts to smooth the stress field on a fault, thereby leading to large events. Very similar effects are seen in the rate-and-state friction models of DIETERICH (1979). Although much of the available data states that there is no significant aseismic slip on many faults, reflecting this condition in simulations yields seismicity patterns with very few large events. Mandating a token amount of creep for each fault gives rise to larger, coordinated event patterns that better reflect observed California seismicity.

The very first versions of the Virtual California model assign an aseismic slip fraction of 0.2 to every fault segment (i.e., 20% of a fault's slip rate is creep). In subsequent versions, field data are used with no minimum slip fraction. Initially, a minimum aseismic slip fraction of 0.2 was then introduced, and later lowered to 0.1. From the 2000 version of the model onward, all faults have aseismic slip values based on field data, when available, with a minimum value of 0.1. Fractions for DENG and SYKES (1997) faults are calculated from their endnotes to Table 1. As the DENG and SYKES (1997) model San Andreas stops at the edge of the creeping section, the bordering segment of the northern San Andreas is designated the creeping section and assigned an aseismic slip fraction of 1.0. The San Andreas segment just north of this section is given a slip fraction identical to that of the DENG and SYKES (1997) segment SACreeping1 to affect a symmetrical tapering of the aseismic slip rate on either side of the creeping section. The northwestern Lenwood fault is given creep because of a non-quantitative note in the Alphabetical Fault Index on the SCEDC website, but the 0.25 fraction is based solely on a comparison of total slip rates from PETERSEN and WESNOUSKY (1994) the SCEDC website, and PETERSEN *et al.* (1996).

The final parameter to set is the mean recurrence interval for large events. The principal difficulty is that reliable and useful seismicity records for California extend back in time no more than 200 years, a shorter period than most of the major California faults' average recurrence intervals. Also, geological records of ruptures are often quite imprecise, even when they exist. All of this makes finding consistent, useful recurrence intervals in the literature a difficult proposition and necessitates a different approach. It is assumed that, as both the coefficient of friction and the normal force at the fault plane are reasonably constant for a strike-slip fault, a segment would be predisposed to have ruptures of a particular magnitude; i.e., a "characteristic earthquake" (SCHWARTZ and COPPERSMITH, 1984). By making reasonable generalizations about properties of earthquakes, the seismic moment of an earthquake is correlated with the coseismic slip it induces. Then, by determining the seismic moment of a segment's characteristic earthquake, a characteristic slip per event is calculated. Dividing this quantity by the segment's coseismic slip rate yields the average interval between ruptures. (For creeping segments, a recurrence interval of 1 year is used.)

The relationship of seismic moment to coseismic slip in each event was developed first. We have the equations $m_o = \mu\langle s\rangle A$ and $\langle s\rangle = f\Delta\sigma\sqrt{A}/\mu$, where m_o is seismic moment, μ is shear modulus, $\Delta\sigma$ is stress drop, $\langle s\rangle$ is average slip, A is fault rupture area, and f is a dimensionless fault shape factor (KANAMORI and ANDERSON, 1975). Combining these, we get

$$\langle s \rangle = \frac{(\Delta\sigma)^{2/3} f^{2/3} m_o^{1/3}}{\mu}. \tag{3}$$

We use reasonably assumed values of $\mu = 3 \times 10^{10}$ Pa, $\Delta\sigma = 5 \times 10^6$ Pa, and $f = 1$ to yield a moment-slip relation that gives realistic values. Numerically, it is quite comparable to the strike-slip moment magnitude regression in Figure 11b of WELLS and COPPERSMITH (1994).

The next step is to determine each segment's characteristic event moment. The first method tried was to simply examine the historic earthquake catalog and attempt to associate a major ($M \geq 6$) historic event with each segment (prior to the segments' being split)[3]. The seismic moment of this event becomes the characteristic major event moment for the fault segment. Since the 1999 fault model was current at the time this method was tried, the catalog used is Table 2 of DENG and SYKES (1997). Known foreshocks and aftershocks are excluded (as they are unlikely to represent a characteristic event), as is any event whose mechanism dip and rake indicated that it is a thrust or reverse fault. Finally, a single event from the catalog is assigned to each segment, based on the known circumstances of the catalog events, or, as a last resort, based on proximity.

This method has serious problems. First, not every segment has an easily associated major historic earthquake, or even one in close proximity relative to other faults. Since every segment needs an event, a number of the event assignments are dubious. A more chronologically extensive catalog would certainly help. Second, this method quite often gives rise to situations in which neighboring, ostensibly continuous fault segments are associated with earthquakes of vastly different moments, and thus possess markedly different frictional properties. It creates a pronounced artificial frictional discontinuity at the shared endpoint, which can impede the realistic propagation of slip during an earthquake.

The problems created by the initial attempts to set characteristic major event moments led to a change in approach. We decided to assume that a fault's characteristic earthquake would be similar to the historic earthquakes that occurred in its vicinity; i.e., that regions, as opposed to individual faults, have characteristic events[4]. Then, a fault segment's characteristic major event moment may become a manner of average of the moments of the historic major events in close proximity. It is most logical to use a distance-weighted average so that faults are most closely

---

[3] The cutoff for characteristic events is set at $M = 6$ because each of the approximately 10-km-long fault segments must fail as a unit. A 10-km rupture length produces an event on the order of $M = 6$.

[4] The definition of *region* is naturally somewhat nebulous. As will become clear, our regions of similar characteristic event moment are essentially defined by the spatial density and moment value diversity of events in the historic earthquake catalogs. There are large regions of similar characteristic event moment where there are few recorded events or where there are many recorded events of similar magnitude. Regions of similar characteristic event moment are much smaller where there are many recorded events of differing magnitude in close spatial proximity.

associated with the nearest possible events. The current method to determine faults' characteristic major event moment is based on this reasoning.

The details of this method are as follows. First, a catalog of all major historic strike-slip events ($M \geq \sim 6$) in the region of the fault system model, excluding foreshocks and aftershocks, is compiled. Each event is characterized by seismic moment and hypocentral latitude and longitude. (The Virtual California model assumes that fault frictional attributes do not change with time, thus the dates of the earthquakes are unnecessary.) Since most earthquake catalogs list events according to magnitude, magnitude values are instead compiled and plugged into a standard moment-magnitude relation

$$m_o = 10^{1.5M+9.0}. \tag{4}$$

Then, the characteristic major event moment m_χ of the j-th fault segment is a heavily distance-weighted average of the moments of all faults in the catalog:

$$m_{\chi j} = \frac{\sum_i m_{oi} r_{ij}^{-3}}{\sum_i r_{ij}^{-3}}, \tag{5}$$

where $m_{\chi j}$ is the characteristic major event moment of the j-th fault segment, m_{oi} is the seismic moment of the i-th catalog event, and $r_{ij} = |\boldsymbol{x}_i - \boldsymbol{x}_j|$ is the distance between the j-th fault segment and the hypocenter of the i-th catalog event.

A few notes about the distance-weighted average method for determining characteristic major event moments: The distance-weighting exponent of –3 is used for two reasons. First, it is large enough to ensure that a fault's characteristic major event magnitude is only significantly influenced by the events closest to it—distant faults contribute to the average in a numerically insignificant way, and a single quake in close proximity dominates the average. Second and more importantly, it mimics the attenuation observed for changes in static stress transfer, so that faults are affected by distant earthquakes in a numerically consistent way. To address the issue of discontinuities between neighboring segments, the distance method is applied after the fault segments are split into ˜10-km segments to create the smoothest possible transitions of frictional properties along fault traces at the system level. Note that all characteristic major event moment values fall between the largest and smallest event moments in the historic catalog. One may think of this method as defining a function, continuous everywhere except at the catalog hypocenters, mapping location in the latitude-longitude domain to a characteristic major event moment—a function completely independent of the locations of fault segments.

In compiling the historic earthquake catalogs, source material was added for each incarnation of Virtual California. The 1999 version of the model uses the catalog in Table 2 of DENG and SYKES (1997), which covers southern California $M \geq 6$ events from 1812–1994. Moment magnitude and/or local magnitude are used wherever possible. The 2000 version adds two more sources. One is Table 1 of JAUMÉ and SYKES (1996), which lists 'moderate' earthquakes in the San Francisco Bay region from 1850–1993. All events with $M < 5.8$ are excluded, and all earthquakes in this region are assumed to be strike-slip events. At some point, the magnitude of the 1906 San Francisco event was changed from 7.8 to 8.2, perhaps to better match the $M = 8.3$ value found in THATCHER (1975), but this change now appears to conflict with prevailing opinion. The other source added in the 2000 model is the UC Berkeley-based CNSS earthquake catalog, via the online search interface (http://quake.geo.berkeley.edu, since changed to the ANSS earthquake catalog). We use all earthquakes from 1900–2000 with $M \geq 5.8$ located between 35° and 42.5°N latitude and 127° and 116° W longitude. There is no easy way to distinguish strike-slip events in this catalog, therefore no further exclusions are made; this obviously presents a problem. The 2001 model adds only the CNSS earthquake catalog events from 1900–2001 with $M \geq 5.8$ located between 31.5° and 35° N latitude and 122° and 114.25°W longitude. In the final version of the model, all events with $M < 6.0$ are excluded.

During initial testing of the 2001 Virtual California model, a problem with the treatment of the largest earthquakes in the catalog was noticed. For earthquakes whose rupture areas fall along more than about 100 km of a fault, there are segments with significant slip far from the epicenter. In slipping, these segments reveal accumulated slip deficits large enough to indicate that their own characteristic events would be similarly large. However, because earthquakes are treated as point sources in our distance-weighted average method, the 'influence' of these large events is diluted for these segments' characteristic major event moments. As a result, their mean recurrence intervals are artificially low. To correct for this, large historic events, defined as events whose length along strike is significantly larger than the depth, are now represented as a set of smaller concatenated events spaced along the rupture length. The magnitude of a concatenated event is based upon the slip observed at its location during the historic large event. Thus the local observed slip is inserted into the moment-slip correlation (Eq. 3) and the moment-magnitude relation (Eq. 4) to yield a magnitude value. In this way, the distance-weighted average method reproduces the observed slip patterns.

The three large earthquakes singled out for splitting into smaller events are the 1812 Wrightwood event, the 1857 Fort Tejon event, and the 1906 San Francisco event. The Wrightwood and Fort Tejon earthquakes are based on the rupture models in Tables 3a and 3b of DENG and SYKES (1997). A smaller event is placed at the location of each listed rupture segment, and each segment is associated with its listed displacement value. In the case of the Fort Tejon event, the surface slip (SS) values are used. The San Francisco event is based on the plot of surface slip vs. distance in Figure 2 of THATCHER (1975), along with the accompanying rupture area map. Seventeen evenly spaced artificial events are created along the rupture between 39.00°N, 123.69°W and 36.82°N, 121.51°W, and the corresponding local slip values are taken from the plot (with an imposed minimum slip of 1 m to enhance slip continuity).

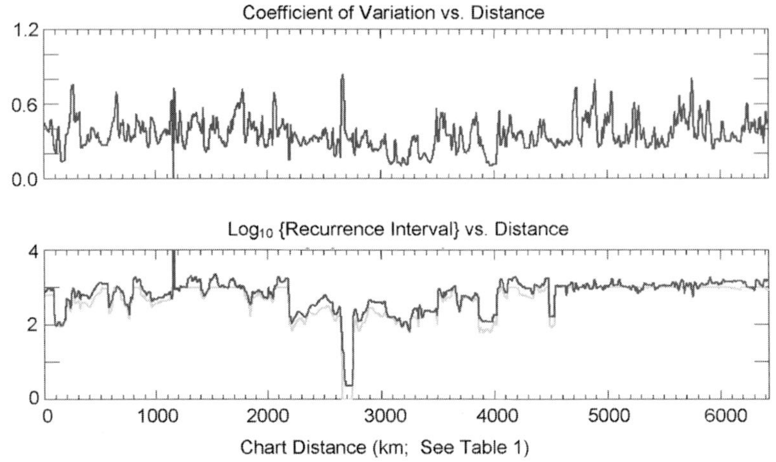

Figure 2
(a) Plot of coefficient of variation along faults. (b) Plot of average recurrence intervals along faults. Light lines: Recurrence intervals as computed from observed data. Dark lines: Recurrence intervals as measured from simulations.

4. *Applications*

We present computations below for a slightly modified version of the 2001 model. In these calculations, we have used a subset of the 650 of the 680 fault segments defined in the full VC 2001 model. Most of the neglected segments are at the extreme northern end of the northern San Andreas fault, where it takes a sharp westward turn north of Cape Mendocino. Scant information is available relative to the long-term slip rate on that section of fault, and its presence seemed to produce unwanted irregularities in the dynamics of the remaining segments. About five other fault segments in southern California on the Manix fault and three segments on the Verdugo fault were found to be interfering with the dynamics of the model for reasons that are presently obscure. This problem is the subject of current investigations. This somewhat reduced model is shown in Figure 1, superposed on a map of California. The fault segments are listed in Table 1 (see also RUNDLE *et al.*, 2004).

We have also assumed a uniform depth of 15 km for all faults in the model, which is in contrast to the "standard depth" of 20 km that was assumed in the construction of the model as described above. The depth value of 15 km was adopted because of a desire to first investigate models with uniform depth, and the value of 15 km is more reflective of the depth of seismicity in California's seismically active regions. In addition, although the data assimilation procedure described above provides our best initial estimate of the friction parameters on the faults, we find that simulated earthquakes often have a somewhat longer time interval between them on a given segment than was computed from the data assimilation procedure (see Fig. 2). This difference is a result of the fault interactions, and their effect on the dynamics. Thus we apply an overall multiplicative factor to the friction values obtained by our data assimilation procedure in order to optimally match the average time intervals between simulated large earthquakes on the San Andreas fault to the intervals observed in nature.

We now discuss an application of our simulations to the problem of earthquake forecasting. The report from the Working Group on California Earthquake Probabilities, WG02 (2003), represents the latest version of a method that has been developed for using observed geological and geophysical data, together with broad assumptions about the appropriate statistical distribution functions that describe the stochastic nature of the process. In the WG02 (2003) method, the models that are used to compute a forecast from the data are statistical in nature, that is, the physics included is typically encoded in 1) the form of the statistical distribution, and 2) the parameters of the distribution, usually the mean and variance. WG02 then uses the statistical distribution for all the fault segments in a region, with means and variances set using the geological and geophysical data, to compute the conditional probability that a large ($M \geq 6.7$) event will occur in the San Francisco Bay region during the period 2002–2031, given that the last such event occurred a time T years ago.

In our approach, we use our Virtual California simulations to generate ("measure") the statistical distribution needed to compute the conditional probability. In addition, we focus on sections of the northern and southern San Andreas fault. The basic idea is to use essentially the same methodology to evaluate the probability for earthquakes on these faults, but to replace the assumptions made about the form of the statistical distribution with measurements of the statistics of earthquake occurrence on the faults of interest that are obtained from our simulations, which include fault interactions.

The primary statistical distributions investigated by the WG02 are the Log-normal distribution (e.g., EVANS *et al.*, 1993), and the Brownian first passage, or Brownian Passage Time (BPT) distribution (RANGARAJAN and DING, 2000; MATTHEWS *et al.*, 2002). The Log-normal distribution is known to describe a number of physical situations where aging, fatigue, and failure processes occur. Then the probability density function (PDF) for failure between time t and $t + \delta t$ is given by:

$$P_{LN}(t) = \frac{1}{\omega t\sqrt{2\pi}} \exp\left\{\frac{-[\log(t/\rho)]^2}{2\omega^2}\right\}. \tag{6}$$

Here the parameters ρ and ω are related to the mean λ and standard deviation θ of the population statistics of waiting times t by the usual relations:

$$\rho \equiv \frac{\lambda}{\sqrt{\left(\frac{\theta}{\lambda}\right)^2 + 1}}, \tag{7}$$

$$\omega \equiv \sqrt{\log\left\{\left(\frac{\theta}{\lambda}\right)^2 + 1\right\}}. \tag{8}$$

Likewise, the Brownian Passage Time PDF for failure between time t and $t+\Delta t$ is given by (RANGARAJAN and DING, 2000; MATTHEWS *et al.*, 2002):

$$P_{BPT}(t) = \sqrt{\frac{\lambda^3}{2\pi\theta^2 t^3}}\exp\left\{-\frac{\lambda(t-\lambda)^2}{2\theta^2 t}\right\}. \tag{9}$$

We also discuss the Weibull PDF (TURCOTTE, 1997), which is the type of statistics often used in Japan (RIKITAKE, 1982). This can be written in the form (EVANS *et al.*, 1993):

$$P_W(t) = \beta\left[\frac{t^{\beta-1}}{\tau^\beta}\right]\exp\left\{-\left(\frac{t}{\tau}\right)^\beta\right\}. \tag{10}$$

The mean λ and standard deviation θ of $P_W(t)$ are given in terms of the parameters τ, β by:

$$\lambda = \tau\Gamma\left[\frac{\beta+1}{\beta}\right], \tag{11}$$

$$\theta = \tau^2\left\{\Gamma\left[\frac{\beta+2}{\beta}\right] - \left(\Gamma\left[\frac{\beta+1}{\beta}\right]\right)^2\right\}. \tag{12}$$

In the WG02 method, the values for the standard deviation of earthquake recurrence time θ_i, and the mean recurrence time interval λ_i are inferred from a variety of observations on the selected Bay Area fault segments. A problem that is sometimes encountered is that for some segments, observations for only the last earthquake involving that segment exist; thus, the values of θ_i and λ_i are estimated by indirect means. Once the PDFs $P(t)$ for the segments are established, the Cumulative Distribution Function $C(t)$, also called simply the "distribution function", is then used to compute the conditional probability (also called the "discrete hazard function") for failure of the segment during the interval $(T, T+\Delta T)$, given that it has been a time T since the last such failure (WG02, 2003):

$$H_{\Delta T}(t,T) = P(T \le t \le T+\Delta T\,|t \ge T) = \frac{C(T+\Delta T) - C(T)}{1 - C(T)}. \tag{13}$$

The WG02 (2003) computes the $\Delta T = 30$ year conditional probabilities (Eq. 13) for the interval 2002–2031, for the occurrence of an earthquake with magnitude $M \ge 6.7$, occurring on the active earthquake faults in the San Francisco Bay region.

Here we compute the entire Discrete Hazard Function $H_{\Delta T}(t, T)$ for the occurrence of an earthquake with magnitude $M \ge 7.5$, for two intervals, 10 years and 30 years, as a function of the time since the last such event anywhere on the fault.

Instead of assuming the PDF $P(t)$ and its associated distribution $C(t)$, we measure these quantities from simulations. Our method uses basically the same type of observational data as does the WG02, but instead of using these data to estimate the statistical parameters θ_i and λ_i for each fault segment, the data are assimilated into physical values for the Virtual California fault segments as described above. We then use the simulations to directly measure $P(t)$ and $C(t)$ for the fault segments of interest, and finally compute $H_{\Delta T}(t,T)$ as in Eq. (8).

The results are shown in Figures 2–7, which are based on 40,000 years of simulated earthquakes on the fault system shown in Figure 1. Figure 2a shows the Coefficient of Variation (CoV) on each fault segment, plotted as a function of "chart distance" along the fault system (see Table 1). Recall that the CoV is defined as the ratio of standard deviation θ to the mean λ of a probability density:

$$CoV \equiv \frac{\theta}{\lambda}. \tag{14}$$

5. *Computational Results*

Figure 3 shows the sections of the San Andreas fault that are considered for our computation of conditional probabilities. Figures 4–7 compare the output of Virtual California simulations to results obtained using WG02 (2003)-type methods employing the Log-normal distribution (dotted lines in the figures), the BPT distribution

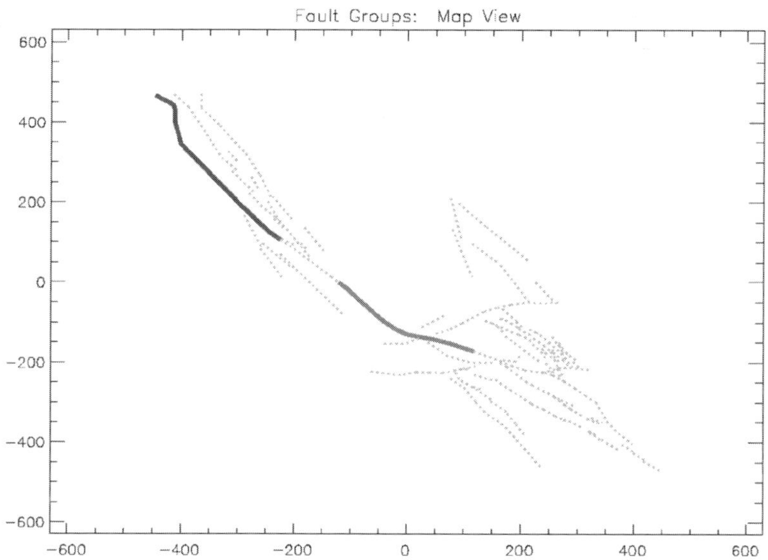

Figure 3
Map of VC 2001 fault model with northern San Andreas fault (NSAF) segments (dark gray) and southern San Andreas fault (SSAF) segments (light gray) indicated. Axes are in km.

(dashed lines in the figures), and the Weibull distribution (solid lines in the figures). The irregular curve is obtained from the simulation data. The Log-normal, BPT and Weibull distributions are chosen to have the same mean and standard deviations as the simulation results (see Table 2). In Figure 4, we show four histograms for inter-event

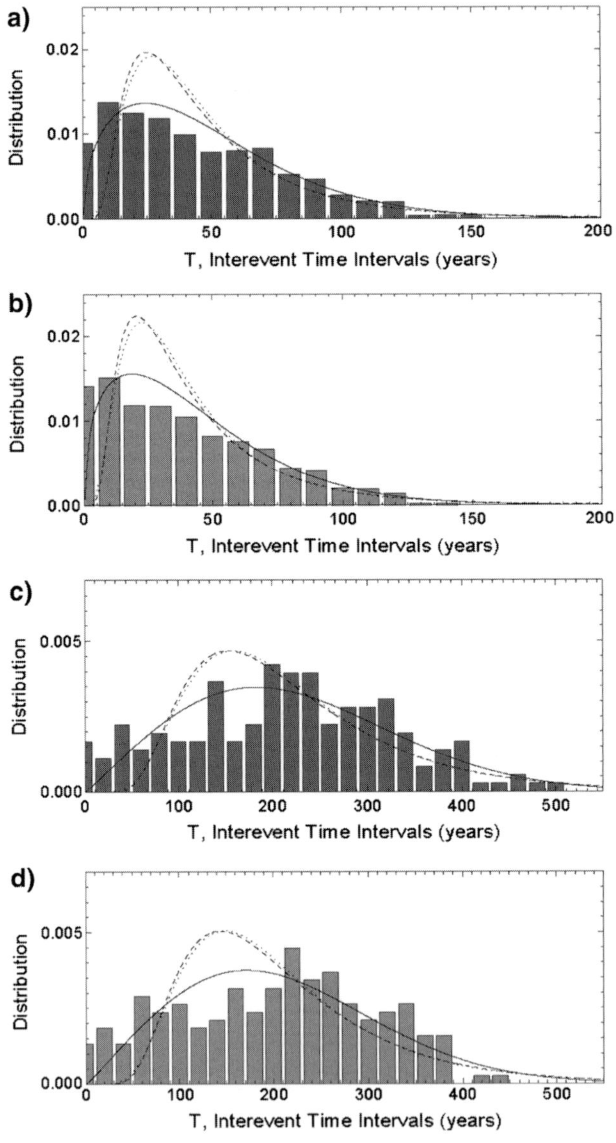

Figure 4

Histograms for interevent time intervals for earthquake data from Virtual California simulations. (a) NSAF events having $M \geq 7.0$. (b) SSAF events having $M \geq 7.0$. (c) NSAF events having $M \geq 7.5$. (d) SSAF events having $M \geq 7.5$.

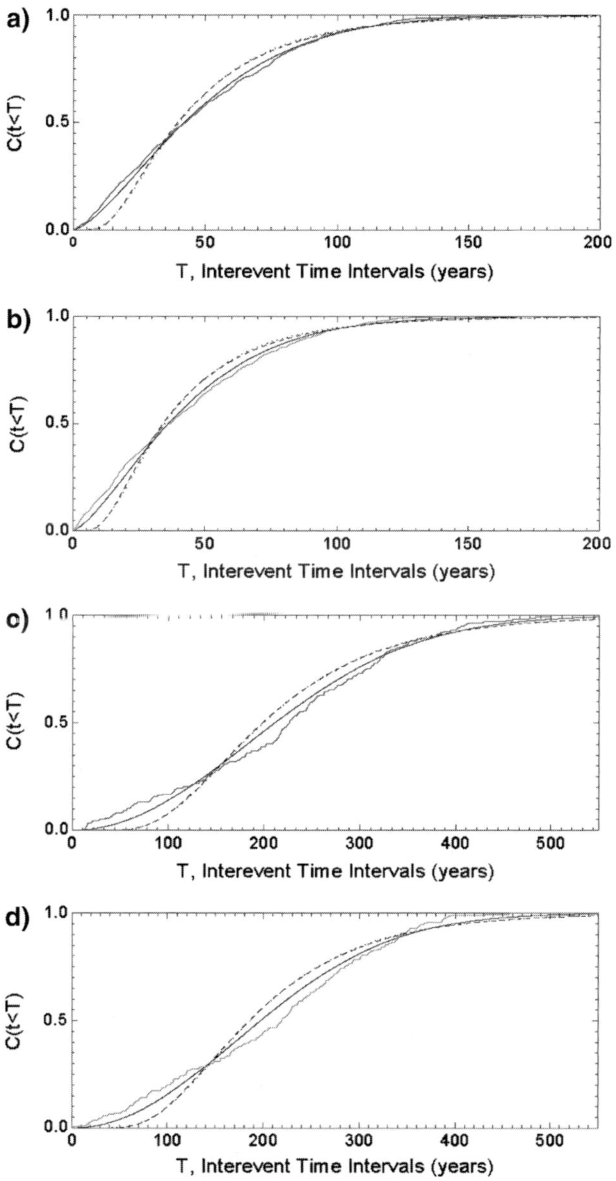

Figure 5

Cumulative Distribution Functions for interevent time intervals for earthquake data from Virtual California simulations. (a) NSAF events having $M \geq 7.0$. (b) SSAF events having $M \geq 7.0$. (c) NSAF events having $M \geq 7.5$. (d) SSAF events having $M \geq 7.5$.

time intervals for a) events having $M \geq 7.0$ on the northern San Andreas fault; b) events having $M \geq 7.0$ on the southern San Andreas fault; c) events having $M \geq 7.5$ on the northern San Andreas fault; and d) events having $M \geq 7.5$ on the southern San Andreas

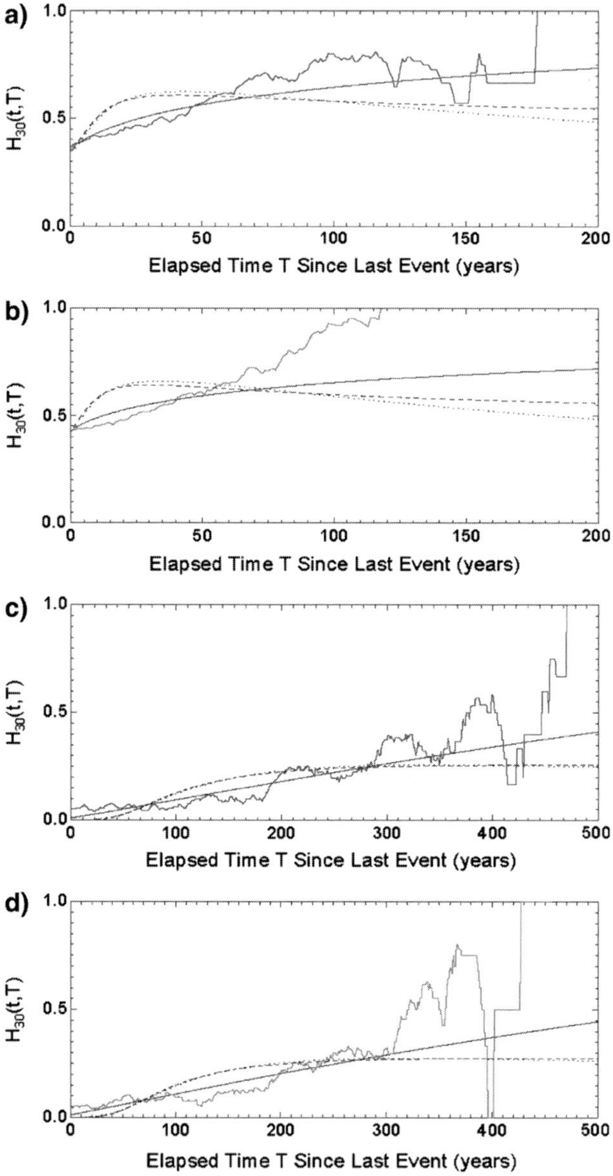

Figure 6
30-year Discrete Hazard Functions (Conditional Probabilities) for earthquake data from Virtual California simulations. (a) NSAF events having $M \geq 7.0$. (b) SSAF events having $M \geq 7.0$. (c) NSAF events having $M \geq 7.5$. (d) SSAF events having $M \geq 7.5$.

fault. In all plots, histograms with the same normalization and the same statistical parameters are shown for the Log-normal (solid) and BPT statistics. When integrated and normalized to 1, the histograms can be used as PDFs to compute $P(t)$. In Figures

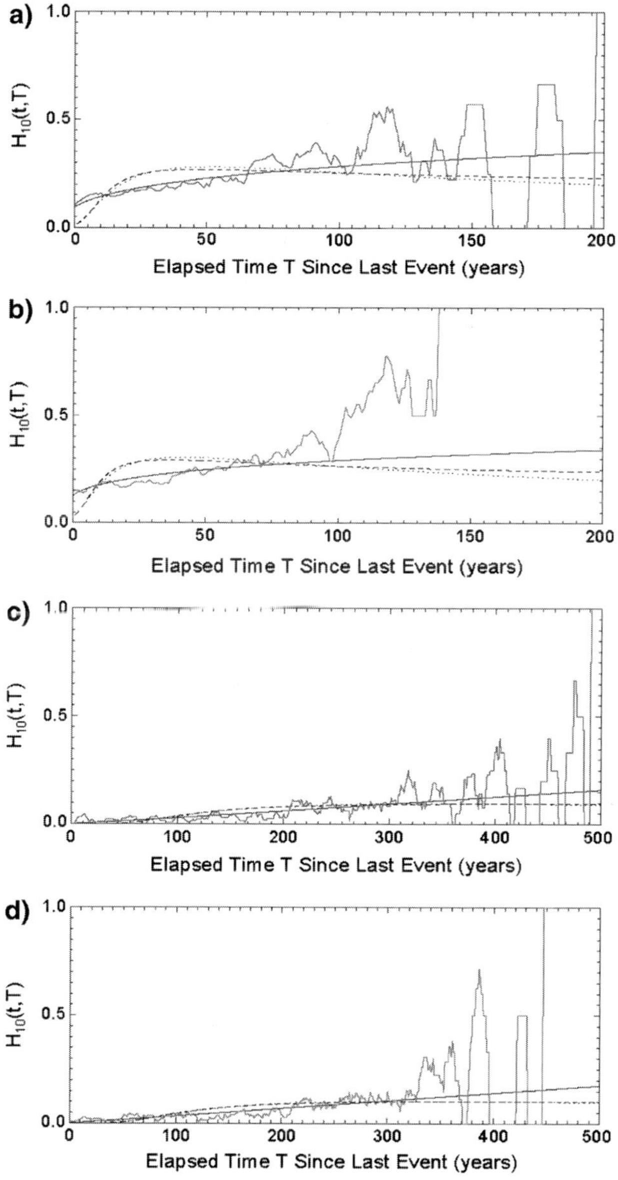

Figure 7
10-year Discrete Hazard Functions (Conditional Probabilities) for earthquake data from Virtual California simulations. (a) NSAF events having $M \geq 7.0$. (b) SSAF events having $M \geq 7.0$. (c) NSAF events having $M \geq 7.5$. (d) SSAF events having $M \geq 7.5$.

4a and 4b, it can be seen that the Log-normal and the BPT curves are characteristically too high at the small-interval end, and too low at the high-interval end. However, the Weibull distribution provides a reasonable, average representation of the simulation data. Figures 5(a–d) show CDFs, or $C(t)$ curves corresponding to the histograms of Figure 4. Differences between the measured data and the theoretical curves for the Log-normal and BPT CDFs are not as apparent to the eye as they are for the histograms; a well-known result that arises due to the fact that the CDFs represent the integration of the histogram or PDF data. However, the Weibull distribution is clearly different and again represents a better fit to the simulation data.

Figures 6(a–d) plot the 30-year values for $H_{30}(t,T)$. Figure 6a is for all northern San Andreas events having $M \geq 7.0$; Figure 6b is for all southern San Andreas events having $M \geq 7.0$; Figure 6c is for all northern San Andreas events having $M \geq 7.5$; and Figure 6d is for all southern San Andreas events having $M \geq 7.5$. It can be seen that the $H_{30}(t,T)$ curves for the simulation data with $M \geq 7.0$ (Figs. 6a,b) are not well-represented by the $H_{30}(t,T)$ curves obtained from either the Log-normal or BPT statistics, but are reasonably well represented by the Weibull statistics. Moreover, the $H_{30}(t,T)$ curves eventually rise to the asymptotic value $H_{30}(t,T) \rightarrow 1$ as t increases, due to the fact that all faults in the simulation eventually fail. By contrast, the $H_{30}(t,T)$ curves for the Log-normal and BPT distributions do not tend asymptotically to $H_{30}(t,T) \rightarrow 1$, which is a well-understood feature of these statistics (WG02 2003). The Weibull distribution produces $H_{30}(t,T)$ curves that do eventually tend asymptotically to 1.

Finally, Figures 7(a–d) plots the 10-year values for $H_{10}(t,T)$. Figure 7a is for all northern San Andreas events having $M \geq 7.0$; Figure 7b is for all southern San Andreas events having $M \geq 7.0$. Figure 7c is for all northern San Andreas events having $M \geq 7.5$; Figure 7d is for all southern San Andreas events having $M \geq 7.5$. Again, the $H_{10}(t,T)$ curves eventually rise to the asymptotic value $H_{10}(t,T) \rightarrow 1$ as t increases, due to the fact that all faults in the simulation eventually fail. The $H_{10}(t,T)$ curves for the Log-normal and BPT distributions also do not tend asymptotically to $H_{10}(t,T) \rightarrow 1$, but the Weibull distributions do. Although the Log-normal and BPT curves approximately represent the simulation data for about 130 years' inter-event time, the $H_{10}(t,T)$ curves thereafter rise abruptly to $H_{10}(t,T) \rightarrow 1$ as all fault segments eventually fail.

6. *Conclusions*

We conclude that topologically realistic simulations such as Virtual California can be easily used to develop forecast statistics for earthquakes. An advantage of simulation data such as that shown here is that interactions among the faults in the system are fully included—they do not have to be assumed or added *a posteriori*. Moreover, all of the faults in the system eventually fail, meaning that there is no paradox that failure becomes increasingly unlikely as the time since the last event recedes, as is the case with renewal models such as the Log-normal distribution.

As an example of how these forecast figures are used, we can consider the Loma Prieta earthquake of 1989 ($T = 15$ years). Note that the southern San Andreas fault depicted in Figure 3 does not include the Parkfield segment. Assuming that the Loma

Prieta earthquake was on the San Andreas fault, Figures 6a and 7a show that the 30-year conditional probability for failure $H_{30}(t,15)$ on the northern San Andreas fault, given the most recent earthquake larger than $M \geq 5.8$, is now approximately 50%. The corresponding 10-year conditional probability for failure $H_{10}(t,15)$ is approximately 18%.

There are two important qualifications on these computations, which are: 1) these hazard probabilities represent lower bounds since we have not included all possible observed (and unobserved) faults in the model; and 2) we have not controlled for the time of earthquake intervals with respect to occurrence time within the earthquake cycle of the greatest events.

The validity of our data assimilation procedure also depends importantly on the characteristics (location, magnitude) of the large historical events upon which it is based. To the extent that these basic parameters are unknown, or that events were not observed or adequately characterized, our procedures for assigning friction values will be in error. We fully expect that knowledge of these parameters will improve as better data are obtained.

Acknowledgements

This work has been supported by a grant from US Department of Energy, Office of Basic Energy Sciences to the University of California, Davis DE-FG03-95ER14499 (JBR); with additional funding from the National Aeronautics and Space Administration under grants to the Jet Propulsion Laboratory and the University of California, Davis (JBR; PBR, GM, DT; AD, PL). This research was also supported in part by the Southern California Earthquake Center (PBR). SCEC is funded by NSF Cooperative Agreement EAR-0106924 and by USGS Cooperative Agreement 02HQAG0008. SCEC contribution number for this paper is 777.

Appendix

Physics of the Virtual California Model

The *Virtual California* model (Rundle *et al.*, 2001, 2004, 2005) is a stochastic, cellular automata instantiation of an earthquake *backslip* model, in that loading of each fault segment occurs via the accumulation of slip deficit $\phi(\mathbf{x},t) = s(\mathbf{x},t) - Vt$, where $s(\mathbf{x},t)$ is slip, V is long-term slip rate, and t is time. At the present time, faults used in the model are exclusively vertical strike-slip faults; the most active faults in California, and upon which most of the seismic moment release is localized. Thrust earthquakes, such as the 1994 Northridge and 1971 San Fernando faults, are certainly damaging, but they occur infrequently and are therefore regarded as perturbations on the primary strike slip fault structures. The Virtual California model also has the following additional characteristics.

1. *Surfaces of discontinuity* (faults) across which slip is discontinuous at the time of an earthquake, and which are subject to frictional resistance. Here we restrict the model to only topologically complex systems of vertically dipping faults mirroring the complexity found on the natural fault networks of southern California.
2. *Stochastic dynamics*. In these models, we are interested in the space-time patterns and correlations that emerge from the underlying stress-strain dynamics. These correlations evolve over many hundreds or thousands of years, time scales much longer than the time scales associated either with rupture or elastic wave periods. Most of the elastic and frictional parameters for faults and earth materials, although known in the laboratory, will likely remain poorly defined in nature. For this reason it makes little sense to attempt a deterministic solution to the equations of motion. Instead, we use a Cellular Automaton (CA) approach, in which the dynamics is parameterized by random variables chosen from well defined probability distributions.
3. *Linear elastic stress transfer* or interactions between fault surfaces. Again, although most of the significant parameters associated with rupture, such as friction coefficients and friction law constants and functions can be defined and measured in the laboratory, current experience indicates they will likely always be poorly known for faults in nature. We therefore use quasistatic stress interaction (Green's function) tensors $T_{ij}^{kl}(\boldsymbol{x} - \boldsymbol{x}')$, which we will write henceforth schematically as $T(\boldsymbol{x} - \boldsymbol{x}')$.
4. *Persistent increase of stresses* on the fault surfaces arising from plate tectonic forcing parameterized via the backslip method. This method has the advantage of matching the long-term rate of offset V in model faults with the geologically known long-term slip rate on faults in nature. Stress increase occurs via the following physics. The stress tensor $\sigma_{ij}(\boldsymbol{x},t)$ is related to the slip $s_l(\boldsymbol{x},t)$ by:

$$\sigma_{ij}(\boldsymbol{x},t) = \int d\boldsymbol{x}_k \; T_{ij}^{kl}(\boldsymbol{x}-\boldsymbol{x}') \; s_l(\boldsymbol{x}',t). \tag{A1}$$

   Now if $\boldsymbol{x} = \boldsymbol{x}'$, a positive slip $s_l(\boldsymbol{x},t) > 0$ results in a *decrease* in stress, $\Delta\sigma_{ij}(\boldsymbol{x},t) < 0$. Therefore, if we write the equation:

$$\sigma_{ij}(\boldsymbol{x},t) = \int d\boldsymbol{x}_k \; T_{ij}^{kl}(\boldsymbol{x}-\boldsymbol{x}') \; \{ s_l(\boldsymbol{x}',t) \; - \; V_l(\boldsymbol{x}')t \}, \tag{A2}$$

   where $V_l(\boldsymbol{x}) = \langle s_l(\boldsymbol{x},t)\rangle$ is the average long-term rate of slip at $\boldsymbol{x}'$, then the second term - $V_l(\boldsymbol{x})t$ leads to an *increase* in the stress, $\Delta\sigma_{ij}(\boldsymbol{x},t) > 0$. Therefore the second term is the stress accumulation term.
5. Parameters for friction laws and fault topology that are determined by assimilating seismic, paleoseismic, geodetic, and other geophysical data from events occurring over the last ~200 years in California (Rundle *et al.*, 2001, 2002).

6. Frictional resistance laws that range from the simplest Amontons-Coulomb stick-slip friction, to heuristic laws such as slip- or stress-rate dependent weakening laws based on recent laboratory friction (TULLIS, 1996) and fracture experiments. These laws are related to rate-and-state and leaky threshold laws (RUNDLE *et al.*, 2002).

In general, several of the friction laws described above can be written in the following representative, equivalent forms on an element of fault surface:

$$\begin{aligned} \frac{\partial \sigma}{\partial t} &= K_L V - f(\sigma V), \\ K_L \frac{\partial s}{\partial t} &= f(\sigma, V). \end{aligned} \tag{A3}$$

Here $s(\boldsymbol{x},t)$ is slip at position $\boldsymbol{x}$ and time t, $\sigma(\boldsymbol{x},t)$ is shear stress, K_L is the self-interaction or "stress drop stiffness" and $f[\sigma,V]$ is the *stress dissipation function* (RUNDLE *et al.*, 2002). For example, the "Amontons" or Coulomb friction law, having a sharp failure threshold, can be written in the form (2) using a Dirac delta function:

$$\frac{\partial s}{\partial t} = \frac{\Delta\sigma}{K_L}\, \delta(t - t_F), \tag{A4}$$

where the stress drop $\Delta\sigma = \sigma - \sigma^R(V)$ and $\sigma^R(V)$ is the velocity-dependent residual stress. For laboratory experiments, K_L is the{machine + sample}stiffness, and for simulations, represents the stiffness of a coarse-grained element of the fault of scale size L. $\delta(\,)$ is the Dirac delta, and t_F is any time at which $\sigma(\boldsymbol{x},t_F) = \sigma^F(V)$. Both σ^F and σ^R can also be parameterized as functions of the normal stress χ by means of coefficients of static μ_S and ("effective") kinetic μ_K coefficients of friction, $\sigma^F = \mu_S\,\chi$, $\sigma^R = \mu_K\,\chi$.

In recent work (RUNDLE *et al.*, 2002), we have introduced another parameter α, which allows for stable stress-dependent aseismic sliding. The process described by α is seen in laboratory friction experiments (TULLIS, 1996), and is expressed by a generalization of equation (A4):

$$\frac{\partial s}{\partial t} = \frac{\Delta\sigma}{K_L}\{\alpha + \delta(t - t_F)\}. \tag{A5}$$

We found that the parameter α, which can be fixed either through laboratory experiments or through field observations (TULLIS, 1996; DENG and SYKES, 1997), acts to smooth the stress field a fault when $\alpha > 0$, and to roughen the fault stress field when $\alpha < 0$.

In the model results that we describe here, we further generalize (A5) to include an additional term which depends on rate of stress increase:

$$\frac{\partial s}{\partial t} = \frac{\Delta\sigma}{K_T}\left\{\alpha + \delta(t - t_F) + \beta\,\delta\!\left(\frac{\partial \sigma}{\partial t} - \eta\right)\right\}, \tag{A6}$$

where β is a constant with appropriate units (stress/time^2) and η is a critical stress rate. Here K_T represents the total spring constant associated with a fault segment. The last term can be considered to be parameterization of effects associated with a dynamic stress intensity factor. It is known that stress rate effects are important in the process of dynamic fracture, such as might be expected during an earthquake. For example, the stress intensity factor K_{I} for mode I tensile fracture is thought to be of the form:

$$K_{ID} = K_{ID}\left(\frac{\partial\sigma}{\partial t}, \; T\right), \tag{A7}$$

where T is temperature. More specifically, for a crack propagating at velocity v, it has been proposed that the time-dependent dynamic stress intensity factor $K_D(t)$ is of the general form:

$$K_D(t) = k(v)\; K_D(0) = k(v)\; K_S, \tag{A8}$$

where K_S is the static stress intensity factor. While not of the exact form of either equation (A7) or (A8), equation (A6) is an expression of the idea that the onset of earthquake sliding depends on the stress rate through a critical threshold value η.

In the simulations described below, we implement equation the physical process described by equation (A6) in our Virtual California CA simulations as follows. We define the Coulomb Failure Function $CFF(\boldsymbol{x},t)$:

$$CFF(\boldsymbol{x},t) \;=\; \sigma(\boldsymbol{x},t) \;-\; \mu_S\; \chi(\boldsymbol{x},t), \tag{A9}$$

According to the first term in the equation, stable slip can occur with amplitude proportional to α for nonzero $\Delta\sigma$. In addition, according to the second term, unstable failure of a fault occurs of when $CFF(\boldsymbol{x},t) = 0$. To implement a failure mechanism in a simple way that demonstrates physics similar to the third term, we allow unstable slip of amplitude:

$$\frac{\Delta\sigma}{K_T} \;=\; \frac{\sigma(\boldsymbol{x},t) \;-\; (\mu_S \;-\; \mu_K)\,\chi(\boldsymbol{x},t)}{K_T}, \tag{A10}$$

when the condition:

$$-\frac{\partial}{\partial t}\;\mathrm{Log}\{\,CFF(\boldsymbol{x},t)\,\} \;>\; \eta \tag{A11}$$

where typically $0 < \eta < 1$, or in discrete terms:

$$\frac{CFF(\boldsymbol{x},t) \;-\; CFF(\boldsymbol{x},t+\Delta t)}{CFF(\boldsymbol{x},t)} \;>\; \eta. \tag{A12}$$

In equation (A12), we interpret Δt as being the time since the beginning of the earthquake at time t. Implicitly, it is assumed in (A6), (A11) and (A12) that:

$$\eta \gg \left. \frac{\partial \sigma(\boldsymbol{x}, t)}{\partial t} \right|_{\text{Interseismic}} = - \int d\boldsymbol{x}_k \, T_{ij}^{kl}(\boldsymbol{x} - \boldsymbol{x}') \, V_l(\boldsymbol{x}'), \tag{A13}$$

i.e., that the η-value for stress-rate triggering is much larger than the stress rate characterizing interseismic stress accumulation.

References

BARNHARD, T. and HANSON, S., compilers (2000–2001 accessed), *Fault parameters used in USGS 1996 Seismic Hazard maps*, http://eqhazmaps.usgs.gov/html/faults.html.

COUNCIL OF THE NATIONAL SEISMIC SYSTEM, in association with the NORTHERN CALIFORNIA EARTHQUAKE DATA CENTER (2000–2001 accessed), *CNSS earthquake catalog*, http://quake.geo.berkeley.edu/cnss/catalog-search.html.

DENG, J. S. and SYKES, L. R. (1997), *Evolution of the stress field in southern California and triggering of moderate-size earthquakes: A 200-year perspective*, J. Geophys. Res. *102*(B5), 9859–9886.

DIETERICH, J. S. (1979), *Modeling of rock friction. 1. Experimental results and constitutive equations*, J. Geophys. Res. *84*, 2161–2168.

EVANS, M., HASTINGS, N., and PEACOCK, B., *Statistical Distributions* (Wiley Interscience, New York. 1993).

JAUMÉ, S. C. and SYKES, L. R. (1996), *Evolution of moderate seismicity in the San Francisco Bay region, 1850 to 1993: Seismicity changes related to the occurrence of large and great earthquakes*, J. Geophys. Res. *101*(B1), 765–789.

KANAMORI, H. and ANDERSON, D. L. (1975), *Theoretical basis of some empirical relations in seismology*, Bull. Seismol. Soc. Am. *65*, 1073–1095.

KARNER, S. L. and MARONE, C., *Effects of loading rate and normal stress on stress drop and stick-slip recurrence interval*, pp. 187–198 in (RUNDLE J. B., TURCOTTE, D. L., and KLEIN, W., eds.) *GeoComplexity and the Physics of Earthquakes*, Geophysical Monograph 120 (American Geophysical Union, Washington, D.C., 2000).

MATTHEWS, M. V., ELLSWORTH, W. L., and REASENBERG, P. A. (2002), *A Brownian model for recurrent earthquakes*, Bull. Seismol. Soc. Am. *92*, 2233–2250.

PETERSEN, M. D. and WESNOUSKY, S. G. (1994), *Fault slip rates and earthquake histories for active faults in Southern California*, Bull. Seismol. Soc. Am. *84*(5), 1608–1649.

PETERSEN, M. D., BRYANT, W. A., CRAMER, C. H., CAO, T., REICHLE, M., FRANKEL, A. D., LIENKAEMPER, J. J., MCCRORY, P. A., and SCHWARTZ, D. P. (1996), *Probabilistic seismic hazard assessment for the state of California*, USGS Open-File Report 96–706, U. S. Govt. Printing Office.

RANGARAJAN, G. and DING, M. (2000), *First passage time problem for biased continuous-time random walks*, Fractals *8*, 139–145.

RIKITAKE, T., *Earthquake Forecasting and Warning* (Center for Acad. Publ. Japan, D. Reidel, Hingham, MA, USA. 1982).

RUNDLE, J. B. (1988), *A physical model for earthquakes, 2, Application to southern California*, J. Geophys. Res. *93*, 6255–6274.

RUNDLE, P. B., RUNDLE, J. B., TIAMPO, K. F., SÁ MARTINS, J. S., MCGINNIS, S., and KLEIN, W. (2001), *Nonlinear network dynamics on earthquake fault systems*, Phys. Rev. Lett. *87*(14), Art. No. 148501 (1–4).

RUNDLE, J. B., TIAMPO, K. F., SÁ MARTINS., and KLEIN, W. (2002), *Self-organization in leaky threshold systems: The influence of near mean field dynamics and its implications for earthquakes, neurobiology and forecasting,* Proc. Nat. Acad. Sci. USA, *99*, Supplement 1, 2514–2521.

RUNDLE, J. B., RUNDLE, P. B., DONNELLAN, A., and FOX, G. C. (2004), *Gutenberg-Richter statistics in topologically realistic system-level earthquake stress-evolution simulations*, Earth Planets Space *56*, 761–771.

RUNDLE, J. B., RUNDLE, P. B., DONNELLAN, A., TURCOTTE, D.L., SHCHERBAKOV, R. LI, P., MALAMUD, B. D., GRANT, L. B., FOX, G. C., MCLEOD, D., YAKOVLEV, G., PARKER, J., KLEIN, W., and TIAMPO, K. F.

(2006, in press), A simulation-based approach to forecasting the next great San Francisco earthquake, Proc. Nat. Acad. Sci.

SCHWARTZ, D. P. and COPPERSMITH, K. J. (1984), Fault Behavior and characteristic earthquakes: Examples from the Wasatch and San Andreas fault zones, J. Geophys. Res. *89*, 5681–5698.

SOUTHERN CALIFORNIA EARTHQUAKE DATA CENTER (2001 accessed), *Clickable Fault Map of Southern California*, http://www.data.scec.org/faults/faultmap.html.

THATCHER, W. (1975), *Strain accumulation and release mechanism of 1906 San Francisco Earthquake*, J. Geophys. Res. *80*(35), 4862–4872.

TULLIS, T. E. (1996), *Rock friction and its implications for earthquake prediction examined via models of Parkfield earthquakes*, Proc. Nat. Acad. Sci. USA *93*, 3803–3810.

TURCOTTE, D.L., *Fractals and Chaos in Geology and Geophysics*, (Cambridge University Press, Cambridge, UK. 1997).

WALD, D. J. and HEATON, T. H. (1994), *Spatial and temporal distribution of slip for the 1992 Landers, California, earthquake*, Bull. Seismol. Soc. Am. *84*(3).

WARD, S.N. (1996), *A synthetic seismicity model for southern California: Cycles, probabilities and hazard*, J. Geophys. Res. *101*, 22393–22418.

WARD, S. N. (2000), *San Francisco Bay Area earthquake simulations: A step toward a standard physical earthquake model*, Bull. Seismol. Soc. Am. *90*(2), 370–386.

WARD, S.N. and GOES, S.D.B. (1993), *How regularly do earthquakes recur?-A synthetic seismicity model for the San Andreas fault*, Geophys. Res. Lett. *20*, 2131–2134.

WELLS, D. L. and COPPERSMITH, K. J. (1994), *New empirical relationships among magnitude, rupture length, rupture width, rupture area, and surface displacement*, Bull. Seismol. Soc. Am. *84*(4), 974–1002.

WORKING GROUP ON CALIFORNIA EARTHQUAKE PROBABILITIES (1999), *Earthquake Probabilities in the San Francisco Bay Region: 2000 to 2030 – A Summary of Findings*, USGS Open-File Report 99–517, U. S. Govt. Printing Office.

(Received December 30, 2004, revised August 10, 2005, accepted September 5, 2005)
Published Online First: August 31, 2006

Pure appl. geophys. 163 (2006) 1847–1865
0033–4553/06/091847–19
DOI 10.1007/s00024-006-0100-8

Pure and Applied Geophysics

Catastrophic Rupture Induced Damage Coalescence in Heterogeneous Brittle Media

FENG RONG,[1] HAIYING WANG,[1] MENGFEN XIA,[1,2] FUJIU KE,[1,3] and YILONG BAI[1]

Abstract—In heterogeneous brittle media, the evolution of damage is strongly influenced by the multiscale coupling effect. To better understand this effect, we perform a detailed investigation of the damage evolution, with particular attention focused on the catastrophe transition. We use an adaptive multiscale finite-element model (MFEM) to simulate the damage evolution and the catastrophic failure of heterogeneous brittle media. Both plane stress and plane strain cases are investigated for a heterogeneous medium whose initial shear strength follows the Weibull distribution. Damage is induced through the application of the Coulomb failure criterion to each element, and the element mesh is refined where the failure criterion is met. We found that as damage accumulates, there is a stronger and stronger nonlinear increase in stress and the stress redistribution distance. The coupling of the dynamic stress redistribution and the heterogeneity at different scales result in an inverse cascade of damage cluster size, which represents rapid coalescence of damage at the catastrophe transition.

Key words: Heterogeneity, damage evolution, FEM, damage coalescence, catastrophic rupture, stress redistribution.

1. Introduction

Damage evolution and rupture of disordered heterogeneous brittle media, such as rocks and the earth's crust, is a key problem in science, especially in earthquake simulations (RUNDLE *et al.*, 2000; BEN ZION and SAMMIS, 2003; MORA *et al.*, 2002). The catastrophe transition occurring in such media is the most important feature in this problem (BAI *et al.,* 1994; XIA *et al.,* 2002; WEI *et al.,* 2000). In rock mechanics (JAEGER and COOK, 1979), it is found that at the beginning of loading, micro-damages appear on mesoscopic heterogeneities. With increasing load, some disordered weakness at mesoscales may be amplified strongly due to strong and trans-scale stress redistribution, and become significant

[1] State Key Laboratory of Nonlinear Mechanics (LNM), Institute of Mechanics, Chinese Academy of Sciences, Beijing 100080, China. E-mail: rongf@hotmail.com
[2] Department of Physics, Peking University, Beijing 100871, China.
[3] Department of Applied Physics, Beihang University, Beijing 100083, China.

effects on the global scale. Beyond the catastrophe point, the responses of the sample, such as damage patterns, energy release, etc., change dramatically with increasing controlling variable. In other words, the system displays trans-scale catastrophic transition.

Though material heterogeneity has been examined in a number of models, such as fiber bundle model (DANIELS, 1945; COLEMAN, 1958; CURTIN, 1997), coupled pattern mapping model (XIA *et al.*, 2000; WEI *et al.*, 2000), network model etc., the mechanisms underlying the catastrophic transition through various spatial scales have not been clearly revealed in computational simulations (CAO *et al.*, 2002). Therefore, it is of the deepest interest to reveal the characteristic features in the catastrophe transition and to seek its underlying mechanism.

Recent works in the study of multiscale coupling (BAZANT and CHEN, 1997; WANG *et al.,* 2004; RONG *et al.,* 2006) suggest that damage evolution and rupture of solid can be regarded as a trans-scale process in a wide range of spatio-temporal scales. Traditionally, researchers working on computational simulations concentrated on bringing practical details into their models, but ignored various spatial scales inherent in heterogeneous media. To obtain proper resolution of small-scale damage effects, finite-element meshes consisting of several million elements are necessary. These often require hours of computation time on even the latest supercomputer and make the analysis of multiscale damage evolution almost impossible. This severely limits the usefulness of computation. Although there are many previous works on the multiscale numerical scheme, such as multilevel method (FISH and CHEN, 2003) and adaptive VCFEM (GHOSH *et al.*, 2001), most of them use the asymptotic homogenization and can be quite effective in analyzing the scale effect, especially in the debonding or crack evolution. However, our concern is the multiscale effects of damage evolution induced catastrophe of heterogeneous media. The detailed stress fields are not enough and the coupling among the micro-damages should be considered to cope with these effects (KRAUSE and RANK, 2003). Therefore, it is necessary to develop a numerical model, which can effectively deal with the coupling of medium heterogeneity and dynamic stress redistribution.

On the other hand, as noted previously that in heterogeneous brittle media, the micro-damages are usually initiated from small scale at some weak locations. Then, micro-damages develop from small to large scales via growth and coalescence, and eventually trigger catastrophe transition. In order to simulate such a phenomena, we must turn to the adaptive scheme. However, in the standard adaptive finite-element algorithm, its constitutive relation most often keeps the same in all meshes (regardless of how large or small), hence it cannot depict heterogeneity properly (RONG *et al.*, 2005b).

In this paper, a Multiscale adaptive Finite-Element Model (MFEM) is proposed, and some 2-D problems, both plane stress and plane strain cases, are simulated with the MFEM. Based on the numerical results, it is found that damage coalescence from mesoscopic to macroscopic scales is crucial in the catastrophe transition. In particular, the underlying mechanism of catastrophe transition, namely the coupling between dynamic stress redistribution and disordered heterogeneity on multiscales, is exposed based on the analysis of the numerical results.

2. Adaptive Multiscale FEM (MFEM)

Since its origin in the early 1960s, FEM is perhaps the most widely applied numerical method across the science and engineering fields (ZIENKIEWICZ and TAYLOR, 1988; JING and HUDSON, 2002). The proposed MFEM is actually a multilevel and self-adaptive one based on the traditional finite-element model. As damage accumulates, the dangerous element will be adaptively refined to a smaller scale. The 4–8 variable nodes serendipity element (KIKUCHI *et al.*, 1999) is adopted to connect a two-scale effect and the polynomial interpolation functions used in the model follow C^0 continuity (SZABO and BABUSKA, 1991).

2.1 Constitutive Relationship and Damage Criterion

For simplification, a linear elastic constitutive relation is used for intact elements, i.e.,

$$\boldsymbol{\varepsilon} = \boldsymbol{E}\boldsymbol{\sigma}, \tag{1}$$

where $\boldsymbol{E}$ is the modulus matrix of elements, $\boldsymbol{\varepsilon}$ and $\boldsymbol{\sigma}$ are the strain and stress tensors, respectively. For damage occurrence, Coulomb criterion is adopted for brittle crack (JAEGAR and COOK, 1979):

$$\tau_{MC} = |\tau| - \mu\sigma \geq \tau_s, \tag{2}$$

where τ_{MC} is the Coulomb stress of an element, μ is the frictional coefficient, τ and σ are the shear and normal stresses (compressive stress is defined to be positive) of the element respectively, and τ_S is the inherent shear strength. The physical meaning of this inequality is as follows. If $\tau_{MC} < \tau_S$ for a particular element, it will remain intact. Otherwise, it will be refined into four smaller elements. If the criterion is also satisfied in the smaller element resulting from the mesh refinement, the smaller element will be further refined into four much smaller elements. This adaptive mesh refinement will continue until the finest elements. The finest element still has $\tau_{MC} \geq \tau_S$, thus, it will be considered damaged and can no longer support shear stress on the corresponding direction.

The material heterogeneity is introduced by Weibull distribution of τ_S:

$$h(\tau_S) = m\left(\frac{\tau_S^{m-1}}{\eta^m}\right)\exp\left[-\left(\frac{\tau_S}{\eta}\right)^m\right], \tag{3}$$

where m and η are the Weibull modulus and scale parameter, respectively. Therefore, each finest element has a given τ_s, according to Equation (2). On the other hand, corresponding spatial distribution in a specimen is assumed to be random. Moreover, for a coarse element, τ_s is supposed to be the average of four smaller elements with τ_s in it. The frictional coefficient μ and the scale parameter η are assumed identical for all elements in the simulations.

Actually, Equation (2) can be satisfied on three different directions within an element (Fig. 1a). In order to simplify the simulation, the response of an element beyond the criterion of (2) is assumed to follow a modified smeared crack model (BAZANT and CHEN, 1997; RONG *et al.*, 2005b) to simulate the post-damage effect. That is, if an element has become damaged and the stress normal to the damage is tension, the failed element is described by a reduced elastic modulus E (Equation (4), R is taken to be 10^3 in the simulation). However, if the normal stress is compressive, i.e., σ_n is positive, here n is the normal to damage, see Figure 1(a), the elastic modulus will remain to be the initial one E_0 (Equation 4).

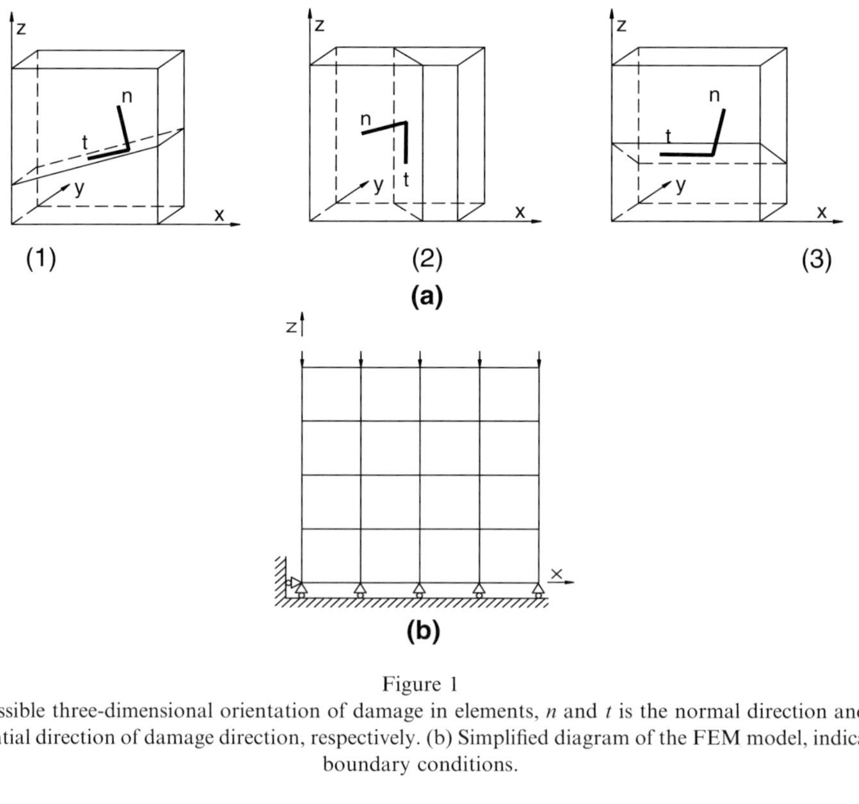

Figure 1
(a) Possible three-dimensional orientation of damage in elements, n and t is the normal direction and the tangential direction of damage direction, respectively. (b) Simplified diagram of the FEM model, indicating boundary conditions.

$$E = \begin{cases} E_0/R & \sigma_n < 0 \\ E_0 & \sigma_n \geq 0 \end{cases}, \tag{4}$$

where E is the Young's modulus in modulus matrix of elements, E_0 is the initial Young's modulus, R is the reduction ratio and σ_n is the stress normal to the damage. Poisson's ratio in the damaged elements is assumed to be identical to the intact one in the simulation. Hence there might be three different damage states in an element and each one will result in a reduction of Young's modulus in corresponding damage directions. If an element cracked on a direction, the damage fraction (defined on the element) is assumed to increase 1/3, and only the element has cracks on all three orientations. The mechanical response of the element will be depicted with a total reduced stiffness matrix. Correspondingly, the element should be regarded as a partially damaged element or fully damaged one depending on how many directions are damaged. More details of the discussion on various partially damaged states are available in RONG *et al.* (2005). In fact, experimental results (PATERSON, 1978) of uniaxial stress in rock mechanics demonstrate that the main crack (plane of failure) occurs on a plane that is inclined to the loading direction. This corresponds to orientations shown in Figure 1a(1) or (3).

The system of Equations (1–4) forms the basic constitutive formation of MFEM in the simulation.

2.2 Case Studies with MFEM

As described in Section 2.1, we examine two-dimensional (2-D) cases: plane stress and plane strain. Note that although the problems are 2-D, damage can form in three different directions according to the nature of the Coulomb criterion.

The boundary conditions used in the simulations are as follows: compressive axial load is acted on upper-surface while the lower surface is fixed, as shown in Figure 1b. The controlling variable in calculations is the upper-surface boundary displacement, namely the displacement increment is applied step by step on the upper surface until the failure of the sample. By solving minimum energy, we let force field reach equilibrium. Note that this equilibrium state of force field does not mean the equilibrium of damage due to the stress redistribution. Thus in one step, the load is held until no more elements become damaged due to stress redistribution.

Hence, in the simulation, the quasi-static loading mode is used. This implies

$$t_P \ll t_D \ll t_{ex}, \tag{5}$$

where t_P is the characteristic time for stress to reach equilibrium, t_D is the characteristic time for damage to nucleate and t_{ex} is the characteristic external loading time. The relationship in (5) implies that the stress field will equilibrate much faster than damage forms. Additionally, damage nucleates faster than the change of

▶

Figure 3
Damage evolution patterns of lane stress case. The labels (A-E, also marked on the σ-ε curve) denote the quasi-static loading steps 2, 202, 502, 531 and 532, respectively. In the patterns, there are three kinds of darkness; black indicates fully damaged, gray indicates partially damaged, and white is still intact.

the external force. These effectively reduce our problems to be a quasi-static one. Therefore, the simulation is effectively a time series of FEM solutions to the elasto-static problem.

Additionally, as a case study, the following parameters (JAEGER and COOK, 1979 and XU *et al.*, 2004) are used in the simulations:

Young's modulus $E_0 = 47.2 \times 10^9$ Pa,
Poisson Ratio $v = 0.250$,
Frictional coefficient $\mu = 0.639$,
Weibull modulus and scale parameter $m = 3$ and $\eta = 3.10 \times 10^5$ Pa.

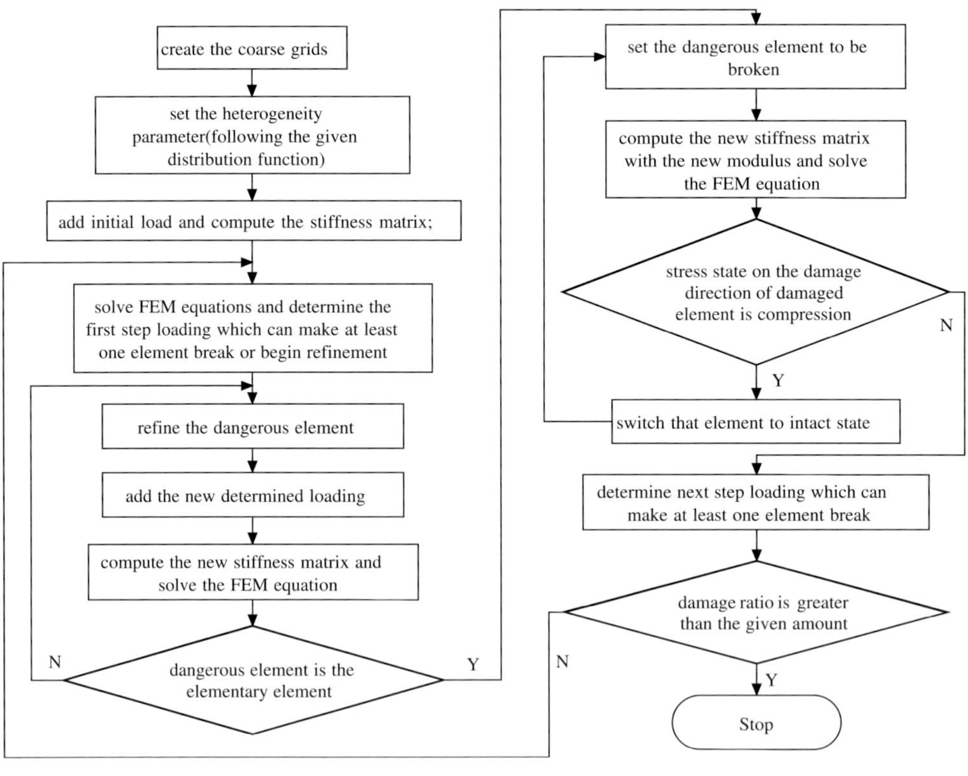

Figure 2
Flow chart of MFEM.

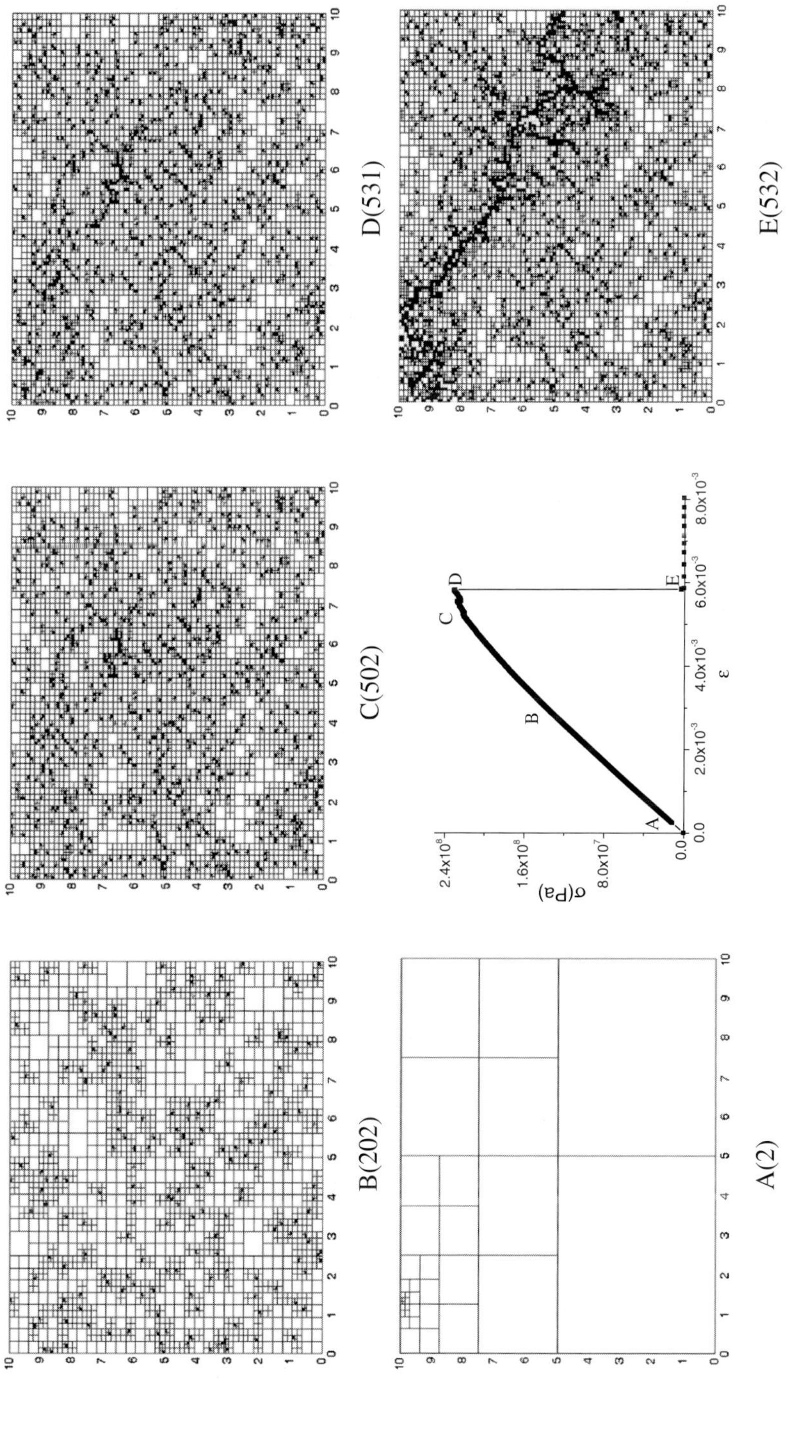
B(202)
C(502)
D(531)
A(2)
E(532)
σ(Pa)
ε
A
B
C
D
E
2.4x10^8
1.6x10^8
8.0x10^7
0.0
2.0x10^-3
4.0x10^-3
6.0x10^-3
8.0x10^-3

▶

Figure 4
Damage evolution patterns of plane strain case. The labels (A-E, also marked on the σ-ε curve) denote the quasi-static loading steps 2, 250, 500, 648 and 649, respectively. In the patterns, for each square element, there are three kinds of darkness; black indicates fully damaged element, gray indicates partially damaged, and white still remains intact.

In the end, for the sake of clarification of the algorithm, the flow chart of MFEM is presented in Figure 2.

3. Damage Evolution and Catastrophe Transition

We examined 2-D problems of plane stress and plane strain using the MFEM. To simulate initial heterogeneity, 15 different spatially random distributions of material strength, following a Weibull distribution, are given over the FEM finest scale elements. This produces different patterns of the inherent shear strength while maintaining the identical probability distribution among the 15 samples. We use the results from both plane stress and plane strain cases to analyze the physical processes that govern the catastrophe transition.

All 15 samples demonstrate a similar evolution trend, namely nonlinear deformation with random damage at an early stage and an abrupt catastrophe transition with trans-scale coalescence. A representative damage evolution pattern for the plane stress case is shown in Figure 3. The central diagram is the curve of nominal stress versus nominal strain in the entire process. At the initial stage, damage appears to be relatively uniform (step 202). The damage occurred at the elements with the lower shear strength in the strength distribution; see Figure 3B and we will discuss this issue in detail later. With increasing loading, more and more elements satisfy Equation (2) and become damaged. This is called damage accumulation; see B and C in Figure 3. Then, more damage clusters are formed (C and D in Fig. 3) but the whole sample still remains stable. Finally, a catastrophic rupture happens at step 532 (from D to E in Fig. 3).

Similarly, a typical damage evolution pattern for plane strain is shown in Figure 4. Although the detail of heterogeneity in this case is identical to that of the previous plane stress case, the final spatial damage pattern is quite different (compare Figs. 3E and 4E). The mean of catastrophe strain in 15 plane stress cases is 6.04×10^{-3}, while that of 15 plane strain cases is 7.95×10^{-3}. The comparison shows that the strain at the catastrophe transition for the plane strain case seems to be greater than that of plane stress, although these two have the identical initial spatial pattern of the inherent shear strength.

Hence, understanding the factors that contribute to these differences is an important part of understanding the catastrophe transition and the process of material failure.

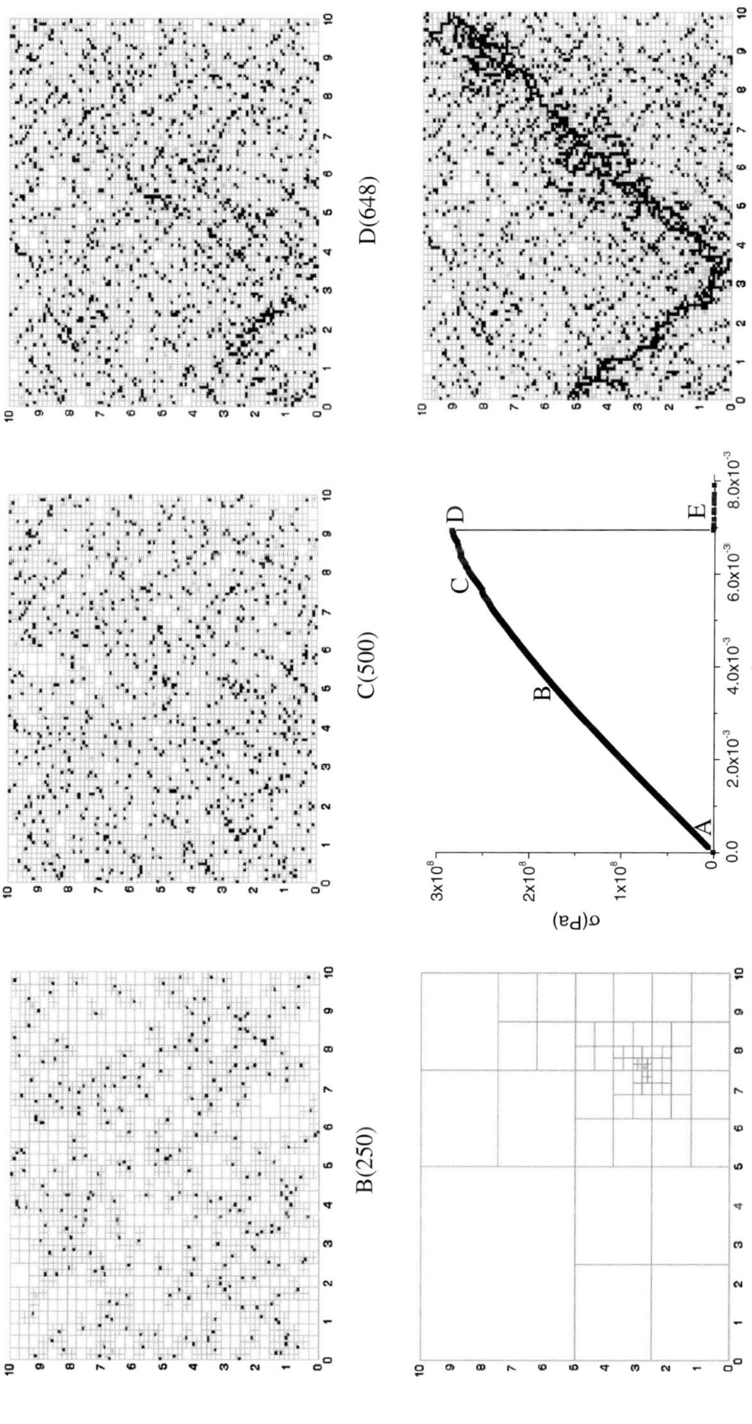
B(250)
C(500)
D(648)
A(2)
E(649)
σ(Pa)
ε
A
B
C
D
E

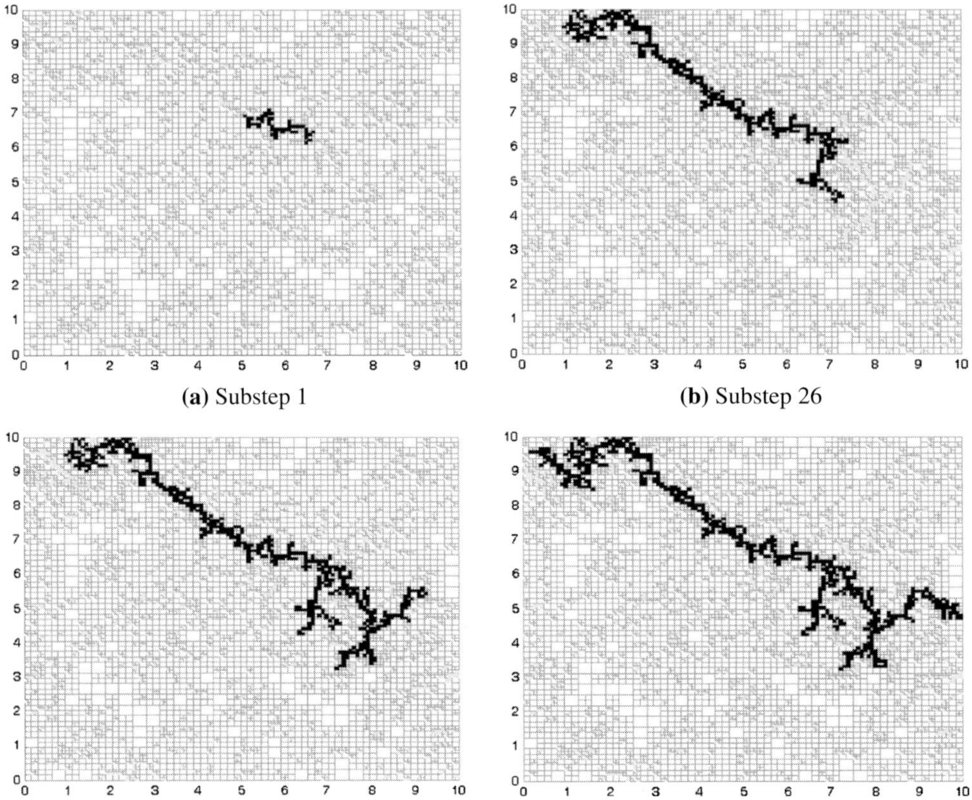

(**a**) Substep 1 (**b**) Substep 26

(**c**) Substep 27 (**d**) Substep 33

Figure 5
The damage coalescence of the plane stress case during the catastrophic transition. In the pattern there are three kinds of elements with different chroma; black indicates the damaged elements forming the largest damage cluster, gray indicates damaged elements elsewhere, and white elements are still intact.

4. Trans-scale Damage Coalescence

It has been suggested that catastrophic rupture can mainly be attributed to trans-scale damage coalescence from mesoscopic to macroscopic, and eventually, to global scales (Wang *et al.,* 2004). Figure 5 shows the nonequilibrium formation of the largest damage cluster during the catastrophic rupture in the plane stress case, i.e., from D (step 531) to E (step 532) in Figure 3. As described before, during stable damage accumulation, both stress and damage reach equilibrium in each step. However, during this catastrophic transition, we should calculate a number of substeps to reveal the trans-scale coalescence. The so-called substeps stand for the procedure, in which even though stress reaches equilibrium, damage still evolves

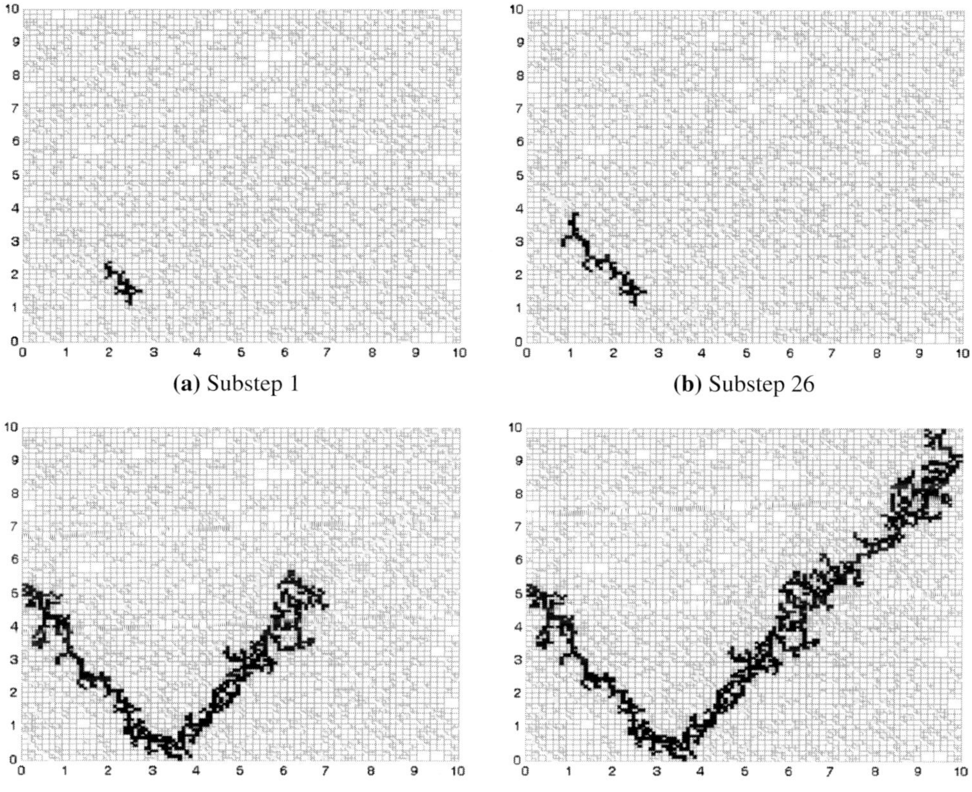

(a) Substep 1 (b) Substep 26

(c) Substep 27 (d) Substep 33

Figure 6
The damage coalescence of the plane strain case during the catastrophic transition. In the pattern, there are three kinds of elements with different chroma; black indicates the damaged elements forming the largest damage cluster, gray indicates damaged elements elsewhere, and white elements are still intact.

and violates the current equilibrium stress state, and then continue on until complete failure. Figure 5 shows the trans-scale damage coalescence of collectively 33 substeps during this non equilibrium catastrophic transition. Moreover, the growth of the cluster in the catastrophe transition is far different from a crack propagation, namely the largest cluster is sometimes formed with the connection of two neighboring ones (for example see (b) and (c) in Figure 5).

Similarly, the nonequilibrium formation of the largest damage cluster during the catastrophic rupture in the plane strain case is given in Figure 6. The growth of the largest damage cluster in the plane stress and plane strain cases is shown in Figures 7 and 8, respectively. In these two figures (7 and 8), we define the characteristic ratio as the ratio between the widths of the damage cluster to the width of the sample. When the characteristic ratio becomes greater than 0.1, we can

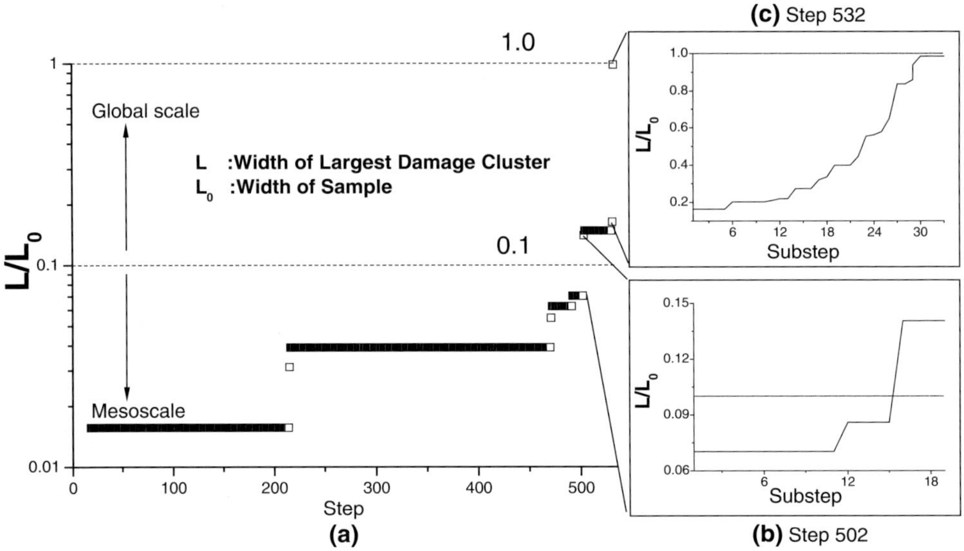

Figure 7
(a) The evolution of the span of the largest damage cluster in the whole damage evolution (from step 0 to step 532) in the plane stress case. (b) The detailed evolution within step 502, and (c) the evolution during eventual catastrophic rupture. The dashed lines stand for the characteristic ratios between the width of the largest damage cluste (L) and the width of the sample (L_0).

see from Figures 5 and 6 that the damage cluster becomes marginal to the catastrophe transition of the entire specimen. Resultingly the largest damage cluster grows as a trans-scale, inverse cascade of damage from small scale to large scale, and eventually to global scale.

The statistics of the growth of the largest cluster for samples are shown in Figures 9 and 10 for the plane stress and plane strain cases, respectively. Although there is variation in the two different stress states, inverse cascade of damage manifests a common feature of catastrophe transition.

5. Underlying Mechanism of Catastrophe

The coalescence of damage which leads to failure is affected by the coupling of nonlinear stress redistribution and heterogeneity. An examination of the evolution of the stress redistribution distance (defined as the distance between newly damaged elements and neighboring elements where the Coulomb stress (τ_{MC} in Equation 2) increases during a specific substep) reveals that the increase in damage results in a concomitant increase in stress and the stress redistribution distance. Figure 11

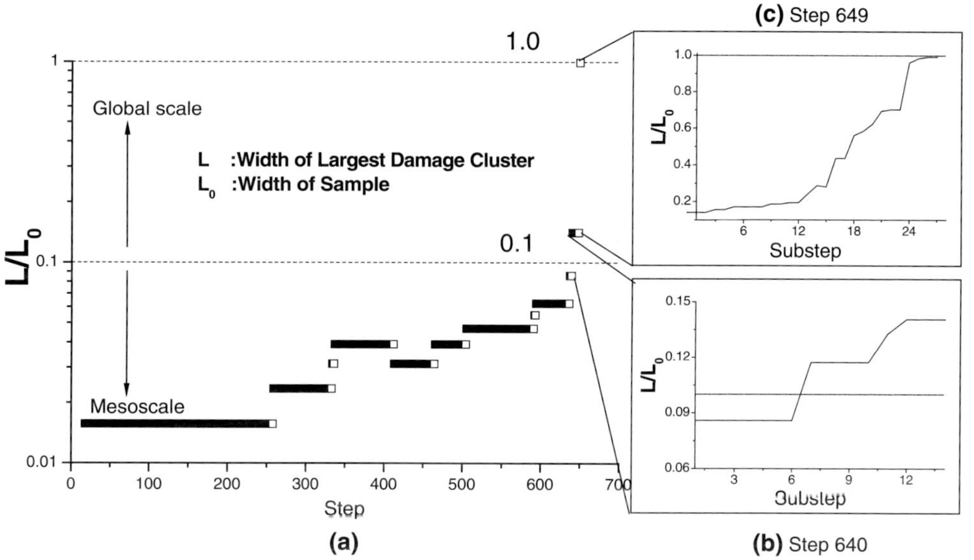

Figure 8
(a) The evolution of the span of the largest damage cluster in the whole damage evolution (from step 0 to step 649) in the plane strain case. (b) The detailed evolution within step 640, and (c) the evolution during eventual catastrophic rupture. The dashed lines stand for the characteristic ratios between the width of the largest damage cluster (L) and the width of the sample (L_0).

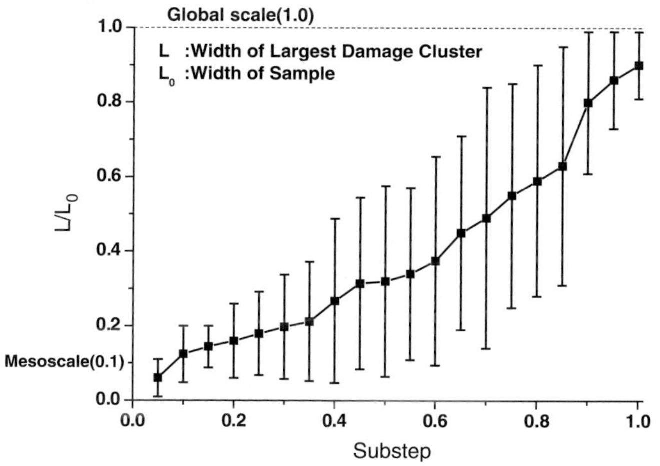

Figure 9
Inverse cascade in the catastrophe transition of the plane stress cases. Fifteen samples were used to calculate the average and the error bars.

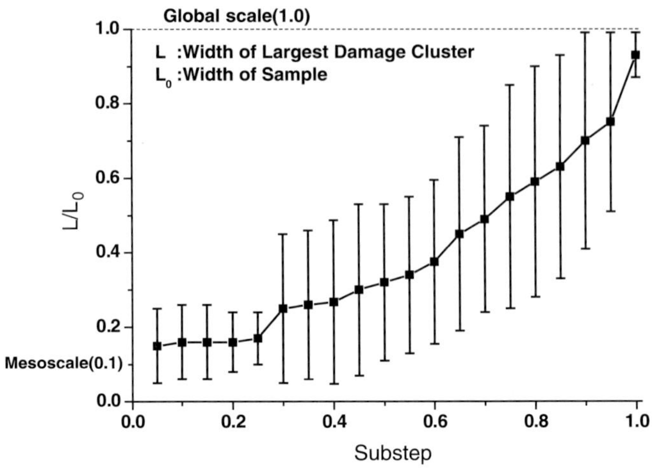

Figure 10
Inverse cascade in the catastrophe transition of the plane strain cases. Fifteen samples were used to calculate the average and the error bars.

shows the stress increase and redistribution distance for the plane stress case. During the early stage of damage accumulation (Fig. 11a), higher stress increment is confined to the elements closer to the damaged elements. As damage accumulates, the stress increment increases and distant elements begin to be affected in addition to those closer to the damaged elements (Fig. 11b). Now, obvious stress redistribution is not only seen near the damaged elements, but also farther away. This phenomenon has a significant effect on the final catastrophic rupture. At the point of catastrophic rupture, elements both proximal and distal to the previously damaged elements experience extreme increases in stress redistribution, indicating complete failure (Fig. 11c). As stress increases, elements with higher strength will be damaged more. This results in a positive feedback mechanism, i.e., stress, damage, and stress redistribution distance become coupled with each other strongly, leading to damage formation at even greater distances. Ultimately, the strong stress fluctuations lead to a breakdown of mean field approximation (ZHANG *et al.,* 2004).

Similar statistics on stress redistribution distance and stress increment in the plane strain case are given in Figure 12. There are also differences in the stress redistribution distance between the plane stress and plane strain cases (Figs. 11 and 12), which make the final damage pattern quite different. Comparing Figures 11(c) and 12(c), the element count reveals that the number of elements where Coulomb stress increases in the plane stain case is greater than that of the plane stress case. And the stress redistribution distance in the plane strain case is lower than that of the plane stress. This statistical result is quite reasonable due to the more

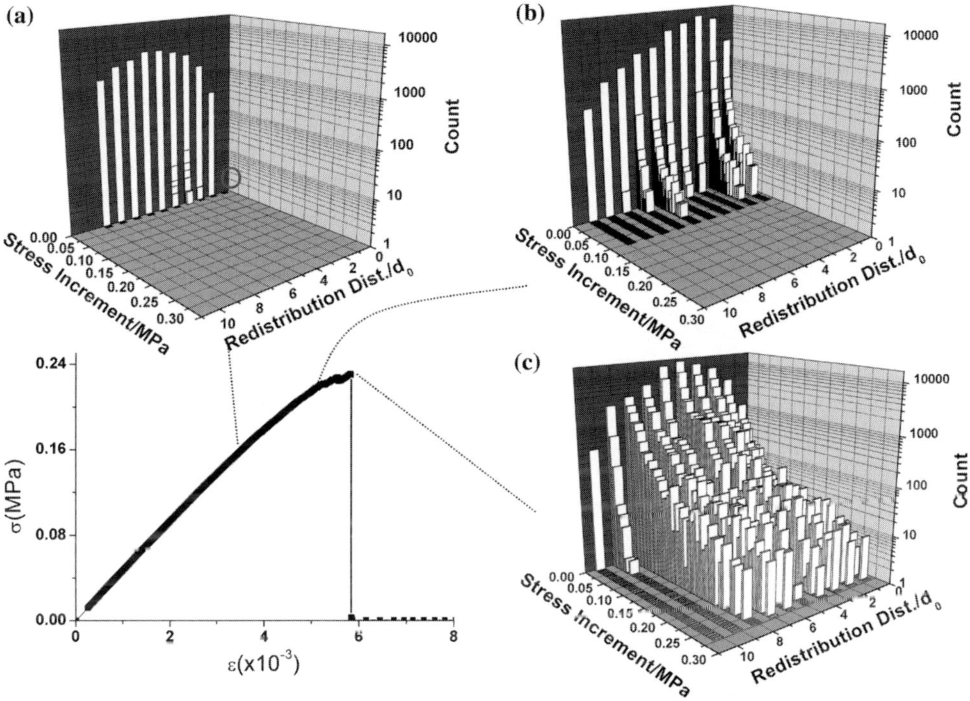

Figure 11
Stress redistribution distance distribution of the plane stress case. Stress increment means the increment of Coulomb stress in elements, redistribution distance indicates the distance between the newly damaged element and the neighboring elements where Coulomb stress increased, and count is the number of elements where the Coulomb stress increased elements. d_0 is the length of finest elements. (a) The early stage of damage accumulation, (b) the later stage of damage accumulation, (c) the catastrophic rupture.

constrained boundary condition in the plane strain case. The different stress redistribution should be the other reasons, which result in the difference in failure pattern in the two cases.

The coupling of heterogeneity with stress redistribution strongly affects the trans-scale inverse cascade of damage evolution. A comparison of the initial distribution representing the heterogeneity and the counts of the damaged elements at different stages of damage evolution show this coupling effect more clearly (Figs. 13 and 14). In order to clarify this effect even more clearly, let us make a comparison with the mean field approximation. The initial heterogeneity in shear strength is Weibull distribution. Based on mean field approximation (Zhang *et al.*, 2004), i.e., all elements have the same stress state, all elements with shear strength below the current Coulomb stress will become damaged, see Figure 13(a). However, in the simulations, some elements with higher shear strength are also damaged due to higher stress resulting from the coupling

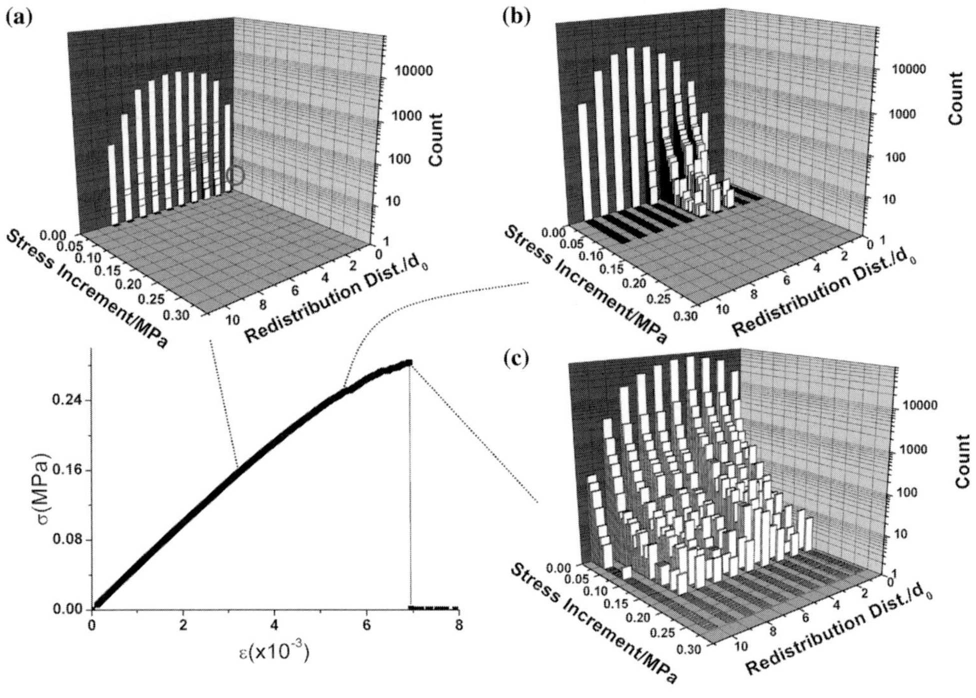

Figure 12
Stress redistribution distance distribution of the plane strain case. d_0 is the length of finest elements. (a) The early stage of damage accumulation, (b) the later stage of damage accumulation, (c) the catastrophic rupture.

between heterogeneity and stress redistribution, whereas some elements with low shear strength remain intact, as shown in Figures 13(b) and 14(a). Although initial Weibull distribution remains the same in plane stress and strain cases, both final catastrophic patterns (Figs. 3 and 4) and the statistics of damaged elements are different (Figs. 13(c) and 14(b)). All these differ from the global mean field approximation and this clearly demonstrates that the mechanism underlying the catastrophe transition is the coupling between dynamical stress distribution and disordered heterogeneity.

6. *Summary*

We use an adaptive multiscale finite-element model (MFEM) to simulate the damage evolution and the catastrophic failure of heterogeneous brittle media. The linear elastic constitutive relationship is used for intact elements. Damage is introduced with Coulomb failure criterion to each element and the heterogeneity is depicted by Weibull distribution of inherent shear strength in Coulomb criterion and

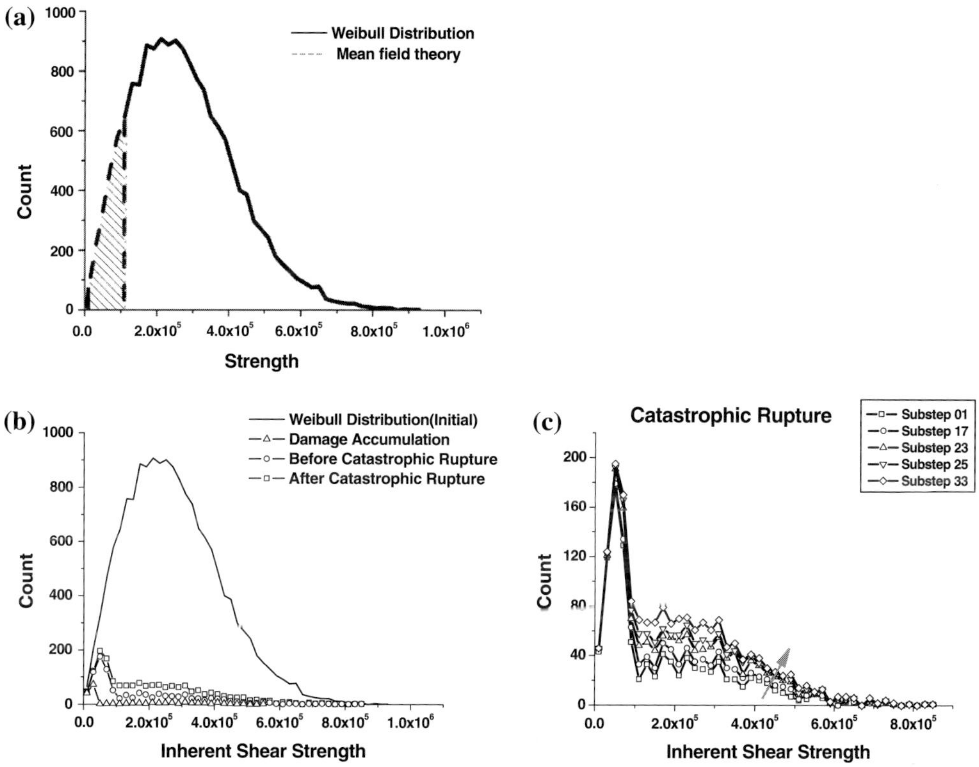

Figure 13
Initial Weibull distribution of shear strength and the counts of damaged elements (the area under corresponding curves) in a plane stress case (a) mean field theory, the area between the Weibull distribution and the vertical dashed line is the total of damaged elements (shaded). (b) The counts for different steps. (c) The counts of zoomed substeps during catastrophic rupture.

random spatial distribution. Large elements should be refined, once the failure criterion is met in them. While the finest elements satisfied the criterion, it will become damaged on three possible orientations where the criterion is satisfied. Consequently, there may be multiple damage in a finest element until complete failure. In order to better understand this catastrophic rupture, a detailed investigation of the damage evolution is carried out, with particular attention to the catastrophe transition. Both plane stress and plane strain cases were investigated. We found that as damage accumulates, there is a strong nonlinear increase in stress and the stress redistribution distance. This is an indication of the mechanism underlying catastrophe transition. The coupling of the dynamic stress redistribution and the material heterogeneity at different scales results in an inverse cascade of damage clusters, which represents the rapid coalescence of damage at catastrophe transition. Therefore, we conclude that the mechanism underlying the catastrophe transition is the coupling between stress redistribution and disordered heterogeneity.

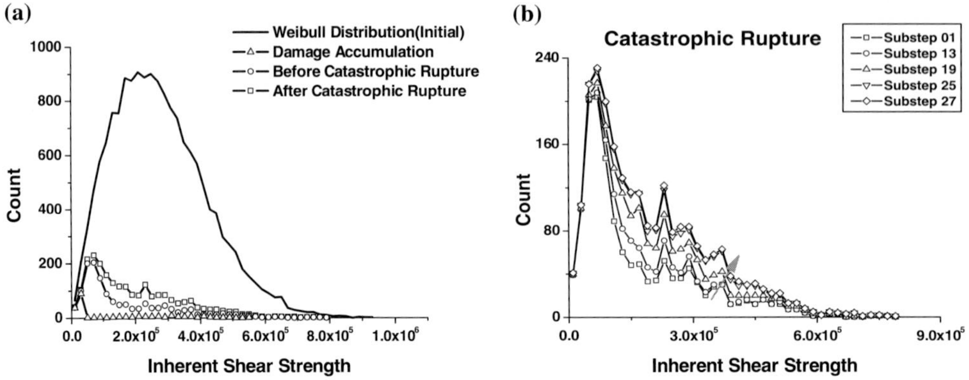

Figure 14

Initial Weibull distribution of shear strength and the counts of damaged elements in the plane strain problem. (a) The counts for different steps. (b) The counts of zoomed substeps during catastrophic rupture.

Acknowledgements

This research is supported by the National Natural Science Foundation of China (Grant No. 10232050, 10302029, 10472118, 10372012 and 10572139) and the Chinese Academy of Sciences. The authors are grateful for the computation on LSSC-II in State Key Laboratory of Scientific and Engineering Computing, Chinese Academy of Sciences.

REFERENCES

BAI, Y.L., LU, C.S., KE, F.J., and XIA, M.F. (1994), *Evolution induced catastrophe*, Phys. Lett. A *185*, 196–200.

BAZANT, Z.P. and CHEN, E.P. (1997), *Scaling of structural failure,* Appl. Mech. Rev. *50*, 593–627.

BEN-ZION, Y. and SAMMIS, C.G. (2003), *Characterization of fault zones*, Pure Appl. Geophys. *160*, 677–715.

CAO, L.Q., CUI, J.Z., and ZHU, D.C. (2002), *Multiscale asymptotic analysis and numerical simulation for the second-order Helmholtz equations with rapidly oscillating coefficients over general convex domains*, SIAM J. Numer. Anal. *40* (2), 543–577.

COLEMAN, B. D. (1958), *On the strength of classical fibers and fiber bundles*, J. Mech. Phys. Sol. *7*, 60–70.

CURTIN, W.A. (1997), *Toughening in disordered brittle materials*, Phys. Rev. *B55*, 11270–11276.

DANIELS, H. E. (1945), *The statistical theory of the strength of bundles of threads*, Proc. Roy. Soc. London *A183*, 405–435.

FISH, J. and CHEN, W. (2003), *RVE-based multi-level method for periodic heterogeneous media with strong scale mixing*. J. Eng. Math. *46*, 87–106.

GHOSH, S., LEE, K., and RAGHAVAN, P. (2001), *A multi-level computational model for multi-scale damage analysis in composite and porous materials*, Int. J. Sol. Struct. *38*, 2335–2385.

JAEGER, J.C. and COOK, N.G.W., *Fundamentals of Rock Mechanics*, 3rd Ed. (Chapman and Hall, London. 1979).

JING, L. and HUDSON, J.A. (2002), *Numerical methods in rock mechanics*, Int. J. Rock Mech. Min. Sci. *39*, 409–427.

KIKUCHI, F., OKABE, M., and FUJIO, H. (1999), *Modification of the 8-node serendipity element*, Comput. Methods Appl. Mech. Engrg. *179*, 91–109.

KRAUSE, R. and RANK, E. (2003), *Multiscale computations with a combination of the h- and p-versions of the finite-element method*, Comput. Methods Appl. Mech. Engrg. *192*, 3959–3983.

MORA, P., WANG, Y.C., YIN, C., PLACE, D., and YIN, X.C. (2002), *Simulation of the load-unload response ratio and critical sensitivity in the lattice solid model*, Pure Appl. Geophys. *159* (10), 2525–2536.

PATERSON, M.S., *Experimental Rock Deformation – The Brittle Field* (Springer-Verlag, New York 1978).

RONG, F., WANG, H. Y., XIA, M.F., KE, F.J., and BAI, Y.L. (2006), *Trans-scale coupling in multi-scale simulation*, Int. J. for Multiscale Numer. Engrg. *4*, 169–182.

RONG, F., XIA, M.F., KE, F.J., and BAI, Y.L. (2005b), *Adaptive mesh refinement FEM for Damage Evolution of Heterogeneous Brittle Media*, Model. Simul. Mater. Sci. Engin. *13*, 771–782.

RUNDLE, J.B., KLEIN, W., TIAMPO, K.F., and GROSS, S. (2000), *Dynamics of seismicity pattern dynamics in systems of earthquake faults*, geocomplexity and the physics of earthquakes, AGU Monograph, Washington, D.C., 127–146.

SZABO, B. and BABUSKA, I., *Finite Element Analysis* (J. Wiley & Sons, New York 1991).

WANG, H. Y., HE, G. W., XIA, M. F., KE, F. J., and BAI, Y. L. (2004), *Multiscale Coupling in complex mechanical systems*, Chem. Engin. Sci. *59*, 1677–1686.

WEI, Y.J., XIA, M.F., KE, F. J., YIN, X.C., and BAI, Y. L. (2000), *Evolution-induced catastrophe and its predictability*, Pure Appl. Geophys. *157*, 1945–1957.

XU, X.H., MA, S.P., XIA, M.F., KE, F.J., and BAI, Y. L. (2004), *Damage evaluation and damage localization of rock*, Theor. Appl. Fract. Mech. *42*, 131–138.

XIA, M. F., KE, F. J., and BAI, Y.L. (2000), *Evolution-induced catastrophe in a nonlinear dynamical model of materials failure*, Nonlin. Dyn. *22*, 205–224.

XIA, M.F., WEI, Y.J., KE, F.J., and BAI, Y.L. (2002), *Critical sensitivity and trans-scale fluctuations in catastrophic rupture*, Pure Appl. Geophys. *159*, 2491–2509.

ZHANG, X. H., RONG, F., XIA, M.F., KE, F.J., and BAI, Y.L. (2004), *Coupling effects of heterogeneity and stress fluctuation on rupture*, Theoreti. Appl. Frac. Mech. *161*, 1931–1944.

ZIENKIEWICZ, O.C. and TAYLOR, R.L., *The Finite-Element Method*, 4th ed., vol. 1 (McGraw-Hill, New York 1988).

(Received December 30, 2004, revised November 17, 2005, accepted December 28, 2005)

Pure appl. geophys. 163 (2006) 1867–1879
0033–4553/06/091867–13
DOI 10.1007/s00024-006-0101-7

Pure and Applied Geophysics

Differences Between Mode I and Mode II Crack Propagation

K. B. Broberg

Abstract—Although mode I and mode II crack propagation show many similarities, in particular when analysed by linear elastic fracture mechanics, they differ significantly in the micro-structural behaviour. These differences, which are clearly noticeable in the fracture surface morphology, lead to fundamental differences in the macroscopic behaviour. Thus, mode II crack expansion under remote loading, appears to obey micro-structural scaling laws, implying that the dimensions of the process region stay essentially constant during crack expansion, rather than increasing with crack length. Therefore, expanding mode II cracks can almost reach the Rayleigh velocity, and actually also intersonic velocities. An expanding mode I crack, on the other hand, seems to obey continuum scaling laws, implying that the dimensions of the process region increase in proportion to crack length, leading to self-similar crack expansion at a velocity significantly below the Rayleigh speed and dependent on the remote load.

Key words: Dynamic crack propagation, mode I, mode II, continuum scaling, micro-structural scaling.

Similarities and Dissimilarities between Modes I and II

Analytical Treatment of Problems in Modes I and II Crack Propagation under Small-scale Yielding

General experience tells us that the small-scale yielding solution for a mode II crack problem can be obtained after only marginal changes from the solution of the corresponding mode I problem, or *vice versa*. Consider, for example, the expressions for the displacement gradients on the upper face of a crack propagating in the x direction:

Mode I:

$$\frac{1}{K_{\mathrm{I}}} \cdot \frac{\partial v_+}{\partial x} = -\frac{V^2/c_S^2\sqrt{1-V^2/c_P^2}}{4\sqrt{(1-V^2/c_P^2)(1-V^2/c_S^2)}-(2-V^2/c_S^2)^2} \cdot \frac{1}{\sqrt{2\pi|x|}},$$

Department of Mathematical Physics, University College Dublin, Belfield, Dublin 4, Ireland.

Mode II:

$$\frac{1}{K_{II}} \cdot \frac{\partial u_+}{\partial x} = -\frac{V^2/c_S^2\sqrt{1 - V^2/c_S^2}}{4\sqrt{(1 - V^2/c_P^2)(1 - V^2/c_S^2)} - (2 - V^2/c_S^2)^2} \cdot \frac{1}{\sqrt{2\pi|x|}}.$$

where K_I and K_{II} are the stress intensity factors, c_P and c_S are the P- and S-wave velocities, and v_+ and u_+ are the upper crack face displacements.

The only difference is that the factor $\sqrt{(1 - V_2/c_P^2)}$ in mode I is changed to $\sqrt{(1 - V_2/c_S^2)}$ for mode II. For sub-Rayleigh velocities, this is a very small difference. On the other hand, it has the fundamental consequence of allowing intersonic velocities for mode II, but not for mode I.

The close similarity that exists in the analytical treatment for modes I and II raises expectations that results from mode I experiments can be carried over to mode II. This would be very desirable, because of the experience that it is generally more difficult to perform experiments in mode II than in mode I. However, differences in the microstructural behaviour, which are very difficult to model analytically, seem to induce fundamental differences between modes I and II in the macroscopic behaviour.

Observations and laboratory experiments on crack propagation reveal conspicuous differences between modes I and II that are not easy to discover from mathematical analysis, using material models for continua, such as LEFM (Linear Elastic Fracture Mechanics), combined with a simple model of the process region. Such models predict that an expanding crack under remote loading will eventually reach the vicinity of the Rayleigh velocity, c_R, however experience shows that this is far from the case for mode I, whereas velocities around 0.9 c_R have often been reported for mode II, and even intersonic velocities (e.g., ARCHULETA, 1982; DUNHAM and ARCHULETA, 2004; ROSAKIS, 2002).

Experimental Difficulties for Mode II

One example of the difficulties with mode II experiments already presents itself by an attempt to realize mode II propagation in the laboratory environment. The main difficulty is to prevent mode I propagation to take over. Figure 1 shows that a high compressive load is needed to prevent kinking under mode I (BROBERG, 1987, 1999). The experiments were performed on PMMA plates, and the loads involved were roughly ten times higher to get mode II propagation instead of kinking under mode I. Special precautions had to be taken to prevent buckling of the plates. The critical stress intensity factor turned out to be about 2.5 higher for mode II than for mode I.

Friction processes, known to create experimental difficulties, are not very common for mode I, but must, of course, usually be considered for mode II. Among

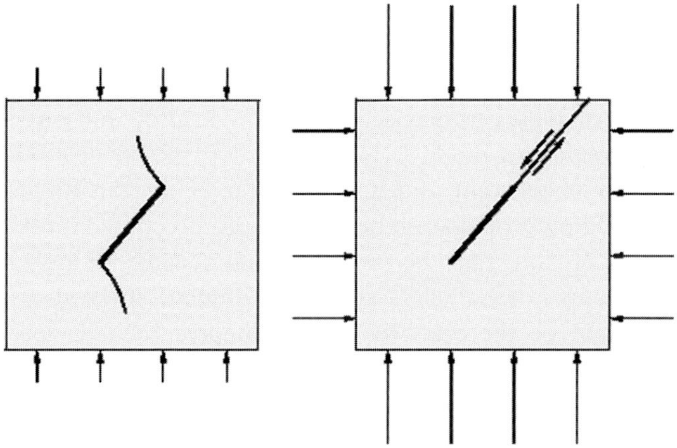

Figure 1
Development of kinks from the edges of a pre-existing crack, in a plate subjected to uniaxial compression (left figure). Such kinks can be suppressed by a high confining pressure, so that unstable mode II crack growth occurs (right figure).

other difficulties, they present scaling problems when trying to extrapolate laboratory results to larger scales.

Fundamentally Different Physical Behaviour between Crack Propagation in Modes I and II

Observations and laboratory experiments reveal conspicuous differences between modes I and II that are not easy to discover from mathematical analysis using material models for continua, such as LEFM (Linear Elastic Fracture Mechanics), combined with a simple model of the process region. Thus, as is well known, such models predict that an expanding crack under remote loading will eventually reach the vicinity of the Rayleigh velocity, c_R, although experience shows that this is far from the case for mode I, whereas velocities around 0.9 c_R have often been reported for mode II, and even transition to intersonic velocities (e.g., ROSAKIS, 2002). Note that it is the front end of the process region whose velocity may approach c_R according to LEFM, not the rear end, where material separation is completed by opening or slip. This distinction is not important if the process region length is negligible compared to the crack length, but otherwise it is.

An expanding mode I crack accelerates to a constant velocity, which is strongly dependent on the preceding history of the acceleration. This history depends both on the magnitude of the remote load and on the length of the pre-existing crack. (Note that the crack propagation can start under different remote loads for pre-existing cracks of the same length but different blunting.) Reported terminal velocities may be as low as 0.1 c_R (e.g., RAVI-CHANDAR and KNAUSS, 1984), but they rarely exceed

0.5 c_R. Such velocities seem to have little to do with the LEFM prediction of acceleration toward the Rayleigh speed. Note that this prediction holds for extensional (bilateral) crack propagation, but not for uni-directional propagation.

Uni-directional (or pulse) propagation with one leading and one trailing edge is hardly possible to realise in mode I, but it seems to be so in mode II. Then the propagation velocity is constant and it may be far below the Rayleigh velocity: it depends on the preceding propagation before the uni-directional motion. However, it does not seem to be obvious that the same constant velocity (if any) will be reached eventually for both edges, nonetheless numerical simulations (JOHNSON, 1990, 1992) indicate that so might be the case. In the first paper, Johnson let one edge of a symmetrically expanding mode II crack be arrested at a barrier, with the result that it reverses direction and changes to a trailing edge with healing. In this way, one single uni-directional crack (or pulse) is created, with a length dependent on the distance from the crack origin to the barrier. In the second paper, he considered the high resistance to slip below the seismogenic layer, which resulted in two uni-directional cracks travelling in opposite directions and originating from the midpoint of the original expanding crack. A third mechanism, also resulting in two symmetrical uni-directional cracks separating from the midpoint of an expanding crack, consists of slip arrest due to velocity-weakening friction, as suggested by ZHENG and RICE (1998).

Micro-structural and Morphological Differences

Micro-structure in Mode I Propagation. The Cell Model. Branching

Post-mortem inspection of the crack faces after a rapid mode I event shows an increase in the surface roughness in the propagation direction. This has often been described in the terms of mirror, mist and hackle. These words may be convenient to use, but it is an illusion that there are sharp boundaries between the three (e.g., RAVI-CHANDAR, 1982, 1998; ARAKAWA and TAKAHASHI, 1991; HULL, 1999). The roughening rather gradually increases almost immediately after crack initiation until the terminal velocity is reached. This increase implies increasing specific fracture energy dissipation, noticed for velocities down to some tenth of the Rayleigh speed. The roughening originates from micro-separations (voids, micro-cracks, etc.) and from further localizations through coalescences of these micro-separations into micro-branches and eventually into (macro-)branches and progressive branching.

The micro-separations nucleate from inhomogeneities in the material, such as voids, particles or grains. The dominating kernels for micro-separations form the basis for the cell model (BROBERG, 1979, 1996), see Figure 2. It is an integrated material model that can adequately describe materials from the continuum level

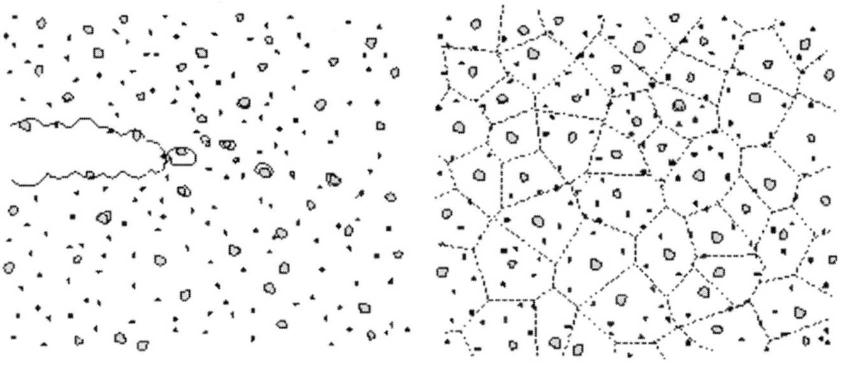

Figure 2
Left: Crack growth as formation, growth and coalescence of micro-separations, opened at certain kernels. The figure shows one example: void growth from two populations of particles, large and small. It is very simplified. For instance, it does not show grain boundaries. Right: A material envisaged as an assembly of cells, each cell containing one micro-separation kernel of dominant type.

down to the cell level. Each cell contains one dominant kernel of micro-separations, and a cell may therefore be considered as the smallest material unit that contains sufficient information on the fracture properties of the material, at least in a statistical sense. Cells that would grow stably under load control are said to be in the cohesive state, and those that would only grow stably under grip control are said to be in the decohesive state. The process region at a crack edge may be viewed as the ensemble of cells that have reached the decohesive state.

For slow mode I crack propagation, only one cell layer will reach the decohesive state, because as soon as it does, cells in the neighbouring layers will be unloaded and continue to stay in the cohesive state, see Figure 3.

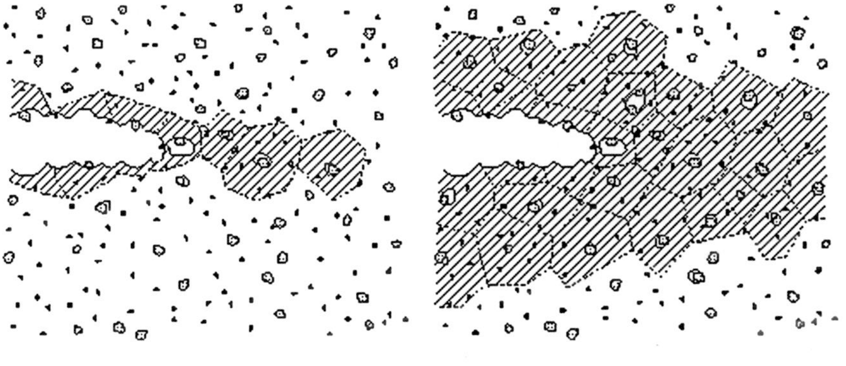

Figure 3
Schematic view of a real process region and its wake (hatched), using the cell model. In three dimensions the process region consists of a wrinkled layer of adjoining cells. The right figure shows a mode I process region for a high crack velocity.

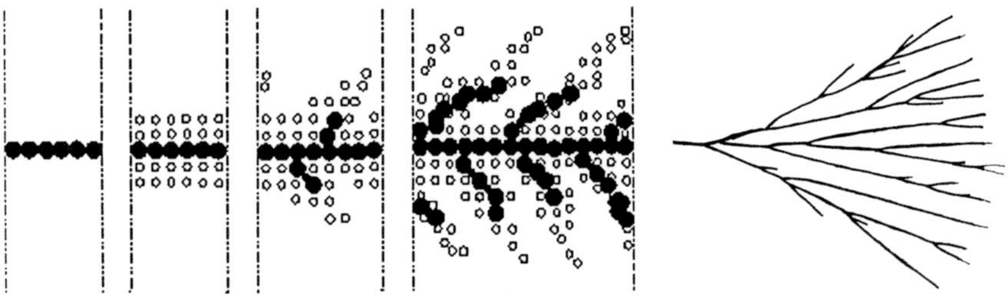

Figure 4

Left part: Progressive thinning of the micro-separation distribution along the crack path, shown at four different positions. Open circles represent cells that have reached the decohesive state, filled circles represent cells that have coalesced with other cells, in particular those in the central layer, forming the main crack. Cells that contain open micro-separations, but have not yet reached the decohesive state, are not shown. Right part shows later macro-branching (on a macro-scale as compared to the micro-scale in the left part) and is sketched after SCHARDIN (1959). Multiple macro-branching forms a fan-like structure if sufficient space is available.

For fast mode I crack propagation, even offside cells will reach the decohesive state, because they do not receive the unloading waves from the central layer in time to prevent continued stretching to decohesion. This is similar to the phenomenon that a tensile piece may exhibit more than one local contraction after a high elongation velocity. With increasing velocity more and more offside cell layers will reach the decohesive state, and the process region increases laterally (BROBERG, 1979), see Figure 3. This implies that eventually the process region also can be considered as a continuum, because the significance of an inherent material length parameter, the distance between the kernels of micro-separations, has disappeared. This is basically the same view as considering an ensemble of more than a few atoms as a continuum.

After further crack growth and increasing velocity, localizations will become prominent, first as small entities of coalescences and then as micro-branches, see Figure 4. Finally (macro-)branches appear and these develop progressively new branches.

Continuum Scaling and Micro-structural Scaling

A beam with either a smooth notch or a sharp crack (Fig. 5) may be used to illustrate the difference between continuum scaling and micro-structural scaling. Elastic-plastic material behaviour and small-scale yielding are assumed. If all linear geometrical dimensions are increased by a factor 2, then continuum mechanics shows that the plastic collapse load for the notched beam increases by a factor 4, whereas LEFM shows that the collapse load for the cracked beam increases by a factor of $2\sqrt{(2)} \approx 2.8$. The reason is that self-similarity prevails for the notched beam, so that the linear size of the plastic region at collapse increases by a factor 2 as the linear geometrical dimensions, whereas the linear size of the dissipative region at the crack edge is the same at collapse. The former scaling may be called continuum scaling, and

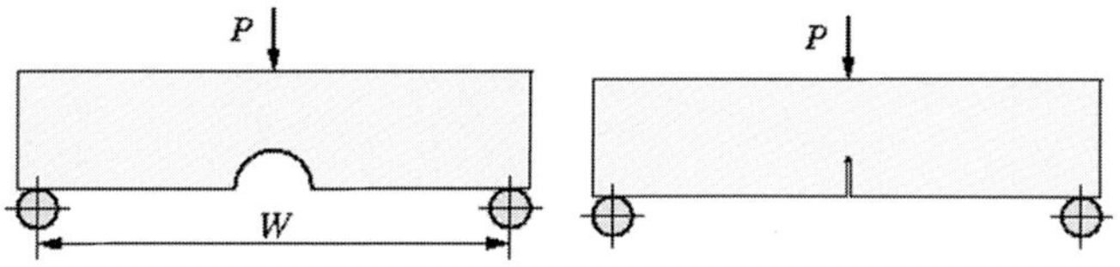

Figure 5
The load is proportional to W^2 for elastic-plastic collapse (left beam) and to $W^{3/2}$ for fracture according to LEFM (right figure).

the latter micro-structural scaling, because the size of the dissipative region is uniquely dependent on the size (and shape) of the process region, which is autonomous at collapse (BARENBLATT, 1959).

The specific energy dissipation (the dissipation per unit of crack front length and crack advance) increases roughly in proportion to the number per unit of crack advance of cells belonging to the process region. Thus, as long as the process region consists of basically one layer of cells, the specific energy dissipation is virtually independent of the crack length, and thus micro-structural scaling prevails. With increasing velocity, the process region of an expanding mode I crack will increase laterally and, correspondingly, the specific energy dissipation will increase with the crack length and velocity. Eventually, at sufficiently high velocity, the process region will encompass so many cells that continuum scaling applies, which leads to self-similar crack expansion.

The fact that the continuum approximation can be used, as it appears from experiments after a velocity of 0.1–0.2 c_R, implies that linear length scaling prevails, i.e., the specific energy dissipation Γ, increases in proportion to the crack length $2a$. There is also a velocity scaling, i.e., the specific energy dissipation increases with the velocity V, although this increase is more complicated than linear scaling. Consideration of the two scaling factors together leads to self-similarity under constant velocity crack expansion, significantly depending on the remote load (BROBERG, 2003), see Figure 6, which includes results from experiments and numerical simulations. The governing equation of motion is

$$G = \Gamma \propto a\, \mathrm{f}(V), \tag{1}$$

where G is the energy flux into the dissipative region at the crack edge and $\mathrm{f}(V) = \mathrm{f}(\mathrm{d}a/\mathrm{d}t)$ is a function that increases linearly, except for very low and very high velocities (cf. SHIOYA and ZHOU, 1995). Note that the equation is satisfied by self-similar propagation under a constant velocity that is dependent on the remote load, because then G is proportional to a and to the remote load. Traditional LEFM would have had

$$G = \Gamma(V), \tag{2}$$

where $\Gamma(V)$ would be independent of a, and consequently result in a velocity that increases toward the vicinity of the Rayleigh velocity.

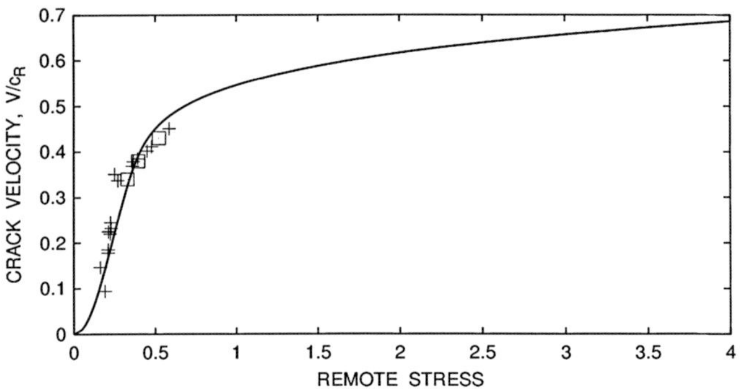

Figure 6
Relation between terminal velocity and remote load for mode I expansion from sharp pre-existing cracks. Crosses show results from experiments by RAVI-CHANDAR and KNAUSS (1984) and squares show results from numerical simulations, using the cell model by JOHNSON (1993). Note that the relation is dependent both on the material and on the history before constant velocity was reached.

Remarkably, the self-similarity extends to the propagation after branching has occurred. Thus, the constant velocity reached before branching remains the same even after branching, including the progressive branching shown in Figure 4. This is well documented, e.g., by SCHARDIN, (1959). The absence of a noticeable velocity drop after branching depends on the fact that even a slight difference of branch length would lead to quick arrest of the (inevitably) shorter branch if the branch velocities were too low (PÄRLETUN, 1979).

It is occasionally argued that the maximum mode I crack velocity should be significantly lower than the Rayleigh velocity because it is the local sound velocities in the highly strained region near the crack edge that matter, and they are substantially lower than farther away for most materials. However, it is actually the global situation that matters (BROBERG, 1973). Moreover, the local theory leads to a terminal velocity that is a material constant and thus it cannot explain its dependence on applied load and acceleration history.

It might be interesting to note that the opposite case, higher sound velocities in the highly strained region near the crack edge than farther away, may lead to crack velocities even in the globally intersonic region, i.e., between the S and P velocities outside the crack edge region (BROBERG, 1995).

Micro-structure in Mode II Crack Propagation

The experience that expanding mode II cracks can accelerate to velocities close to the Rayleigh speed, and even to intersonic velocities, indicates strongly that the governing equation of motion is of the type Equation (2) rather than Equation (1). This profound difference appears to depend on differences in the micro-structural behaviour, which, in turn, are due to different stress distribution around the crack edge.

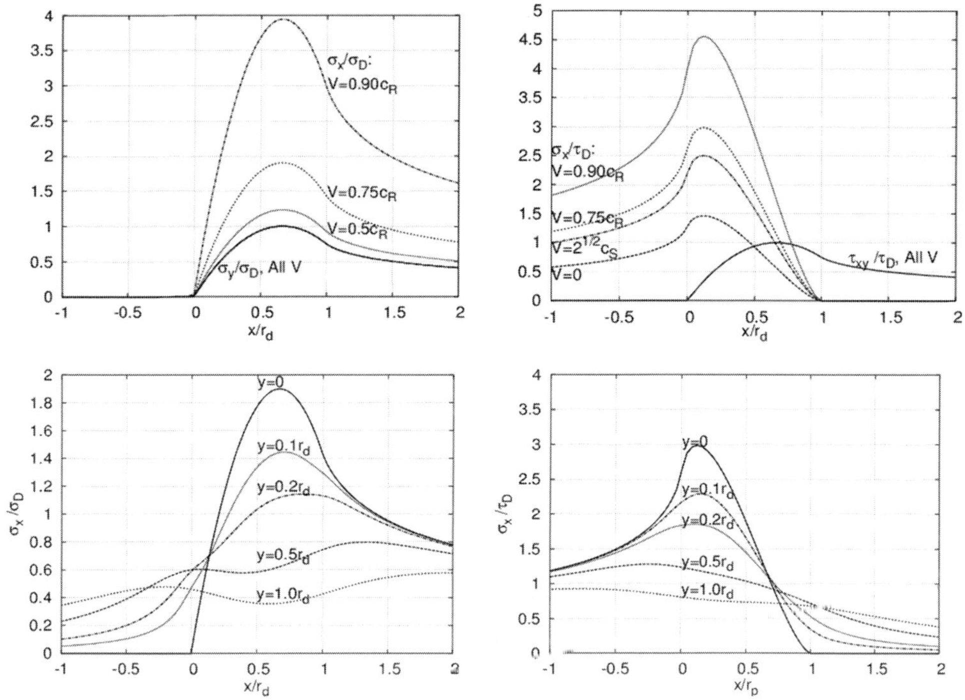

Figure 7

Upper left figure: Stresses σ_y and σ_x in the crack plane for different mode I crack velocities. For the velocity $V = 0$, $\sigma_x = \sigma_y$ Upper right figure: Stresses τ_{xy} and σ_x near the crack plane for different mode II crack velocities. In both figures, r_d is the forwards extension of the dissipative region. σ_D and τ_D are the cohesive strength in tension and shear, respectively. Plane strain and Poisson's ratio $= 0.25$ is assumed. Lower figures: The stress σ_x at some different distances from the plane of a crack, moving with velocity $0.75\ c_R$. Notations and data as for the upper figures. Left figure: Mode I. Right figure: Mode II.

Propagation in the x direction is driven by the opening stress σ_y for a mode I crack, and by the shear stress τ_{yx} for a mode II crack. These stresses are shown for the crack plane in Figure 7, assuming a smooth dissipative region model (BROBERG, 2002). However, there is also a strong σ_x component shown for the side where it is positive, potentially capable of opening up micro-cracks or voids, but weakened by compressive stresses. As the figure shows, this component becomes large already in front of the crack in mode I, whereas it develops later in mode II, implying a reduced potential of counteracting the localization of material separations to the crack plane. On the other hand, its magnitude increases dramatically as the Rayleigh velocity is approached. Figure 7 also shows the stress σ_x at offside locations.

Figure 8 shows the principal normal stress distribution around the edge of a mode I crack and the principal shear-stress distribution around a mode II crack. For mode II, there is a significant confinement of high shear stresses to the crack plane vicinity, whereas for mode I, high normal stresses exhibit a considerable lateral extension.

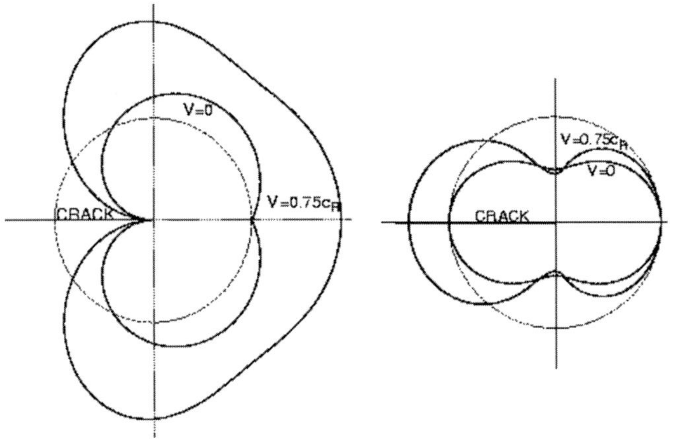

Figure 8
Comparison between the angular distribution of principal normal stresses in mode I (left figure) and principal shear stresses in mode II (right figure) at crack velocities 0 and 0.75 c_R. The stresses (mode I) and τ_{xy} (mode II) straight ahead of the cracks equal those for $V = 0$, and they also determine the radii of the reference circles.

Very important is the direction of offside micro-separations. In mode I, the stress directions are such that micro-separations tend to grow in lateral directions, whereas there is no such preference for mode II, if the compressive stress is sufficiently strong to prevent kinking. Thus, formation, growth and coalescence of micro-separations lead to lateral patterns of the kind shown in Figure 4 in mode I, but to slip-bands along the crack direction in mode II.

The confinement of high stresses to the crack plane vicinity in mode II and the preference for micro-separation extension along the crack direction, constitute a characteristic of mode II crack propagation that is very different from mode I. It may prevent substantial lateral growth of the dissipative region so that self-similar crack expansion under low constant velocities, as known from mode I, cannot take place in mode II. However, because of the dramatic increase of the magnitude of normal stresses as the velocity comes close to the Rayleigh velocity, it cannot be excluded that continuum scaling also exists for mode II, although that it takes place only after very high velocities have been reached, say 0.7–0.9 c_R. In such a case, there might be a constant terminal velocity, perhaps about 0.9 c_R, although this view is difficult to reconcile with the fact that intersonic propagation is possible.

It should also be noted that for the vicinity of the Rayleigh velocity it is important to make the distinction of measuring the crack length to the front or the rear of the process region at an expanding crack. It is the front velocity that is limited to the Rayleigh speed. Furthermore, it should be recalled that uni-directional mode II crack propagation may proceed at virtually any constant velocity, depending on how it is obtained.

Detailed discussions about offside slip and branching during mode II propagation have been undertaken by KAME *et al.* (2003) and by RICE *et al.* (2005). They considered several different situations, but one conclusion was that the likelihood for offside ruptures increases strongly when the Rayleigh velocity is approached (or the S-wave velocity for mode III crack propagation, which KAME *et al.* also considered).

Intersonic Crack Expansion

There are a few possibilities for a mode II crack to jump over the forbidden region between the Rayleigh and the S-wave velocities. Some depend on a damaged region on the crack path or a curved crack front, however for extensional (bilateral) crack propagation the jump may take place because of a curious stress peak travelling with S-wave speed. (Note that there is no such peak for steady uni-directional propagation.) This peak becomes "fatter" with increasing velocity as compared with the peak at the crack front, and at crack velocities close to the Rayleigh speed it will have the potential of opening a crack in front of the main crack. Both peaks are shown in Figure 9, which is based on a calculation for the crack velocity 0.94 c_R. The solid curve shows the LEFM result for linear elasticity throughout, whereas the dotted curve shows the more realistic case in which consideration is taken of the dissipative region that limits the stresses to the cohesive stress.

When the two cracks coalesce, the result is a sudden velocity jump of the main crack, which then may continue travelling at intersonic speed and accelerate towards the P-wave velocity. This mechanism has been shown to operate in numerical simulations by ABRAHAM and GAO (2000) and in experiments by XIA *et al.* (2003).

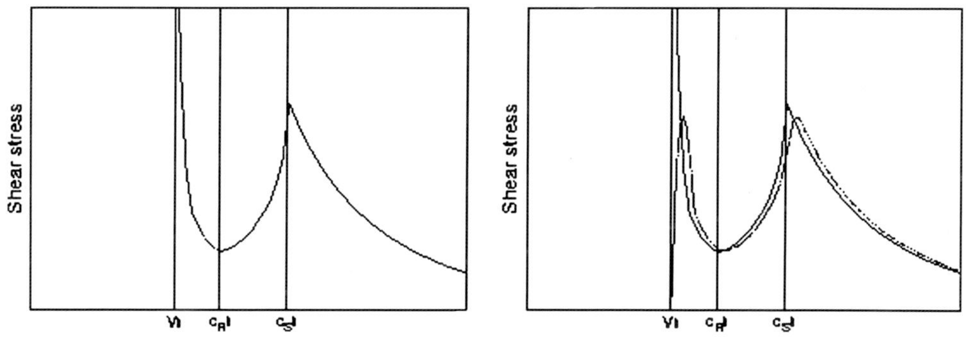

Figure 9
Shear stress near the Rayleigh and S-wave velocities for a mode II crack, expanding at 94% of the Rayleigh speed. The left figure shows the idealization to infinitesimal small-scale yielding, and the right figure compares this with a case of finite, but small, scale of yielding. Note that the S-wave peak may tend to be both higher and fatter than the crack front peak. If they are about equal, a particle in front of the S-wave peak will experience about the same stress history as one in front of the crack.

Now, the distance between the main crack and the peak travelling with S-wave speed is quite substantial, of the order of $0.1a$ where a is the half-length of the main crack. This implies that there is some uncertainty as to whether the two cracks actually will meet. If they miss each other, the jump to intersonic velocity might fail. It is well known that two coplanar mode I cracks regularly miss each other, even if later coalescence occurs through cross fracture (MELIN, 1983). This fact motivates an investigation about the directional stability of mode II cracks.

Consider, for simplicity, a periodic array of stationary mode II cracks, $-a + 2nd < x < a + 2nd,\ y = y(x) = y(-x), n$ integer, and assume that the remote loads are $\sigma_x^\infty = \sigma_y^\infty$. From MELIN (1983), for $y'(a) \ll 1$, it follows that

$$\sqrt{\frac{2}{d}} \cdot \frac{K_{\rm II}}{|\sigma_y^\infty|} = \sqrt{\tan\frac{\pi a}{2d}}\left[2\frac{\tau_{xy}^\infty}{|\sigma_y^\infty|} - y'(a) + \frac{\pi}{d} \cdot \frac{y(a)}{\sin(\pi a/d)}\right].$$

For straight cracks, $y(a) = 0$:

$$K_{\rm II} = \tau_{xy}^\infty \sqrt{2d\tan\frac{\pi a}{2d}} \to \tau_{xy}^\infty \sqrt{\pi a} \quad \text{as } a/d \to 0 \text{ (isolated cracks)}$$

But $K_{\rm II}$ can also have this value, if

$$y\prime(a) - \frac{\pi}{d} \cdot \frac{y(a)}{\sin(\pi a/d)} = 0 \Rightarrow y(a) = \delta \cdot \tan\frac{\pi a}{2d},$$

which expression is valid for δ and a/d small enough to satisfy the assumption $y'(a) \ll 1$. The result indicates that crack coalescence is not uniquely guaranteed: The cracks may shoot beyond each other, as they do (invariably) in mode I. If they do or coalesce too late after their fronts have missed each other, there will be no jump, but new attempts may be made repeatedly, resulting in a series of daughter-cracks at the side of a mother crack.

REFERENCES

ABRAHAM, F.F. and GAO, H. (2000), *How fast can cracks propagate?* Phys. Rev. Lett. *84*, 3113–3116.

ARAKAWA, K. and TAKAHASHI, K. (1991), *Relationships between fracture parameters and fracture surface roughness of brittle polymers*, Int. J. Fract. *48*, 103–114.

ARCHULETA, R.J. (1982), *Analysis of near source static and dynamic measurements from the 1979 Imperial Valley earthquake,* Bull. Seismol. Soc. Am. *72*, 1927–1956.

BARENBLATT, G.I. (1959), *The formation of equilibrium cracks during brittle fracture. General ideas and hypotheses. Axially-symmetric cracks,* J. Appl. Math. Mech, *23*, 622–636, English translation from PMM *23* (1959), 434–444.

BROBERG, K.B. *On dynamic crack propagation in elastic-plastic media.* In *Dynamic Crack Propagation* (ed. Sih G.C.) (Noordhoff International Publishing, Leyden 1973) pp. 461–499.

BROBERG, K.B., On the behaviour of the process region at a fast running crack tip. In *High Velocity Deformation of Solids* (ed. Kawata, K. and Shioiri, J.) (Springer-Verlag, Berlin Heidelberg 1979) pp. 182–194.

BROBERG, K.B. (1987), *On crack paths,* Engng Fract. Mech. *28*, 663–679.

BROBERG, K.B. (1995), *Dynamic crack propagation in a layer,* Int. J. Solids Struct. *32*, 883–896.

BROBERG, K.B. (1996), *The cell model of materials,* Comp. Mech. *19*, 447–452.

BROBERG, K.B., *Cracks and Fracture* (Academic Press, London 1999).

BROBERG, K.B. (2002), A smooth dissipative region model, Int. J. Fract. *113*, L3–L8.

BROBERG, K.B. (2003), *Constant velocity crack propagation – A microstructural or a continuum event? The Frithiof Niordson Volume.* Proc. *DCAMM Internat. Symp. on Challenges in Applied Mechanics*, 25–27 July 2002 (ed. Pedersen, P. and Olhoff, N.) (Kgs. Lyngby, Denmark) pp. 69–78.

DUNHAM, E.M. and ARCHULETA, R.J. (2004), *Evidence for a supershear transient during the 2002 Denali Fault Earthquake,* Bull. Seismol. Soc. Am. *94(6B)* S256–S268.

HULL, D. *Fractography* (Cambridge University Press, Cambridge, UK 1999).

JOHNSON, E. (1990), *On the initiation of unidirectional slip,* Geophys. J. Int. *101*, 125–132.

JOHNSON, E. (1992), *The influence of the lithospheric thickness on bilateral slip,* Geophys. J. Int. *108*, 151–160.

JOHNSON, E. (1993), *Process region influence on energy release rate and crack-tip velocity during rapid crack propagation,* Int. J. Fract. *61*, 183–187.

KAME, N., RICE, R.J., and DOMOWSKA, R. (2003), *Effects of prestress state and rupture velocity on dynamic fault branching,* J. Geophys. Res. *108(B5)*, 2265, doi:10.1029/2002JB002189.

MELIN, S. (1983), *Why do cracks avoid each other?* Int. J. Fract. *23*, 37–45.

PÄRLETUN, L.G. (1979), *Determination of the growth of branched cracks by numerical methods,* Eng. Fract. Mech. *11*, 343–358.

RAVI-CHANDAR, K., *An experimental investigation into the mechanics of dynamic fracture* (Ph.D. Thesis, California Institute of Technology, Pasadena, California 1982).

RAVI-CHANDAR, K. (1998), *Dynamic fracture of nominally brittle materials,* Int. J. Fract. *90*, 83–102.

RAVI-CHANDAR, K. and KNAUSS, W.G. (1984), *An experimental investigation into dynamic fracture: III On steady-state crack propagation and crack branching,* Int. J. Fract. *26*, 141–154.

RICE, J.R., SAMMIS, C.G., and PARSONS, R. (2005), *Off-fault secondary failure induced by a dynamic slip pulse*, Bull. Seismol. Soc. Am. *95*, 109–134.

ROSAKIS, A.J. (2002), *Intersonic shear cracks and fault ruptures.* Advances in Physics *51*, 1189–1257.

SCHARDIN, H., *Velocity effects in fracture.* In *Fracture* (eds. Averbach, B.L., Felbeck, D.K., Hahn, G.T. and Thomas, D.A.) (John Wiley & Sons, New York 1959) pp. 297–329.

SHIOYA, T. and ZHOU, F., *Dynamic fracture toughness and crack propagation in brittle material.* In *Constitutive Relation in High/Very High Strain Rates* (eds. Kawata K. and Shioiri, J.) (Springer-Verlag, Tokyo 1995) pp. 105–112.

XIA, K., ROSAKIS, A. and KANAMORI, H., (2003), *Laboratory earthquakes: The sub-Rayleigh-to-supershear rupture transition*, Science *303*, 1859–1861.

ZHENG, G. and RICE, J.R. (1998), *Conditions under which velocity-weakening friction allows a self-healing versus a crack-like mode of rupture*, Bull. Seismol. Soc. Am. *88*, 1466–1483.

(Received December 30, 2004, revised August 18, 2005, accepted September 5, 2005)

Pure appl. geophys. 163 (2006) 1881–1892
0033–4553/06/091881–12
DOI 10.1007/s00024-006-0102-6

Pure and Applied Geophysics

Dynamic Rupture in a 3-D Particle-based Simulation of a Rough Planar Fault

STEEFEN ABE,[1,2,3] SHANE LATHAM,[1,2] and PETER MORA[1,2]

Abstract—An appreciation of the physical mechanisms which cause observed seismicity complexity is fundamental to the understanding of the temporal behaviour of faults and single slip events. Numerical simulation of fault slip can provide insights into fault processes by allowing exploration of parameter spaces which influence microscopic and macroscopic physics of processes which may lead towards an answer to those questions. Particle-based models such as the Lattice Solid Model have been used previously for the simulation of stick-slip dynamics of faults, although mainly in two dimensions. Recent increases in the power of computers and the ability to use the power of parallel computer systems have made it possible to extend particle-based fault simulations to three dimensions. In this paper a particle-based numerical model of a rough planar fault embedded between two elastic blocks in three dimensions is presented. A very simple friction law without any rate dependency and no spatial heterogeneity in the intrinsic coefficient of friction is used in the model. To simulate earthquake dynamics the model is sheared in a direction parallel to the fault plane with a constant velocity at the driving edges. Spontaneous slip occurs on the fault when the shear stress is large enough to overcome the frictional forces on the fault. Slip events with a wide range of event sizes are observed. Investigation of the temporal evolution and spatial distribution of slip during each event shows a high degree of variability between the events. In some of the larger events highly complex slip patterns are observed.

Key words: Lattice Solid Model, fault simulation, slip complexity.

1. *Introduction*

Over the last decade a large number of numerical models of earthquake faults have been investigated. Amongst the dynamic or quasi-dynamic models it is possible to distinguish two broad classes: Models with and models without static heterogeneity such as heterogeneous distribution of model parameters like strength or static stress drop. Both classes of models have shown to be able to show complexity both in long-term behaviour of the fault model and in the dynamics of individual slip event. Thus, as NIELSEN *et al.* (2000) remark, there is a debate about

[1] ACcESS MNRF, The University of Queensland, Brisbane, Australia.
E-mail: {steffen, slatham, morap}@esscc.uq.edu.au
[2] Earth Systems Computational Centre, The University of Queensland, Brisbane, Australia.
[3] now at School of Geological Sciences, University College Dublin, Ireland.

the origin of the complexity, whether it is due to the frictional instabilities or fixed geometric irregularities.

The "smooth" models, i.e., those without static heterogeneity, (COCHARD and MADARIAGA, 1996; MADARIAGA and COCHARD, 1996; RICE and BEN-ZION, 1996; LANGER *et al.*, 1996) use rate-dependent friction laws to generate dynamical heterogeneities which give rise to the observed complexities. In most of those models the emergence of slip complexity is highly dependent on the friction law used. RICE and BEN-ZION (1996) suggest that, on smooth faults, friction laws with an inherent characteristic distance that lead to a finite nucleation size will result in slip complexity. MADARIAGA and COCHARD (1996) find that on a homogeneous fault, with rate- and state-dependent friction, high rate dependency will lead to complex slip while rate-independent friction will suppress instabilities.

Fault heterogeneity has been shown to be an important property in several earthquake processes. A heterogeneous fault model containing barriers has been suggested by DAS and AKI (1977) to explain several observations pertaining to earthquakes. Based on simulations, numerous of authors suggest that fault heterogeneity has a significantly influences both rupture propagation (BEN-ZION and RICE, 1995; GROSS, 1996) and temporal behaviour of the fault and the resulting event size distribution (BEN-ZION and RICE (1993); HEIMPEL (2003); ZÖLLER *et al.*, 2006). RICE (1993) suggests that geometric disorder is necessary to provide a physically viable description of the seismic complexity. BEROZA and MIKUMO (1996) find in a kinematic model of the 1984 Morgan Hill, California, earthquake that the short rise times observed in this earthquake can be fully explained by a strongly heterogeneous spatial distribution of stress drop and strength excess and that a self-healing mechanism such as a strongly velocity-dependent friction is not required. Additional argument for including geometric heterogeneity into fault models comes from observations of roughness in natural faults (POWER *et al.*, 1988; POWER and TULLIS, 1991).

In this paper an application of the Lattice Solid Model (MORA and PLACE, 1994) to the simulation of a rough planar fault embedded in a three dimensional elastic medium is presented. In the simulations the fault is represented as a discrete arrangement of particles. Particles interact with each other by imparting elastic and frictional forces when in contact. The particle obeys Newton's laws of motion, making the model fully dynamic. While this approach is computationally expensive, the discrete model has significant advantages. In particular, fault roughness comes immediately from the arrangement of different sized spherical particles (Fig. 1). In contrast to the models mentioned previously, this model introduces the heterogeneity not by varying some material property along the surface of a smooth fault but by using a rough geometry on the fault surface. The friction law employed in the model is pure Coulomb friction and the intrinsic coefficient of friction is the same for all contacts between particles along the fault. However, due to the fault roughness, the contacts are inclined relative to the fault plane (Fig. 1).

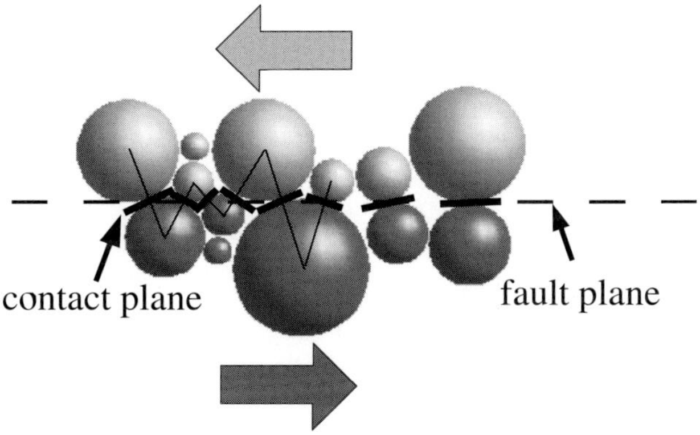

Figure 1
Schematic drawing of the fault surface roughness. The red and blue particles form the two fault surfaces. The direction of the contacts between particles is determined by the relative position and size of each particle and can therefore change over time if the particles move.

Also, because the two sides of the fault move past each other, the geometry of the fault surface will change over time, so the heterogeneity is not spatially or temporally static. Particle-based models such as the Lattice Solid Model have been shown to be a valuable tool for the simulation of stick-slip processes on rough faults (MORA and PLACE 1994; WEATHERLEY and ABE, 2006). Recent advances in the development of computer technology, in particular parallel computer systems, and the development of a parallel implementation of the Lattice Solid Model by ABE *et al.* (2004) have made it possible to extend the study of the stick-slip dynamics from 2-D to 3-D models. In particular, the good scaling behaviour of the parallel implementation (LATHAM *et al.*, 2004) allows the number of particles in the simulation model to increase with the number of available CPUs, without a major increase in total computing time. However, the long simulations necessary to generate a sufficient number of slip events for thorough statistical analysis are still largely restricted to 2-D models. Therefore the work presented here concentrates mainly on the investigation of the dynamics of single slip events.

2. Lattice Solid Model Overview

The Lattice Solid Model (MORA and PLACE, 1994; PLACE and MORA, 1999) is a particle-based model similar to the Discrete Element Model (DEM) developed by CUNDALL and STRACK (1979). The model consists of spherical particles which are characterised by their radius, mass, position and velocity.

The particles interact with their nearest neighbours by imparting elastic and frictional forces. The particles can be linked together by elastic bonds or springs

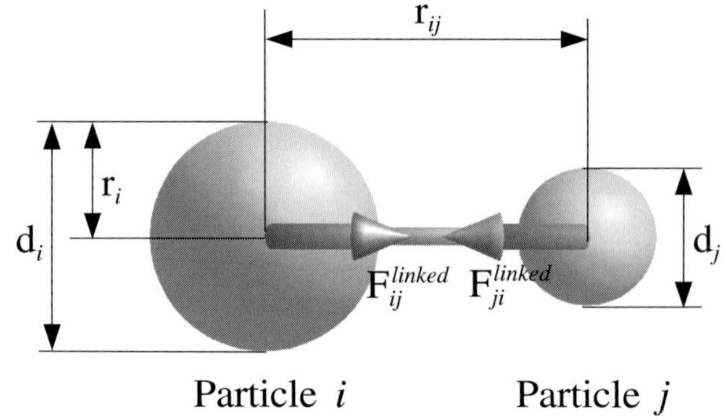

Figure 2
Elastic forces between bonded particles.

(Fig. 2), in which case the elastic forces are attractive or repulsive, depending on whether the particles are closer or more distant than the equilibrium distance $(r_0)_{ij}$

$$\mathbf{F}_{ij}^{\text{linked}} = \begin{cases} k_{ij}(r_{ij} - (r_0)_{ij})\mathbf{e}_{ij} & r_{ij} \leq (r_{\text{cut}})_{ij} \\ \mathbf{0} & r_{ij} > (r_{\text{cut}})_{ij} \end{cases}, \tag{1}$$

where k_{ij} is the spring constant for the elastic interaction between the particles, r_{ij} is the distance between the particles i and j, $(r_{\text{cut}})_{ij}$ the breaking distance for the link between the particles and $\mathbf{e}_{ij}$ is a unit vector in the direction of the interaction.

Links are broken if the distance between the particles exceeds the threshold breaking distance $(r_{\text{cut}})_{ij}$. If two particles are not linked together (Fig. 3) the elastic force $\mathbf{F}_{ij}^{\text{free}}$ between the particles i and j is purely repulsive

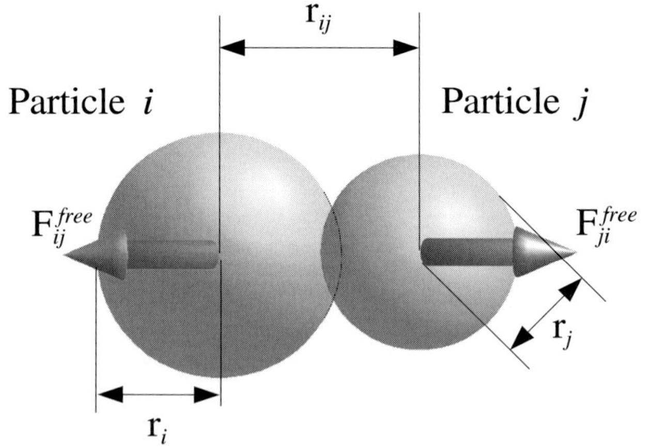

Figure 3
Elastic forces between unbonded particles.

$$\mathbf{F}_{ij}^{\text{free}} = \begin{cases} k_{ij}(r_{ij} - (r_0)_{ij})\mathbf{e}_{ij} & r_{ij} \leq (r_0)_{ij} \\ \mathbf{0} & r_{ij} > (r_0)_{ij} \end{cases}. \tag{2}$$

An intrinsic friction between particles has been incorporated in the model (PLACE and MORA, 1999). Two unbonded interacting particles can be in static or dynamic frictional contact (Fig. 4). The force on particle i due to the dynamic frictional contact with particle j is given by

$$\mathbf{F}_{ij}^{D} = -\mu F_{ij}^{n} \mathbf{e_{ij}}^{T}, \tag{3}$$

where μ is the coefficient of friction between the particles, F_{ij}^{n} is the magnitude of the normal force and $\mathbf{e_{ij}}^{T}$ is a unit vector in the direction of the relative tangential velocity between the particles (CUNDALL and STRACK, 1979; MORA and PLACE, 1998; PLACE and MORA, 1999).

Both Coulomb friction with a constant coefficient of friction and velocity weakening friction have been implemented, however, in the current study only constant friction with no difference between static and dynamic friction has been used.

3. *The Model*

The model consists of a rough planar fault embedded in elastic 3-D medium (Fig. 5). The particles are arranged in a dense random packing close to the fault to

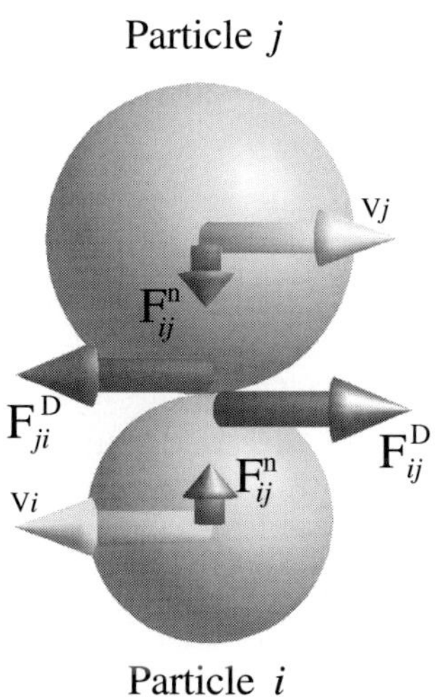

Figure 4
Frictional forces between particles.

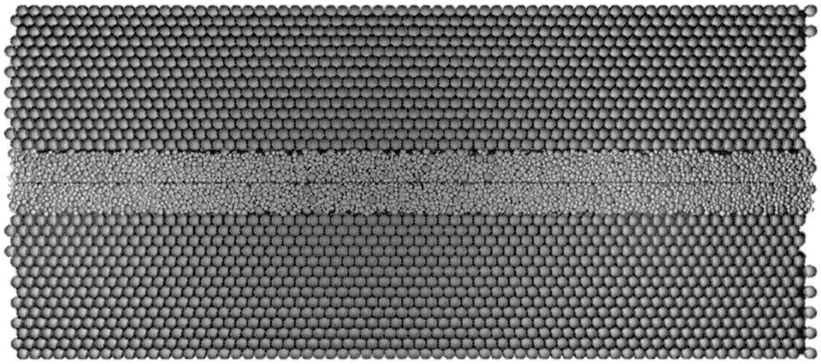

Figure 5
3-D model of a rough planar fault embedded between two elastic blocks consisting of approximately 10^5 particles, looking sideways onto the fault trace. The fault area is 135 × 20 model units, the thickness of the model perpendicular to the fault 60 model units and the radii of the particles are between 0.2 and 1.0 model units.

enable random roughness at the fault and in a regular triangular arrangement away from the fault in order to reduce the total number of particles and still maintain a sufficiently large volume of elastic material around the fault. The random particle packing is generated by first randomly placing particles in the space and then filling in particles so that they always touch either four other particles or they touch three particles and the fault plane (PLACE and MORA, 2001). This algorithm produces a planar fault with a random small scale roughness due to the spherical shape and size range of the particles.

In the simulations presented here particles with a radius of 0.2–1.0 model units have been used. In the models used the fault plane had a size of 20 × 135 model units, while the model thickness perpendicular to the fault plane was 60 model units. Those dimensions resulted in approximately 10^5 model particles. A friction coefficient $\mu = 0.6$ between particles has been used. The particles within the elastic blocks are fully linked together (Fig. 2), while the particles on one side of the fault are in elastic and frictional interactions (Fig. 4) with the particles on the other side of the fault. During the simulations a normal loading perpendicular to the fault plane is applied to the model, while the edges of the elastic bulk parallel to fault plane are moved with a constant velocity of 5×10^{-5} model units parallel to the long dimension of the fault. The shear velocity is equivalent to 0.005% of the P-wave speed in the model (or 0.25 m/s assuming a P-wave speed of 5 km/s). This is significantly faster than tectonic loading rates, but the use of real tectonic velocities of a few cm/year, i.e. $\approx 10^{-9}$m/s would require simulations with 10^{13}–10^{14} time steps which is currently computationally infeasible. The main simulations were run on 24 CPUs of an SGI Altix3700, while smaller, preliminary tests with fault sizes ranging from 16 × 16 to 20 × 54 model units were run on 4-12 CPUs. The simulation software "LSMearth" described by ABE *et al.* (2004) has been used.

4. Results

During the simulation the behaviour of the model is characterised by long stick phases, interrupted by slip events. During the stick phases the potential, (elastic) energy stored in the model builds up due to the applied shear loading. During the slip events the stored elastic energy drops and part of it is released as kinetic energy. Figure 6 shows the evolution of the elastic energy in the model during a simulation lasting 10^6 time steps. The total displacement at the end of the simulation is approximately 1.7 models units, i.e., ≈1.3% of the fault length. A wide range of different slip events was observed, 17 of which were further investigated. The elastic energy drops corresponding to those events are marked by the event number (1...17) in Figure 6.

A measure to quantify the size of seismic events is the product of average slip $\bar{u}$ and fault area A, called potency by BEN-MENAHEM and SINGH (1981) and geometric moment by KING (1978). This quantity can easily be calculated in the model from the displacement of the particles along the fault, i.e.,

$$P_0 = \int_A u \, dA. \tag{4}$$

where u is the local slip displacement and A the slip area. The more common measure in seismology is the seismic moment M_0 (e.g., AKI and RICHARDS, 1980)

$$M_0 = \mu \bar{u} A = \mu P_0 \tag{5}$$

where μ is the shear rigidity at the source. However, the calculation of the seismic moment in the model would require an estimate of μ at the source. BEN-ZION (2001) points out that the potency is directly observable in real earthquakes while the determination of the seismic moment relies on an assumed rigidity. Another measure

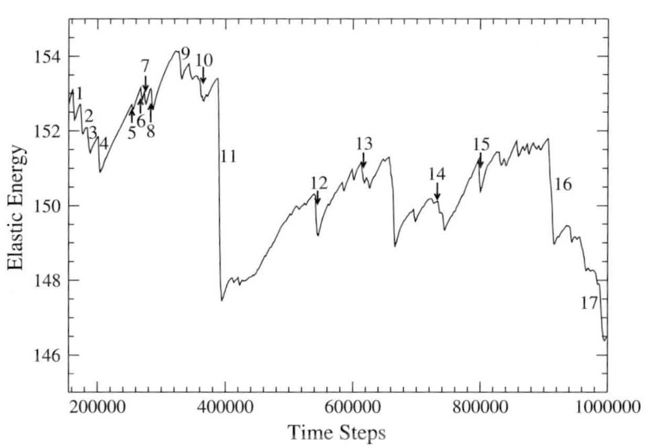

Figure 6
Elastic energy stored in the model during stick-slip cycles.The numbers denote the slip events which have been further analysed in this paper.

of event size which can be readily calculated is the amount of stored elastic energy lost from the model during the event.

For the chosen 17 simulated events, the potency P_0 and the drop in elastic energy stored in the model have been calculated. The result (Fig. 7) shows that the event sizes span a range of $\approx 22 \ldots 395$ model units of potency and $\approx 0.3 \ldots 7$ model units in energy drop.

In order to investigate the temporal evolution of the slip within the events, the potency rate $P(t) = dP_0/dt$ has been calculated for the events. To enable a better

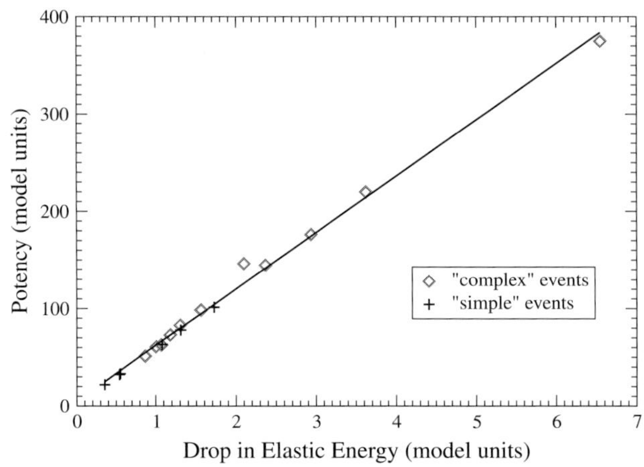

Figure 7
Event potency and drop in elastic energy stored in the model for 17 selected slip events. The straight line is a least-squares linear fit to the data.

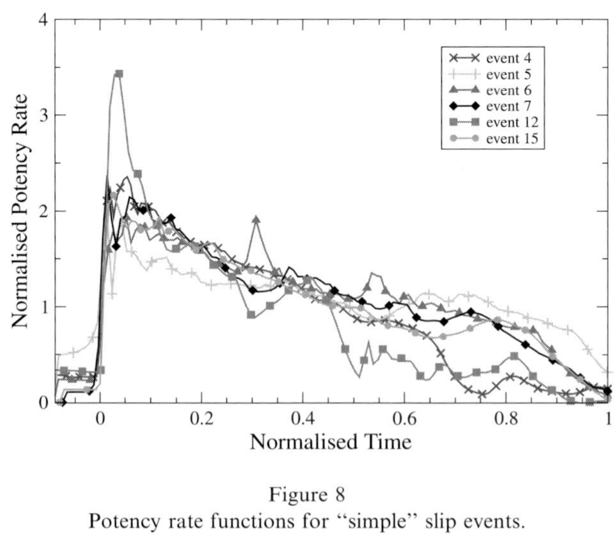

Figure 8
Potency rate functions for "simple" slip events.

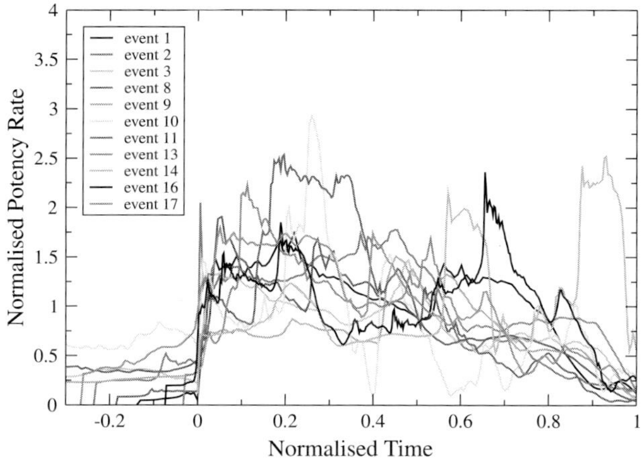

Figure 9
Potency rate functions for "complex" slip events.

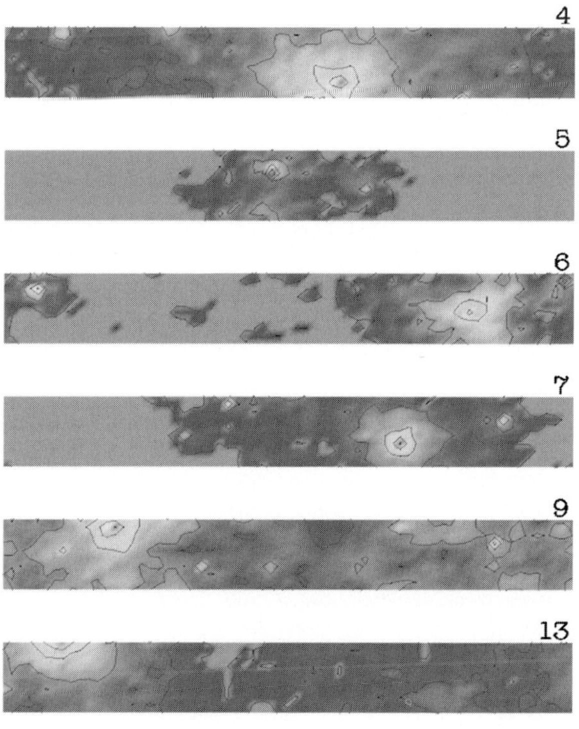

Figure 10
Slip distributions for the "simple" events. The colors show the amount of slip at the fault plane, with red representing the highest slip, blue lower slip and gray slip of less than 0.01 model units. The black iso-lines are drawn at 0.01, 0.02, 0.04, 0.08 and 0.16 model units of displacement. The numbers correspond to the event numbers in Figures 6 and 8.

comparison of the shape of the potency rate functions for different events a scaled potency rate $P_{\mathrm{scaled}}(t) = P(t/t_0)/P_0$ where t_0 is the event duration and P_0 the total potency of the event has been plotted (Figs. 8 and 9).

The variations in both the potency rate functions and slip distributions (Figs. 10 and 11) suggest that there are two different types of slip events. The first type, mainly restricted to smaller events, exhibits relatively simple potency rate functions (Fig. 8) and slip concentrated in a roughly circular or elliptic area around the point of maximum displacement (Fig. 10). The events which show a more complicated potency rate function (Fig. 9) also display a more irregular distribution of slip in the fault region. Also, unlike the "simple" events, the scaled potency rate functions for those events show very little similarity to each other.

A possible explanation for the different appearance of the two types of events is that the "simple" events consist of the slip of a single, point-like asperity and the surrounding area. In contrast, the events which exhibit complicated potency rate functions and slip distributions consist of a sequence of slips on multiple asperities.

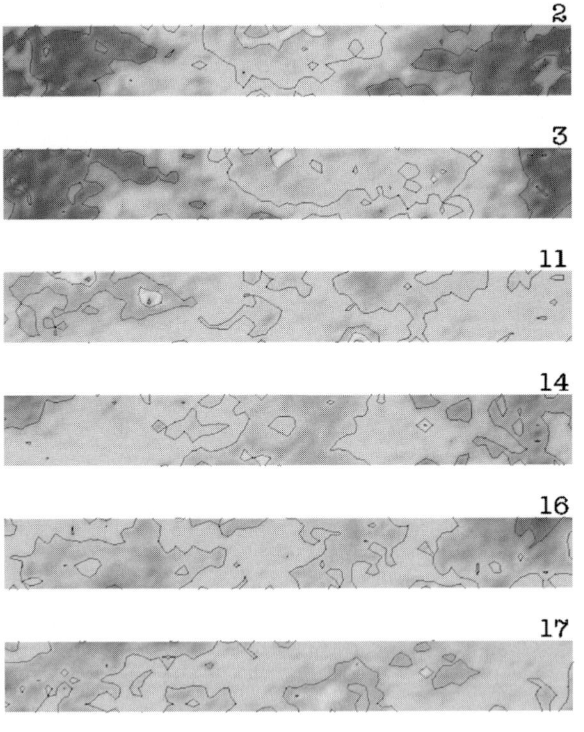

Figure 11
Slip distributions for some of the "complex" events. The colors show the amount of slip at the fault plane, with red representing the highest slip, blue lower slip and gray slip of less than 0.01 model units. The numbers correspond to the event numbers in Figures 6 and 9.

Another interesting observation is that, if the the moment rate functions for the "simple" events are scaled by event time and total potency, they are very similar to each other (see Fig. 8) despite the fact that the absolute potency for those events varies from 22 to 101 model unit, i.e., by a factor of ≈ 5. However, there is currently insufficient data to confirm a "universal" shape of the moment rate function for those small events.

5. *Conclusions*

The results obtained from the simulations presented here have shown that three-dimensional particle-based models can be a useful tool for the study of the stick-slip dynamics of faults. The observation of a wide variety of dynamic slip events in a model using simple Coulomb friction indicates that the presence of geometric roughness is sufficient to cause complex stick-slip dynamics in a fault model. No variation in the intrinsic constitutive parameters or complex friction laws appears to be necessary.

6. *Acknowledgements*

Support is gratefully acknowledged by the Australian Computational Earth Systems Simulator Major National Research Facility, the Queensland State government, The University of Queensland, and SGI. The ACcESS MNRF is funded by the Australian Commonwealth Government and participating institutions (Univ. of Queensland, Monash Univ., Melbourne Univ., VPAC, RMIT, ANU) and the Victorian State Government. Computations were made using the ACcESS MNRF supercomputer, a 208 processor 1.1 TFlops SGI Altix 3700 which was funded by the Queensland State Government Smart State Research Facility Fund and SGI.

References

Abe, S., Place, D., and Mora, P. (2004), *A parallel implementation of the lattice solid model for the simulation of rock mechanics and earthquake dynamics*, Pure Appl. Geophys. *161*, (9/10).

Aki, K. and Richards P.G., *Quantitative Seismology, Theory and Methods* (Freeman and Company, San Francisco 1980).

Ben-Menahem, A. and Singh, S.J., *Seismic Waves and Sources*, (Springer-Verlag, New York 1981).

Ben-Zion, Y. (2001), *On Quantification of the Earthquake Source*, Seis. Res. Lett. *72*, 151–152.

Ben-Zion, Y. and Rice, J.R. (1993), *Earthquake failure sequences along a Cellular Fault Zone in a Three-Dimensional elastic Solid Containing Asperity and Nonasperity Regions*, J. Geophys. Res. *98*, 14109–14131.

Ben-Zion, Y. and Rice, J.R. (1995), *Slip patterns and earthquake populations along different classes of faults in elastic solids*, J. Geophys. Res. *100*, 12959–12983.

Beroza, G.C. and Mikumo, T. (1996), *Short slip duration in dynamic rupture in the presence of heterogeneous fault properties*, J. Geophys. Res. *101* B10, 22449–22460.

COCHARD, A. and MADARIAGA, R. (1996) *Complexity of seismicity due to highly rate dependent friction* J. Geophys. Res. *101* B11, 25321–25336.

CUNDALL, P.A. and STRACK, O.D.A. (1979), *A discrete numerical model for granular assemblies*, Geótechnique, *29,* 47–65.

DAS, S. and AKI, K. (1977), *Fault plane with barriers: A versatile earthquake model*, J. Geophys. Res. *82*, 5658–5670.

GROSS, S. (1996), *Magnitude distribution and slip scaling of heterogeneous seismic source*, Bull. Seismol. Soc. Am. *86*, 498–504 .

HEIMPEL, M.H. (2003), *Characteristic scales of earthquake rupture from numerical models*, Nonlinear Processes in Geophys. *10*, 573–584.

KING, G.C.P. (1978), *Geological faults: fracture, creep and strain*, Phil. Trans. R. Soc. Lond. A *288*, 197–212.

LATHAM, S., ABE S., and DAVIES, M. (2004), *Scaling evaluation of the lattice solid model on the SGI Altix 3700*, HPCAsia2004, Proceed. 7th Internat. Conf. on High Performance Computing and Grid in the Asia Pacific Region, 226–233.

LANGER, J.S., CARLSON, J.M., MYERS, C.R., and SHAW, B.E. (1996), *Slip Complexity in dynamic models of earthquake faults*, Proc. Nat. Acad. Sci. USA, *93*, 3825–3829.

MADARIAGA, R. and COCHARD, A. (1996) *Dynamic friction and the origin of the complexity of earthquake sources*, Proc. Nat. Acad. Sci. USA, *93*, 3819–3824.

MORA, P. and PLACE, D. (1994), *Simulation of the stick-slip instability*, Pure Appl. Geophys. *143*, 61–87.

MORA, P. and PLACE, D. (1998), *Numerical simulation of earthquake faults with gauge: towards a comprehensive explanation for the low Heat Flow*, J. Geophys. Res. *103*, 21067–21089.

NIELSEN, S.B., CARLSON, J.M., and OLSEN, K.B. (2000), *Influence of friction and fault geometry on earthquake rupture*, J. Geophys. Res. *105 B3*, 6069–6088.

PLACE, D. and MORA, P. (1999), *The lattice solid model to simulate the physics of rocks and earthquakes: incorporation of friction*, J. Comp. Physics, *150*, 332–372.

PLACE, D. and MORA, P. (2001), *A random lattice solid model for simulation of fault zone dynamics and fracture processes.* In: *Bifurcation and Localisation Theory for Soils and Rocks'99* (eds. H. B. Mühlhaus, A. Dyskin and E Pasternak) (AA Balkema, Rotterdam/Brookfield 2001).

POWER, W.L., TULLIS, T.E., and WEEKS, J.D. (1988), *roughness and wear during brittle faulting*, J. Geophys. Res. *93*, 15268–15278.

POWER, W.L. and TULLIS, T.E. (1991), *Euclidean and fractal models for the description of rock surface roughness*, J. Geophys. Res. *96,* 415–424.

RICE, J.R. (1993), *Spatio-temporal complexity on a fault*, J. Geophys. Res. *98*, 9885–9907.

RICE, J.R. and BEN-ZION, Y. (1996), *Slip complexity in earthquake fault models*, Proc. Nat. Acad. Sci. USA, *93*, 3811–3818.

WEATHERLEY, D. and ABE, S. (2004), *Earthquake statistics in a block slider model and a fully dynamic fault model*, Nonlinear Processes in Geophys., *11*, 553–560.

ZÖLLER, G., HOLSCHNEIDER, M., and BEN-ZION, Y. (2005), *The role of heterogeneities as a tuning parameter of earthquake dynamics*, Pure Appl. Geophys., *162*, 1077–1111.

(Received November 1, 2004, revised September 10, 2005, accepted December 30, 2005)

Pure appl. geophys. 163 (2006) 1893–1913
0033–4553/06/091893–21
DOI 10.1007/s00024-006-0103-5

Pure and Applied Geophysics

Numerical Simulation of the Effect of a DC Electric Field on Seismic Wave Propagation with the Pseudospectral Time Domain Method

LANBO LIU,[1] LEI XIAO,[1] HONG LIU,[2] and HONGRUI YAN[3]

Abstract—We have modeled the effect of a direct current (DC) electric field on the propagation of seismic waves by the pseudospectral time domain (PSTD) method, based on a set of governing equations for the poroelastic media. This study belongs to the more general term of the seismoelectric coupling effect. The set of physical equations consists of the poroelastodynamic equations for the seismic waves and the Maxwell's equations for the electromagnetic waves; the magnitude of the seismoelectric coupling effect is characterized by the charge density, the electric conductivity, the Onsager coefficient, a function of the dielectric permittivity, the fluid viscosity, and the zeta potential. The poroelastodynamic vibration of a solid matrix generates an electric oscillation with the form of streaming current via the fluctuation of pore pressure. Meanwhile, fluctuating pore pressure also causes oscillatory variation of the electric resistivity of the solid matrix. The simulated poroelastic wave propagation and electric field variation with an existing background DC electric field are compared with the results of a physical experiment carried out in an oilfield. The results show that the DC electric field can significantly affect the propagating elastic energy through the seismoelectric coupling in a wide range of the seismic frequency band.

Key words: Seismoelectric coupling, pseudospectral time domain method, DC current, seismic exploration.

1. *Introduction*

Elucidations on the electric, magnetic, and electromagnetic disturbances associated with different stages of a seismogenic cycle can be traced back to well beyond a century ago. Nevertheless, the international earthquake research community, as a whole, is still very skeptical of the actual physical mechanism, or mechanisms, of the mechanic-electromagnetic energy conversion. The credibility of case reports on seismo-electric, seismo-magnetic and seismo-electromagnetic conversion measurement in literature up-to-date varies widely, from very credible to very suspicious. Especially, the soundness of the electric, magnetic, and electromagnetic precursors

[1] Department of Civil and Environmental Engineering, University of Connecticut, Storrs, CT 06269, U.S.A. E-mail: lanbo.liu@uconn.edu
[2] Institute of Geophysics, Chinese Academy of Science, Beijing 100101, P. R. China.
[3] Daqing Petroleum Administrative Bureau, Heilongjiang 163453, P. R. China.

prior to a great earthquake is still under hot worldwide debate among geophysicists (VAROTSOS *et al.*, 1994; UYEDA *et al.*, 2004). Nevertheless, physical experiments at laboratory scale did observe very credible rupture-electromagnetic (EM) emissions (BRADY and ROWELL 1986; YAMADA *et al.*, 1989).

From a very different perspective, the geophysical exploration community has also been interested in the seismoelectric coupling effect for more than half a century. BLAU and STATHAM (1936) made the first documented observations of electric current modulation, crossing a pair of electrodes when seismic waves passed through the electrodes. THOMPSON (1936) published a paper on the seismoelectric effects in the first issue of *Geophysics*, the primary international journal on exploration geophysics. There has been a resurgence of interest in seismoelectric effects since the early 1990s (NEEV and YEATTS, 1989; THOMPSON and GIST, 1991, 1993). PRIDE (1994) systematically reviewed the theory and developed a set of the governing equations based upon complete Maxwell's equations. HAARTSEN (1995) solved the macroscopic governing equations numerically for the case of a layered poroelastic media and used this technique to interpret two data sets with analytic Green's functions. PRIDE and HAARTSEN (1996) derived nine Green's tensors for various wavefields in a general anisotropic and heterogeneous porous medium.

Understanding the fundamental physics of the coupling between the poroelastodynamic and the electric, magnetic, and electromagnetic aspects associated with a great earthquake from the first principles is the only path to gain insight of the electromagnetic consequences which resulted from the considerably complicated and highly nonlinear earthquake processes. The rapid development and global availability of high-performance computational capacities (hardware, software, middleware, and high-speed network, etc.) has been facilitating our understanding of the seismo-electromagnetic coupling through multi-scale and multi-physics numerical modeling. MIKHAILENKO and SOBOLEVA (1997) modeled the seismomagnetic effects arising in the seismic wave motion in a constant geomagnetic field by the finite Fourier transform. HAN and WANG (2001) modeled the coupling of the elastic SH wave and the TE mode electric field by a fast finite-element time-domain method. This paper presents a numerical modeling scheme that models the fully coupled seismoelectric wave propagation using the pseudo-spectral time domain (PSTD) method (LIU, 1997, LIU and ARCONE, 2003). We are particularly interested in the case of a seismic wave propagating through a porous medium with the presence of an externally exerted electric field, by the stimulation of observed results from a physical experiment conducted at a petroleum exploration site.

2. The Governing Equations

In a porous medium, the equations that govern the elastic and electromagnetic behavior in both the fluid and the solid skeleton are the following set of equations.

2.1 Continuity Equation for Material Conservation

$$\frac{\partial \rho_b}{\partial t} + \nabla \cdot (\rho_b) = 0,$$

where ρ_b is the bulk density of the porous medium, $\mathbf{v}$ is the particle velocity vector. This equation states that the porous medium is continuous during the entire course of the motion.

2.2 Elastodynamic Equation for Momentum Conservation

Application of the dynamic equation to continuum media results in the following equation that governs the elastic field

$$\rho_b \frac{\partial \mathbf{v}}{\partial t} = \nabla \cdot \boldsymbol{\sigma} + \mathbf{F},$$

where $\boldsymbol{\sigma}$ is the stress tensor, and $\mathbf{F}$ is the body forces, which includes the gravity force, the electromagnetic force (coupled from the electromagnetic field), and the elastic source terms. For a porous medium, consider the elastic fields in the fluid and the solid skeleton separately, we can write the individual momentum conservation equations for the fluid as

$$\phi \rho_f \frac{\partial \mathbf{v}_f}{\partial t} = \phi \nabla P + \phi \rho_{ef} \mathbf{E} + A_f(\mathbf{v}_s - \mathbf{v}_f)$$

and the solid skeleton as

$$(1-\phi)\rho_s \frac{\partial \mathbf{v}_s}{\partial t} = (1-\phi)\nabla \cdot \boldsymbol{\sigma} + (1-(1-\phi)C)\nabla P + (1-\phi)\rho_{es}\mathbf{E} + A_s(\mathbf{v}_s - \mathbf{v}_f) + (1-\phi)\mathbf{S}$$

where the subindices f and s denote the fluid and solid skeleton, respectively; so that ρ_f is the density of the fluid and ρ_s is the density of the solid skeleton; ρ_e is the electric charge density; ϕ is the porosity; $(1-\phi)$ is the specific volume of the solid skeleton; C is the degree of consolidation; $\mathbf{S}$ is the elastic source; P is the pore pressure; $\mathbf{v}_f$ and $\mathbf{v}_s$ are the particle velocity of the fluid and the solid skeleton, respectively. Apparently, the right-hand side of the above two equations are in the dimension of force per volume, i.e., they stand for all the forces in a unit volume. The constants, A_f and A_s, are defined as

$$A_f = \frac{\phi^2 \eta}{k},$$

$$A_s = \frac{\phi(1-\phi)\eta}{k},$$

where k is the hydraulic permeability; and η is the fluid viscosity. From the constitutive relationship for the porous medium, we also have

$$-\frac{\partial P}{\partial t} = D_f \nabla \cdot \mathbf{v}_f + D_s \nabla \cdot \mathbf{v}_s,$$

where D_f and D_s are constants of diffusivity related to the bulk modulus of the fluid and the solid skeleton:

$$D_f = \frac{K_s + K_f b}{1 + b},$$
$$D_s = \alpha \frac{K_f}{1 + b},$$

where K_f and K_s are the bulk modulus of the fluid and solid phase, parameter b is defined as

$$b = \frac{K_f}{\phi K_s^2}((1 - \phi)K_s - K_{dr}),$$

where K_{dr} is the drained bulk modulus of the solid skeleton.

The seismic wavelength ranges generally from several tens of meters to several kilometers; then the size of the grid cells in the PSTD algorithm is around meters to hundreds of meters, as governed by numerical accuracy and stability conditions (LIU, 1997). Meanwhile, the grain size or the pore size is around the orders of submillimeter to submicrometers. Obviously, the pore sizes are much smaller than the size of the grid cells and the seismic wavelength. The size relationship enables us to make a volumetric average of all the material properties of a porous medium over a grid cell. For a transient passing of the seismic wave, it is reasonable to assume the 'un-drained' condition is applicable to the porous medium, i.e., only pore pressure varies but the pore fluid does not flow between the adjacent pores. Thus, the particle velocities of the fluid and the solid skeleton are virtually identical, i.e., $\mathbf{v}_f = \mathbf{v}_s$. Then, the volume-averaged equation for momentum conservation becomes

$$\rho_b \frac{\partial \mathbf{v}}{\partial t} = (1 - \phi)\nabla \cdot \boldsymbol{\sigma} + (1 - C')\nabla P + \rho_e \mathbf{E} + (1 - \phi)\mathbf{S}, \quad (1)$$

where $\rho_b = \phi\rho_f + (1-\phi)\rho_s$ is the bulk density; $\rho_e = \phi\rho_{ef} + (1-\phi)\rho_{es}$ is the bulk electric charge density, $C' = \phi + (1-\phi)C$ is also a constant. Meanwhile, we also have

$$\frac{\partial P}{\partial t} = -D\nabla \cdot \mathbf{v} \quad (2)$$

with $D = D_f + D_s$.

2.3 Maxwell's Equations for Governing the Electromagnetic Field

$$\begin{aligned}\nabla \times \mathbf{E} &= -\mu \frac{\partial \mathbf{H}}{\partial t} \\ \nabla \times \mathbf{H} &= \varepsilon \frac{\partial \mathbf{E}}{\partial t} + \mathbf{j} \\ \nabla \cdot (\mu \mathbf{H}) &= 0 \\ \nabla \cdot (\varepsilon \mathbf{E}) &= \rho_e\end{aligned}$$

where $\mathbf{E}$ is the electric field, and $\mathbf{H}$ is the magnetic field; μ is the magnetic permeability; ε is the electric permittivity. In the second equation, $\varepsilon \partial \mathbf{E}/\partial t$ represents the displacement current and $\mathbf{j}$ is the total conductive current. For the problem considered here, the frequency of seismic signals generated in seismic exploration is usually around or lower than 100 Hz. The frequency band of the seismic waves generated by a natural earthquake is even lower. In these frequency bands, the electromagnetic fields associated with the propagation of the seismic wave can be considered as static. That is, all the time derivatives of the electromagnetic fields tend to be vanishing. Then the original Maxwell equations can be rewritten as

$$\nabla \times \mathbf{E} = 0 \tag{3a}$$

$$\nabla \times \mathbf{H} = \mathbf{j} \tag{3b}$$

$$\nabla \cdot (\mu \mathbf{H}) = 0 \tag{3c}$$

$$\nabla \cdot (\varepsilon \mathbf{E}) = \rho_e. \tag{3d}$$

2.4 The Seismic-electric Coupling Mechanisms

In Equation (3) the conductive electric current density $\mathbf{j}$ consists of currents generated by the electric field, and the streaming current generated from the coupling with the elastic field in a porous medium, i.e.,

$$\mathbf{j} = \sigma \mathbf{E} + \mathbf{j}_s, \tag{4}$$

where σ is the electric conductivity; $\mathbf{j}_s$ is the streaming current density. The value of σ is associated with the pore pressure, and $\mathbf{j}_s$ is associated with the gradient of the pore pressure as

$$\sigma = \sigma_0(1 + \gamma P), \tag{5a}$$

$$\mathbf{j}_s = -\frac{\varepsilon \zeta}{\eta} \nabla P, \tag{5b}$$

where ζ is the zeta potential, a measure of the strength of the electric double layer at the solid-fluid interface; γ is a constant ranging from $10^{-5} \sim 10^{-10}$ Pa^{-1}, depending on the physical properties of the porous medium. Equation (5a) is the mathematical expression of the current effect or the so-called "I-effect;" while Equation (5b) is the mathematical expression of the electrokinetic effect, or the "E-effect." Apparently, in Equation (5b), the dielectric permittivity ε, the zeta potential ζ, and the viscosity η can also vary as functions of the pore pressure and/or the electric field. Nevertheless, their variations should be a secondary effect and we assume they are constants. The relationships stated in Equations (5a) and (5b) are the only mechanisms for the seismoelectric coupling discussed in this paper. They state the effect of pore pressure on the electric field through two channels: 1) changing the conductivity; and 2) building up the streaming current.

2.5 Constitutive Equation and the Strain Rate-velocity Relationship for Poroelastic Deformation

Biot (1956, 1962a, b) systematically derived a group of equations to describe the static and dynamic deformation of a porous medium. Using the concurrent notation convention, the constitutive relationships for a poroelastic medium can be written as:

$$\sigma_{ij} = 2\mu\left[\varepsilon_{ij} + \frac{v}{1-2v}\varepsilon_{kk}\delta_{ij}\right] - \alpha P\delta_{ij}, \tag{6}$$

where σ_{ij}, ε_{ij}, are the stress and strain tensor, respectively; μ is the shear modulus; v is the Poisson's ratio; δ_{ij} is the Kronecker delta. The Einstein summation convention is understood. The coefficient α is directly related to the original definition of the skeleton parameter H defined by Biot (1956, 1962a, b) as

$$\alpha = \frac{2(1+v)\mu}{3(1-2v)H}.$$

The strain rate-velocity relation keeps the same as for the elastic case

$$\dot{\varepsilon}_{ij} = \frac{1}{2}(v_{i,j} + v_{j,i}), \tag{7}$$

where $\dot{\varepsilon}_{ij}$ is the time derivative of ε_{ij}, and $v_{i,j} = \partial v_i/\partial x_j$.

The governing equations listed here form the base for the numerical simulations, which is the theme of this paper. A more rigorous derivation of the governing equations for the seismoelectric coupling can be found from the theoretical work done by Pride (1994).

3. Seismic Wave Propagation with the Existence of an External Electric Field

Under natural circumstances, the extremely weak signal from the seismoelectric effects makes it very difficult to distinguish the electrokinetic signal from the pure electric signal. This technical difficulty hinders the application of seismoelectric effects from practical deployment in the field. One way to overcome this technical barrier is to boost the background electric field by injecting a directional current (DC) to build up a temporary background electric field. Observation of the seismoelectric effect subjected to an imposed electric field may lead to a better understanding of the nonlinear mechanic-electrical coupling, thus reaching a more feasible separation of these two kinds of signals. If an external electric field is exerted, the total current density expressed in Equation (4) should be written as

$$\mathbf{j} = \sigma \mathbf{E} + \mathbf{j}_s + \mathbf{j}_0$$

where $\mathbf{j}_0$, is the external current source density, and $\mathbf{E}$ should also include the electric field caused by the external current source. Put (4) and (5) into (3b) we get

$$\nabla \times \mathbf{H} = \sigma \mathbf{E} - \frac{\varepsilon\zeta}{\eta} \nabla P + \mathbf{j}_0. \tag{8}$$

Taking divergence on both sides of Equation (8) and rearrange the terms, we obtain

$$\nabla \cdot (\sigma \mathbf{E}) = \frac{\varepsilon\zeta}{\eta} \nabla^2 P - \nabla \cdot \mathbf{j}_0, \tag{9}$$

with the understanding that $\nabla \cdot (\nabla \times \mathbf{H}) = 0$.

4. *The Numerical Algorithms*

Considering a two-dimensional case, we can rewrite Equations (1), (2), (6) and (7) into component form:

$$\begin{cases} \rho \dfrac{\partial v_x}{\partial t} = \dfrac{\partial \sigma_{xx}}{\partial x} + \dfrac{\partial \sigma_{xy}}{\partial y} - (1 - C') \dfrac{\partial P}{\partial x} + \rho_e E_x + S_x \\ \rho \dfrac{\partial v_y}{\partial t} = \dfrac{\partial \sigma_{xy}}{\partial x} + \dfrac{\partial \sigma_{yy}}{\partial y} - (1 - C') \dfrac{\partial P}{\partial y} + \rho_e E_y + S_y \end{cases} \tag{10}$$

$$\begin{cases} \dfrac{\partial \sigma_{xx}}{\partial t} = (\lambda + 2\mu) \dfrac{\partial v_x}{\partial x} + \lambda \dfrac{\partial v_y}{\partial y} \\ \dfrac{\partial \sigma_{yy}}{\partial t} = (\lambda + 2\mu) \dfrac{\partial v_y}{\partial y} + \lambda \dfrac{\partial v_x}{\partial x} \\ \dfrac{\partial \sigma_{xy}}{\partial t} = \mu \left(\dfrac{\partial v_x}{\partial y} + \dfrac{\partial v_y}{\partial x} \right) \end{cases} \tag{11}$$

$$\frac{\partial P}{\partial t} = -D \left(\frac{\partial v_x}{\partial x} + \frac{\partial v_y}{\partial y} \right) \tag{12}$$

Let U represent the electric potential, then equation (9) becomes

$$\frac{\partial}{\partial x}\left(\sigma\frac{\partial U}{\partial x}\right)+\frac{\partial}{\partial y}\left(\sigma\frac{\partial U}{\partial y}\right)=\frac{\varepsilon\zeta}{\eta}\left(\frac{\partial^2 P}{\partial x^2}+\frac{\partial^2 P}{\partial y^2}\right)-\left(\frac{\partial j_{0x}}{\partial x}+\frac{\partial j_{0y}}{\partial y}\right) \tag{13}$$

where

$$E_x=-\frac{\partial U}{\partial x}, \tag{14}$$

$$E_y=-\frac{\partial U}{\partial y}. \tag{15}$$

4.1 Pseudospectral Time Domain Solutions of the Poroelastic Fields

Using a finite-difference model to simulate the real world, we first need to spatially discretize the model into finite cells, each cell (i, j) has a cell length of Δx and Δy. All the field components and the medium properties are measured at the center of the cell, and they represent their volumetric average value throughout the cell. The spatial derivatives can be calculated by the pseudospectral algorithm. For example, the spatial derivative of a generic field component W at cell (i, j) with respect to x direction can be formulated as

$$\frac{\partial W(i,j)}{\partial x}=\tilde{D}_x(W(i,j))=F_x^{-1}(ik_xF_x(W(i)))_j,$$

where $\tilde{D}_x$ means the spatial derivative done in wave number domain, and F_x and F_x^{-1} denote for the forward and inverse Fourier transform along the x direction, respectively. In time domain, we also need to discretize it into time steps; each step has a time span of Δt. The time derivatives of the field components can then be approximated by the central finite difference. That is, for each field component W at the n-th time step, we have

$$\frac{\partial W^{(n+1/2)}}{\partial t}=\frac{W^{(n+1)}-W^{(n)}}{\Delta t}.$$

Then the pseudospectral time domain solutions for Equations (10–12) are

$$\begin{cases} v_x^{(n+1)}(i,j)=v_x^{(n)}(i,j)+\frac{\Delta t}{\rho}\Big[\tilde{D}_x\Big(\sigma_{xx}^{(n+\frac{1}{2})}(i,j)\Big)+\tilde{D}_y\Big(\sigma_{xy}^{(n+\frac{1}{2})}(i,j)\Big) \\ \qquad -(1-C')\tilde{D}_x\Big(P^{(n+\frac{1}{2})}(i,j)\Big)+\rho_eE_x^{(n+\frac{1}{2})}(i,j)+S_x(i,j)\Big] \\ v_y^{(n+1)}(i,j)=v_y^{(n)}(i,j)+\frac{\Delta t}{\rho}\Big[\tilde{D}_x\Big(\sigma_{xy}^{(n+\frac{1}{2})}(i,j)\Big)+\tilde{D}_y\Big(\sigma_{yy}^{(n+\frac{1}{2})}(i,j)\Big) \\ \qquad -(1-C')\tilde{D}_y\Big(P^{(n+\frac{1}{2})}(i,j)\Big)+\rho_eE_y^{(n+\frac{1}{2})}(i,j)+S_y(i,j)\Big] \end{cases} \tag{16}$$

$$\begin{cases} \sigma_{xx}^{(n+\frac{1}{2})}(i,j) = \sigma_{xx}^{(n-\frac{1}{2})}(i,j) + \Delta t[(\lambda+2\mu)\tilde{D}_x(v_x^{(n)}(i,j)) + \lambda\tilde{D}_y(v_y^{(n)}(i,j))] \\ \sigma_{yy}^{(n+\frac{1}{2})}(i,j) = \sigma_{yy}^{(n-\frac{1}{2})}(i,j) + \Delta t[(\lambda+2\mu)\tilde{D}_y(v_y^{(n)}(i,j)) + \lambda\tilde{D}_x(v_x^{(n)}(i,j))] \\ \sigma_{xy}^{(n+\frac{1}{2})}(i,j) = \sigma_{xy}^{(n-\frac{1}{2})}(i,j) + \Delta t\mu[\tilde{D}_y(v_x^{(n)}(i,j)) + \tilde{D}_x(v_y^{(n)}(i,j))] \end{cases} \tag{17}$$

$$P^{(n+\frac{1}{2})}(i,j) = P^{(n-\frac{1}{2})}(i,j) - \Delta t D[\tilde{D}_y(v_x^{(n)}(i,j)) + \tilde{D}_x(v_y^{(n)}(i,j))]. \tag{18}$$

The next step is to solve the electric potential U.

4.2 Two-dimensional Finite-difference Solution for the Electric Field

4.2.1 The cell equation

Equation (13) is actually a Poisson's equation for the electric field with the right side representing the effective electric current sources, which include the effect of both the injected electric current and the seismically induced streaming current.

Since the left-hand side of Equation (13) is a sparse matrix, instead of using the pseudospectral algorithm, it is more efficient to use a central difference algorithm. The equation for each cell (i, j) at time step n can then be expressed as

$$\frac{1}{\Delta x}\left(\frac{U^{(n)}(i+1,j)-U^{(n)}(i,j)}{\bar{\rho}^{(n)}(i+\frac{1}{2},j)\Delta x} - \frac{U^{(n)}(i,j)-U^{(n)}(i-1,j)}{\bar{\rho}^{(n)}(i-\frac{1}{2},j)\Delta x}\right)$$

$$+\frac{1}{\Delta y}\left(\frac{U^{(n)}(i,j+1)-U^{(n)}(i,j)}{\bar{\rho}^{(n)}(i,j+\frac{1}{2})\Delta y} - \frac{U^{(n)}(i,j)-U^{(n)}(i,j-1)}{\bar{\rho}^{(n)}(i,j-\frac{1}{2})\Delta y}\right) = s^{(n)}(i,j) \tag{19}$$

where $\bar{\rho}_e$ is the average resistivity, and $s(i, j)$ is the effective source including both the external current source and the streaming current (caused by the 'E-effect').

$$s^{(n)}(i,j) = \frac{\varepsilon\zeta}{\eta}[\tilde{D}_x(\tilde{D}_x(P^{(n)}(i,j))) + \tilde{D}_y(\tilde{D}_y(P^{(n)}(i,j)))] - [\tilde{D}_x(j_{0x}^{(n)}(i,j)) + \tilde{D}_y(j_{0y}^{(n)}(i,j))].$$

The resistivity is spatially interpolated with the known conductivity on the integer nodes.

$$\bar{\rho}^{(n)}(i+\tfrac{1}{2},j) = \frac{1}{2}\left(\frac{1}{\sigma^{(n)}(i+1,j)} + \frac{1}{\sigma^{(n)}(i,j)}\right),$$

$$\bar{\rho}^{(n)}(i-\tfrac{1}{2},j) = \frac{1}{2}\left(\frac{1}{\sigma^{(n)}(i,j)} + \frac{1}{\sigma^{(n)}(i-1,j)}\right),$$

$$\bar{\rho}^{(n)}(i,j+\tfrac{1}{2}) = \frac{1}{2}\left(\frac{1}{\sigma^{(n)}(i,j+1)} + \frac{1}{\sigma^{(n)}(i,j)}\right),$$

$$\bar{\rho}^{(n)}(i,j-\tfrac{1}{2}) = \frac{1}{2}\left(\frac{1}{\sigma^{(n)}(i,j)} + \frac{1}{\sigma^{(n)}(i,j-1)}\right).$$

The effect of pore pressure on conductivity, i.e., the I-effect (Equation (5a)), has been incorporated here for obtaining the average resistivity to solve for the electric potential U. It is noteworthy to point out that both the seismoelectric coupling 'I-effect' and 'E-effect' have been fully considered in the simulation by our PSTD algorithm.

4.2.2 The formation of a linear system of equation

In order to form a system of equations to be solved by linear algebra, we use the index

$$k = i' + j' * nx \quad (i' = 0 \dots nx; \quad j' = 0 \dots ny)$$

for cell (i', j'), where nx is the maximum number of i' (total cell numbers along x direction), ny is the maximum number of j' (total cell numbers along y-direction). Equation (19) can be rewritten as

$$\sum_{k=0}^{nx*ny-1} a^{(n)}(k) U^{(n)}(k) = s^{(n)} \tag{20}$$

where

$$a^{(n)}(k) = \begin{array}{l} -\dfrac{1}{\bar{\rho}_e^{(n)}(i+\frac{1}{2},j)\Delta x^2} - \dfrac{1}{\bar{\rho}_e^{(n)}(i-\frac{1}{2},j)\Delta x^2} \\ -\dfrac{1}{\bar{\rho}_e^{(n)}(i,j+\frac{1}{2})\Delta y^2} - \dfrac{1}{\bar{\rho}_e^{(n)}(i,j-\frac{1}{2})\Delta x^2} \end{array}, \quad \text{if} \quad i' = i \text{ and } j' = j$$

$$a^{(n)}(k) = \frac{1}{\bar{\rho}_e^{(n)}(i+\frac{1}{2},j)\Delta x^2}, \quad \text{if} \quad i' = i+1 \text{ and } j' = j$$

$$a^{(n)}(k) = \frac{1}{\bar{\rho}_e^{(n)}(i-\frac{1}{2},j)\Delta x^2}, \quad \text{if} \quad i' = i-1 \text{ and } j' = j$$

$$a^{(n)}(k) = \frac{1}{\bar{\rho}_e^{(n)}(i,j+\frac{1}{2})\Delta y^2}, \quad \text{if} \quad i' = i \text{ and } j' = j+1$$

$$a^{(n)}(k) = \frac{1}{\bar{\rho}_e^{(n)}(i,j-\frac{1}{2})\Delta y^2}, \quad \text{if} \quad i' = i \text{ and } j' = j-1$$

$$a^{(n)}(k) = 0, \quad \text{otherwise}$$

Assembly Equation (20) for each cell ($l = i + j*nx$), we obtained a system of equations

$$\sum_{k=0}^{nx^*ny-1} a^{(n)}(k,l) U^{(n)}(k) = s^{(n)}(l) \qquad (l = 0 \dots nx*ny - 1).$$

In matrix form, it is

$$\mathbf{AU} = \mathbf{S}. \tag{21}$$

A is a linear sparse matrix with a dimension of $(nx \times ny, nx \times ny)$, **U** and **S** are both vectors. Obviously, this is a linear system.

There exist numerous approaches to solve this system of linear equations (21). In view of the sparse nature of matrix **A**, the most efficient approach is the conjugate gradient (CG) method. The CG approach is an iterative method, in which the solution is an approximation. The efficiency or the convergence speed of this method largely depends on the singularity property of matrix **A**. To improve this property, we applied the Cholesky preconditioner (ZHANG and XIAO, 2003) in the iterations, which largely improve the convergence property of matrix **A**. Finally, with the use of Equations (14–15) we have

$$E_x^{(n+\frac{1}{2})}(i,j) = -\tilde{D}_x\left(U^{(n+\frac{1}{2})}(i,j)\right) \tag{22}$$

$$E_y^{(n+\frac{1}{2})}(i,j) = -\tilde{D}_y\left(U^{(n+\frac{1}{2})}(i,j)\right) \tag{23}$$

5. *Numerical Implementation*

We used the pseudospectral time domain (PSTD) algorithm to simulate the poroelastic field and finite-difference time-domain method to solve the electric field. For each time step, the numerical algorithm is subject to the following steps sequentially:

i. Calculating the velocity fields **v** by Equation (16);
ii. calculating the stress fields σ by Equation (17);
iii. calculating the pore pressure P by Equation (18);
iv. solving Equations (21) to obtain electric potential U;
v. calculating electric fields **E** from electric potential by Equation (22) and (23);
vi. output results.

The flowchart of the program in Figure 1 outlines the process in greater detail.

6. *Simulation Results*

6.1 Brief Description of the Field Experiment

This study is stimulated by a field experiment on seismoelectric effect conducted by YAN *et al.* (1998). The field experiment results are briefly described here. The data

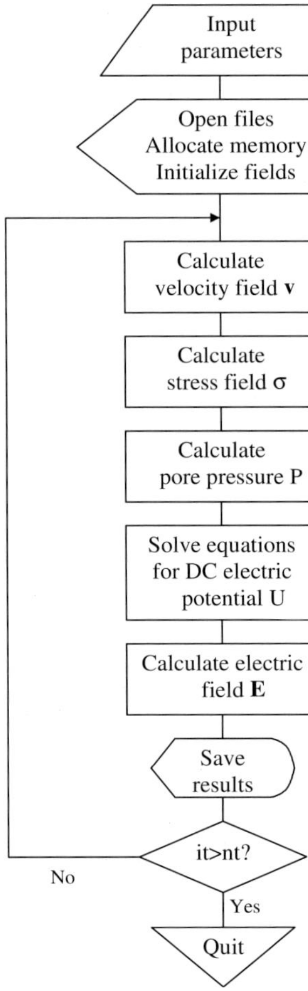

Figure 1
The flow chart of the PSTD numerical algorithm for simulation of the seismoelectric effects.

acquisition was in a two-dimensional profile fashion (Fig. 2). The seismic source was an explosion generated by a 4-kg TNT explosive fired in an 8-meter deep borehole. A total number of 97 shots were fired between Location No. 1998 to Location No. 1806, with 50-m shot interval (Fig. 2). The distance between two adjacent survey Locations is 25 meters. A receiver array consisting of 60 geophones was planted at a 25-meter spacing on the right-hand side of the shot point (Fig. 2). The offset between the first geophone and the shot point was 50 meters. The recording system has a sampling interval of 1 millisecond, with a total recording time of 3 seconds. The DC electric current source was one used for magnetotelluric surveys. The DC electric field was exerted by two non-polarization electrodes with 4125-meter spacing. For each shot, seismic data were acquired without injecting the DC current, then with a

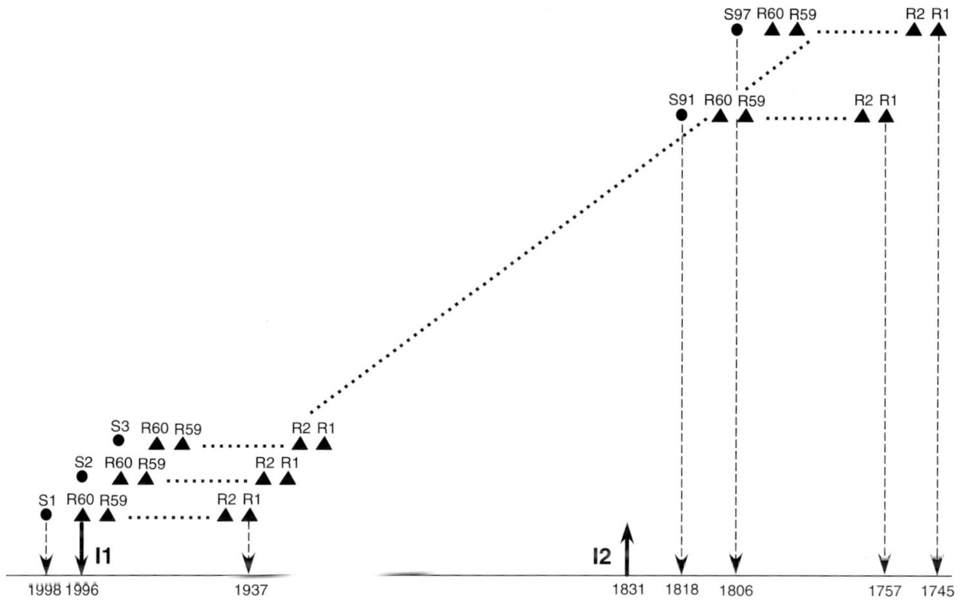

Figure 2

Illustration of the seismoelectric effect field experiment setup. The electrodes that established the DC electric field were set at Location numbers 1996 and 1831. The first shot was located at Location 1998, and the 60-channel geophone spread located from 1996 to 1937. The last shot (shot no. 97) was located at Location number 1806, and the geophone spread located from 1804 to 1745. The geophone spread was always located at the right-hand side of the shot. The first geophone (Channel R60) was always two location numbers away from the shot. Every location number equals 25 meters, the geophone spacing used in the survey. The distance between the two electrodes then was 4125 m.

DC electric field by injecting 25A current at Location 1996 as the positive electrode (noted as I1 in Fig. 2, and the case of +25 A current), and Location 1831 as the negative electrode (noted as I2 in Fig. 2). After waiting one hour for the start of current injection, the source shot was fired to acquire the seismic data. The same procedure was repeated, exchanging the polarity of the electrodes (with I1 as the negative electrode and I2 as the positive electrode, and the case of −25 A current).

Figure 3 shows the seismic records for shot No. 91 for the cases without the DC electric field inserted and with a DC field generated by a +25A current. The difference (the bottom panel) clearly shows that the existence of the DC electric field does significantly effect the magnitude of seismic vibrations.

To examine the effect of the DC electric field on seismic wave dynamics in detail, Figure 4a shows the single channel record acquired by Geophone R56 for shot no. 60. The seismic vibration amplitude was enhanced by a DC field generated from a +25 A current, and suppressed by a field caused by a −25A current. Figure 4b displays the similar results in frequency domain, for Channels 1–10 for shot 60.

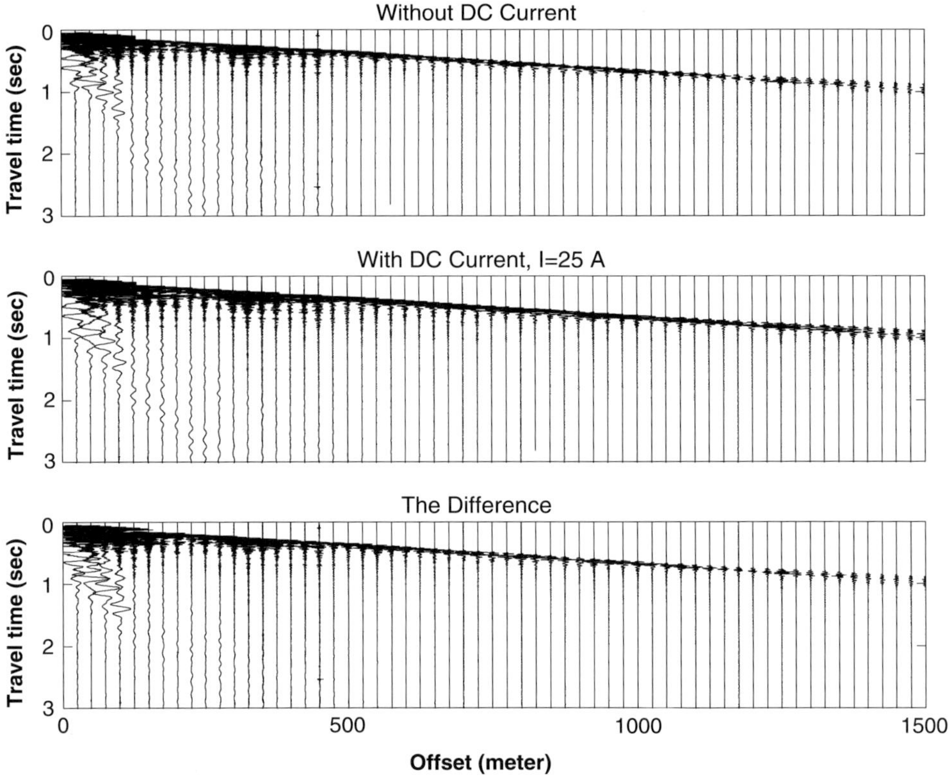

Figure 3
The observed seismic record for shot number 91 (shot located at 1818, geophone spread from 1816 to 1757) for the case of no DC electric field (top), with an injected DC electric field (center) of 25 A DC current from location 1996 (positive electrode) and 1831 (negative electrode) and the difference for the two cases (bottom).

Again, it is clear that at the locations near shot 60 (Location No. 1880), the positive DC field boosts the amplitude of the seismic vibration, and the field with opposite polarity suppressed the seismic vibration. In summary, the field experiments demonstrated that the DC electric field can significantly affect seismic wave dynamics.

6.2 The Modeling Results

With the use of the PSTD algorithm, the elastic wave propagation in a poroelastic media under the influence of a DC electric field can be simulated. The two electrodes imbedded in the media provide a strong background electric field. Figure 5 shows the geometry. The electric resistivity and seismic velocity values used are those best estimated from the local region based on geophysical explorations. A 25A DC

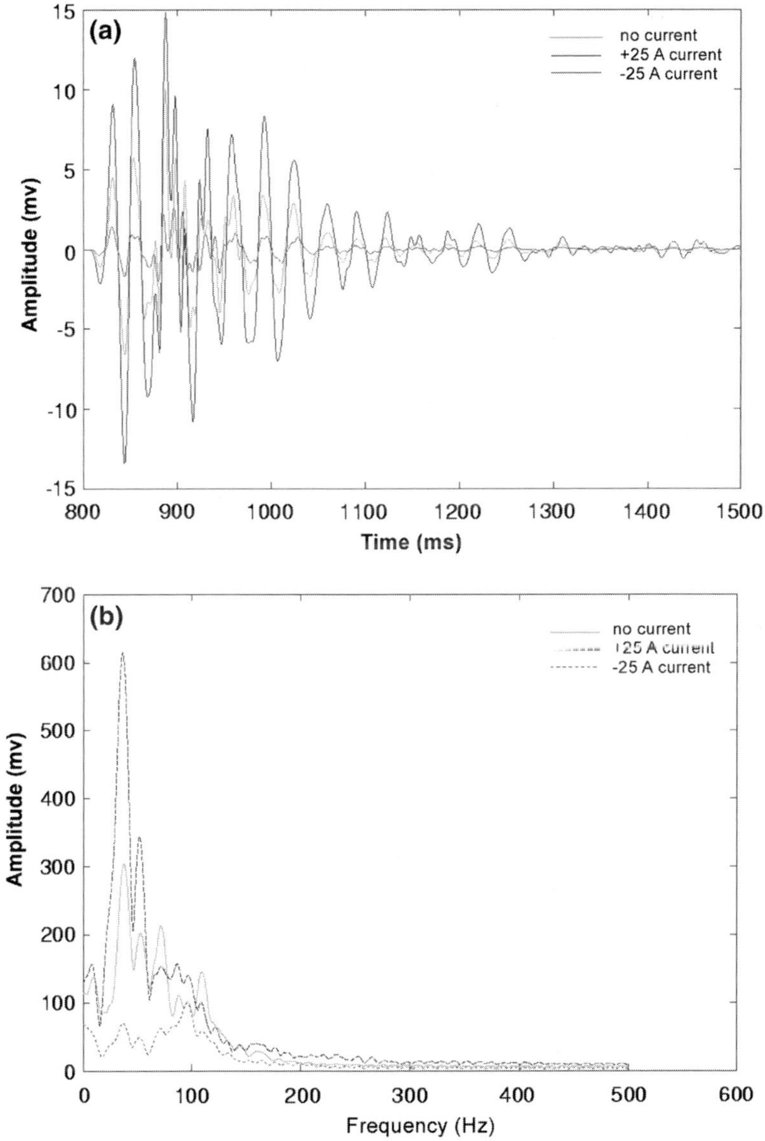

Figure 4

(a) A single channel time record extracted from Channel 56 (Location no. 1874) for shot no. 60 (Location no. 1880). Time records were acquired without a DC electric field, one hour after starting the injection of a +25A DC current from the electrode at Location 1996, and one hour after starting the injection of a +25A DC current from the electrode at Location 1831. (b) The averaged amplitude spectra from Channels 1–10 for shot no. 60.

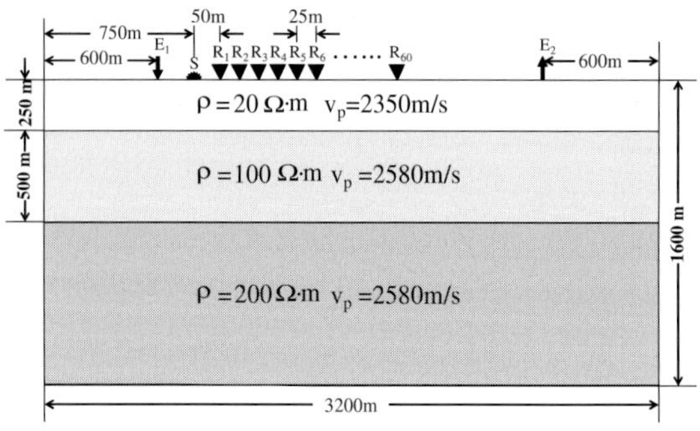

Figure 5

A layered model for modeling the seismic wave propagation under the influence of a DC electric current. The model is 3.2 km in width and 1.6 km in depth with three layers. The first layer is 250-m thick, the second layer is 500-m thick, and the third layer is a half space. The two electrodes are placed 2000 m apart. The elastic source and receiver array is located between the two electrodes. The source is 150-m from the positive electrode. The elastic source signal is a Ricker wavelet with a central frequency of 10 Hz. Sixty receivers are located on the right-hand side of the source with a spatial interval of 25 m.

electric current is applied to a pair of electrodes. The DC electric field in the ground is shown as Figure 6.

We used 256 × 128 grids to simulate this model. The dimension of each grid is 12.5 meters square. The time interval is 0.0001 sec with 15000 total time steps and a recording time window of 1.5 seconds. The DC current is injected from the 2-electrode position to the model from zero to 25 A within 100 time steps (0.01 sec); an isotropic

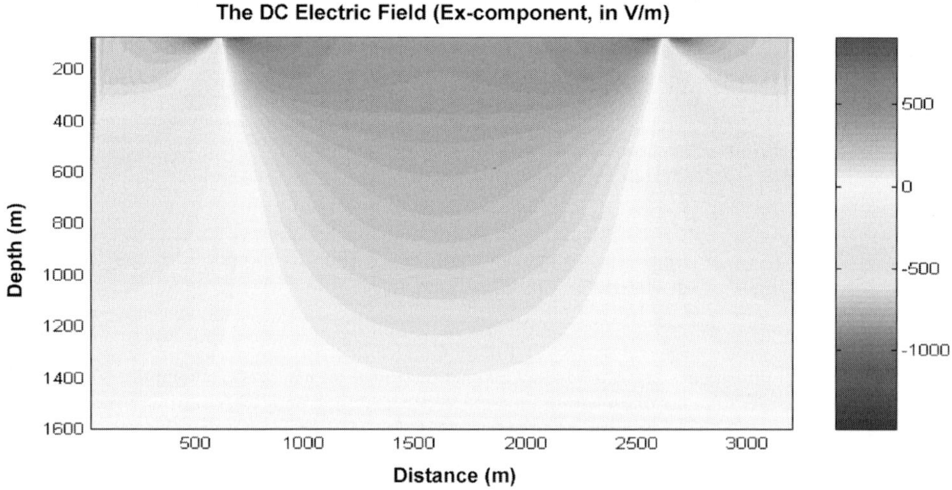

Figure 6

The horizontal component Ex of the electric field generated by injecting DC current at $X_{11} = 600$ m and $X_{12} = 2600$ m.

pressure source of 1000 Pa is fired 50 meters right to the positive electrode at 0.2 second. Besides the parameters annotated in Figure 5, other elastic and electric properties of the formation used in the modeling are some typical values and kept as constants over the entire domain of the model. They are: porosity 0.2, charge density 0.2 C/m^3, effective bulk modulus of the solid skeleton 2.3×10^9 Pa, effective bulk modulus of the pore fluid 1.7×10^9 Pa, pressure-conductivity coefficient 1.0×10^{-8} Pa^{-1}, and the electrokinetic Onsager coefficient $\varepsilon\zeta/\eta$ 1.0×10^{-9} [Am^{-2} Pa^{-1}]. The simulation results are shown in Figures 7 to Figure 10.

Figure 7 shows the synthetic seismograms for horizontal particle velocity (top) and electric field (bottom) records without exertion of the DC electric field. Since only the elastic source is used, the events shown in the electric field can only be the products of the seismoelectric coupling, here both the pressure-resistivity effect (the 'I-effect') and the electrokinetic effect (the 'E-effect') play roles generating the electric

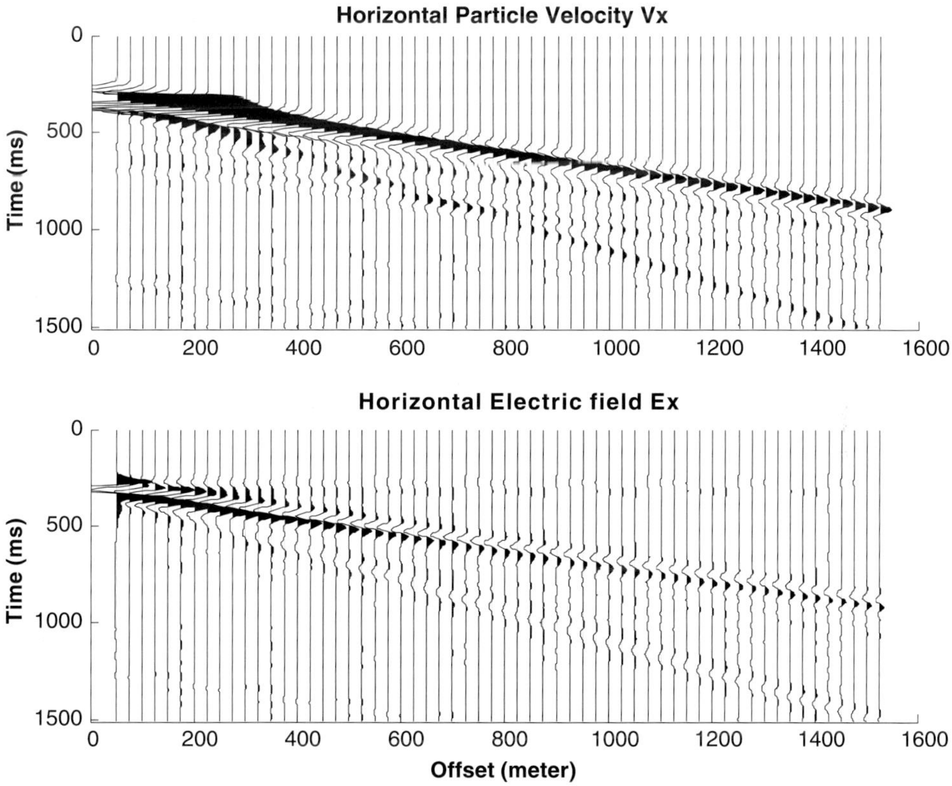

Figure 7
The synthetic seismograms for horizontal particle velocity (top) and electric field (bottom) records without exertion of the DC electric field. The origin time is arbitrarily chosen. An automatic gain (AGC) is applied to the amplitude in the seismic particle velocity and the electric field to show the features over the entire profiles.

events. The events in the electric records are quite similar to those in the seismic record, which are the responses to formation interfaces. Figure 7 demonstrates that the surface and interfaces are the important secondary sources to generate the seismically induced electric signals.

Figure 8 compares the horizontal particle velocity field V_x for the case of no DC electric field and with exertion of the DC electric field at receiver location R1 (50 m right to the elastic pressure source). The seismically induced vibration arrives at R1 at 0.43 sec. The low-frequency uprising before this point is caused by the transition of the exertion of the DC electric field.

Figure 9 compares the horizontal electric field E_x for the case of no DC electric field and with exertion of the DC electric field at receiver location R1. The seismically-induced electric signal also arrives at R1 at 0.43 sec. It is apparent that changing the polarity of the DC current results in an opposite effect on the seismically-induced electric signals. The magnitude of the changes in the seismically-induced electric fluctuation is on the same order of magnitude with previous published results (e.g., HAARTSEN and PRIDE, 1997).

Figure 10 shows the seismoelectric coupling effect by comparing the elastic record with full consideration of the seismoelectric coupling and the one without any coupling at receiver R1. Clearly a significant amount of electric energy has been transformed to elastic energy thus enhancing the seismic vibration at R1 to a certain degree.

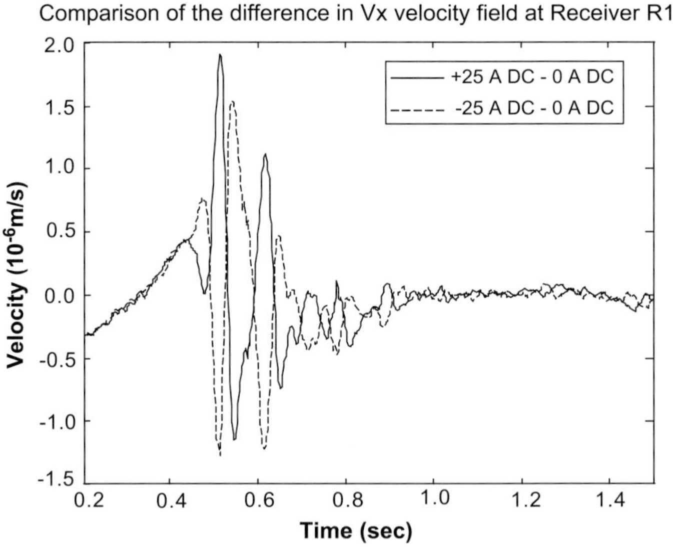

Figure 8
Comparison of the seismoelectric coupling effect on the simulated horizontal particle velocity field Vx at receiver location R1. The solid line represents the differential Vx of the exerted DC field and without the DC field. The broken line represents the differential Vx of the reversely exerted DC field and without the DC field. The seismically induced vibration arriving R1 at 0.43 sec. The low-frequency vibration before this point is caused by the exertion of the DC electric field.

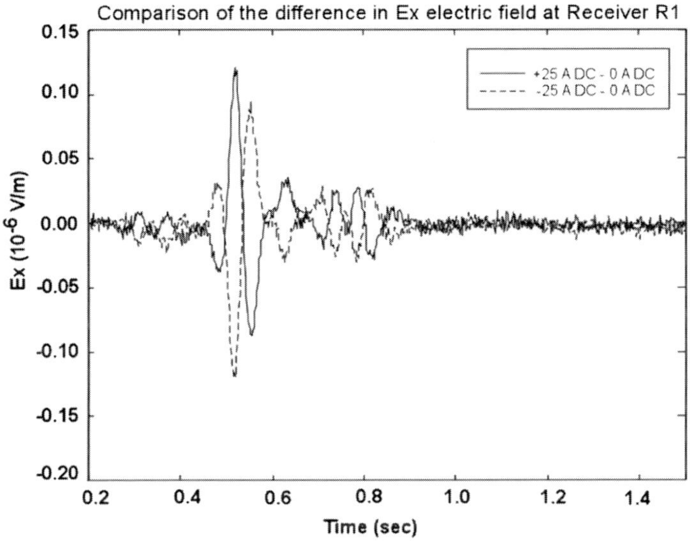

Figure 9

Comparison of the seismoelectric coupling effect on the simulated horizontal electric field Ex at receiver location R1. The solid line represents the differential Ex of the exerted DC field and without the DC field. The broken line represents the differential Ex of the reversely exerted DC field and without the DC field.

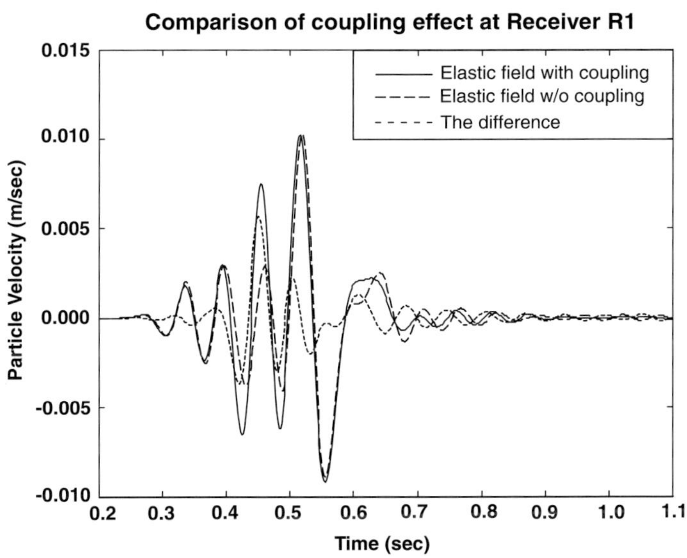

Figure 10

Comparison of the seismoelectric coupling effect for the ground vibration particle velocity at receiver location R1. The solid line represents the coupled elastic field with the the contribution of the background DC electric field. The broken line represents the original elastic field in which no DC current was injected. The dotted line is the difference in the elastic field between received signals under those two conditions.

7. Conclusion

A set of governing equations for seismoelectric coupling in poroelastic media is used to simulate the elastic wave propagation in the porous media with an externally exerted DC electric field. Assuming that there is no relative motion between the pore fluids and the solid skeleton, which is the 'undrained case' suitable for describing a seismic wave propagating through a porous medium, we used a volume average representation for all the physical parameters. In the frequency band of seismic waves, the electromagnetic fields can be approximated by a static electric field. Both the formation resistivity and the current source are time dependent; they vary along with the variation of the pore pressure generated by the passing seismic waves. The pore pressure importantly influences seismoelectric coupling. It generates streaming current (the 'E-effect') and causes variations in formation resistivity (the 'I-effect'). In the future, variations in the dielectric permittivity, the zeta potential, and the fluid viscosity as the pore pressure could also be considered as the additional seismoelectric coupling mechanisms. The PSTD method is applied in the numerical simulation. The electric field is solved by the conjugate gradient method. It is proved that the Cholesky preconditioner for accelerating the convergence is very accurate and efficient. The numerical simulation of elastic wave propagating in the poroelastic media with a strong DC electric field successfully displayed that a DC field significantly influences the propagation of seismic waves. If the condition is loosened to include a drained condition to allow a relative motion of pore fluids comparative to the solid skeleton, the simulation technique may be extended to the case for slow deformations in a static electric or magnetic field, which is more direct to the study of seismogenic processes.

Acknowledgments

The authors would like to thank Professor X. Yin for his encouragement to submit the research results to the special issue on Seismic Simulation in *Pure and Applied Geophysics*. The lead author is grateful to Rob Mehl for his editorial review of an early version of the manuscript.

References

Biot, M. A. (1956), *Theory of propagation of elastic waves in a fluid-saturated porous solid,* J. Acoust. Soc. Am. *28,* 168–191.

Biot, M. A. (1962a), *Mechanics of deformation and acoustic propagation in porous media*, J. Appl. Phys. *33,* 1482–1498.

Biot, M. A. (1962b), *Generalized theory of acoustic propagation in porous dissipative media*, J. Acoust. Soc. Am. *34,* 1254–1264.

BLAU, L. and STATHAM, L. (1936), *Method and apparatus for seismic electric prospecting*, Technical Report 2054067, U.S. Patent.

BRADY, B. T. and ROWELL, G. A. (1986), *Laboratory investigation of the electrodynamics of rock fracture*, Nature *321*, 488–492.

HAARTSEN, M.W. *Coupled electromagnetic and acoustic wavefield modeling in poroelastic media and its applications in geophysics exploration* (Ph.D. Thesis, MIT, 1995).

HAARTSEN, M. W. and PRIDE, S. (1997), *Electroseismic waves from point sources in layered media*, J. Geophys. Res. *102*, 24745–24769.

HAN, Y. and WANG, Z. (2001), *Time domain simulation of SH-wave-induced electromagnetic field in heterogeneous porous media: A fast finite-element algorithm*, Geophysics *66*, 448–461.

LIU, L. and ARCONE, A. S. (2003), *Numerical simulation of the wave-guide effect of the near-surface thin layer on radar wave propagation*, J. Environ. Eng. Geophys. *8*, 133–141.

LIU, Q. (1997), *The PSTD algorithm: a time domain method requiring only two cells per wavelength*, Microwave Opt. Tech. Lett. *15*, 158–165.

MIKHAILENKO, B. G. and SOBOLEVA, O.N. (1997), *Mathematical modeling of seismo-magnetic effects arising in the seismic wave motion in the earth's constant magnetic field*, Appl. Math. Lett. *10*, 47–51.

NEEV, J. and YEATTS, F. R. (1989), *Electrokinetic effects in fluid-saturated poroelastic media*, Phys. Rev. *B40*, 9135–9141.

PRIDE, S. (1994), *Governing equations for the coupled electromagnetic and acoustics of porous media*, Phys. Rev. B*50*, 15678–15696.

PRIDE, S. R. and HAARTSEN, M. W. (1996), *Electroseismic wave properties*, J. Acoust. Soc. Am. *100*, 1301–1315.

THOMPSON, A. H. and GIST, G. A. (1991), *Electroseismic prospecting*, Soc. Expl Geophys. 61th Ann. Internat. Mtg, Expanded Abstracts 425–427.

THOMPSON, A. and GIST, G. (1993), *Geophysical applications of electrokinetic conversion*, The Leading Edge *12*, 1169 1173.

THOMPSON, R. (1936), *The seismic-electric effect*, Geophysics *1*, 327–335.

UYEDA, S., NAGAO, T. and MEGURO, K., *Report of the RIKEN International Frontier Research Project on Earthquakes (IFREQ) and the future of the short-term earthquake prediction research*, 3rd Internat. Conference on Continental Earthquakes, Abstract, July, 2004.

VAROTSOS, P., UYEDA, S., ALEXOPOULOS, K., NAGAO, T., and LAZARIDOU, M. (1994), *Prediction of recent destructive seismic activities in Greece based on seismic electric signals*, In *Electromagnetic Phenomena Related to Earthquake Prediction* (eds. Hayakawa M. and Fujinawa Y.) (Terra Scientific Publishing Company, Tokyo 1994) pp. 13–24.

YAMADA, I., MASUDA, K., and MIZUTANI, H. (1989), *Electromagnetic and acoustic emission associated with rock fracture*, Phys. Earth Planet. Int. *57*, 157–168.

YAN, H., LIU, H., and LI, Y. (1998), *Seismic survey in electric field: A new seismo-electric prospecting method*, Soc. Expl Geophys. 68th Ann. Internat. Mtg., Expanded Abstracts, pp. 160–163.

ZHANG, J. and XIAO, T. (2003), *A multilevel block incomplete Cholesky preconditioner for solving normal equations in linear least-squares problems*, J. Appl. Math. Computing *11*, 59–80.

(Received December 30, 2004, revised November 28, 2005, accepted December 30, 2005)
Published Online First: August 31, 2006

Pure appl. geophys. 163 (2006) 1915–1932
0033–4553/06/091915–18
DOI 10.1007/s00024-006-0104-4

Pure and Applied Geophysics

Quartz Rheology and Short-time-scale Crustal Instabilities

KLAUS REGENAUER-LIEB[1] and DAVID A. YUEN[2]

Abstract—We present numerical results of thermal-mechanical feedback in crustal quartz rheology and contrast this behavior to the vastly different character of an olivine mantle. In the numerical experiments quartz is found to have a very strong tendency for short-time-scale instabilities, while our numerical experiments show that olivine has a decisive tendency for a stable thermally lubricated slip. At the same time, olivine can also go through a transitional period of creep bursts, which are physically caused by multiple interacting ductile faults at various length and time scales which collocate quickly into a major shear zone. Since olivine has this strong propensity to self organize in a large apparently stable fault system, it lacks the dynamics of interacting ductile faults evident in other minerals. Quartz behaves totally different and keeps its jerky slip behavior for prolonged deformation. An example is shown here in which a 30 × 50 km piece of a wet quarzitic crust is extended for about 2 Ma. The associated total displacement field clearly shows the unstable slipping events, which have a characteristic time frame of one to several years, In contrast, olivine is very stable and has a much longer time scale for thermal instability of 100 kyrs.

Key words: Quartz, rheology, slow earthquake, thermal-mechanical mode, instability.

1. Introduction

Up to now, most previous work on thermo-mechanical feedback has been focussed on the olivine rheology, characteristic of the mantle (REGENAUER-LIEB and YUEN, 2000, 2003, 2004) yet the influence of crustal rheology on the dynamics has not been explored. YUEN and SCHUBERT (1979) have shown that, for the dynamics of large ice masses, the activation energy plays an important role in the shear-heating instabilities of ice sheet dynamics, cast within the framework of a one-dimensional model. There an activation energy contrast of a factor of 4 was examined. It is well known (EVANS and KOHLSTEDT, 1995) that the activation energies for quartz are considerably smaller than those for olivine, therefore, we will now explore the differences in the rheologic paths taken between olivine and quartz. This issue is

[1] Earth and Geographical Sciences, University of Western Australia; CSIRO Exploration and Mining, P.O. Box 1130, Bentley WA 6102, Australia and Institut für Geowissenschaften, Johannes Gutenberg University, D-55099 Mainz, Germany. E-mail: klaus@cyllene.uwa.au
[2] Department of Geology and Geophysics and Minnesota Supercomputing Institute, University of Minnesota, Minneapolis, MN 55455-0219, U.S.A.

especially critical in time scales, as what YUEN and SCHUBERT (1979) found for ice-sheet dynamics. Our objective in this work is to compare the dynamical behavior in the instabilities between quartz and olivine. Of particular importance is the difference in the time scales of the instabilities, which would have strong implications for earthquakes with a long time scale, such as the recent Sumatran event, whose source spectrum does not saturate at low frequencies (STEIN and OKAL, 2005). This implies that the physics of the process is governed by slow slip not detectable from surface waves (ISHII *et al.,* 2005; KRUGER and OHRNBERGER, 2005; NI *et al.,* 2005) but rather bears the characteristics of a slow earthquake.

First, we will discuss the pioneering experiments carried out by EVANS (1984) on quartz. In particular, their experiments revealed a sharp difference in the yield strength between quartz and olivine. Then we will display the differences in the numerical results of the two-dimensional numerical models. We will discuss the ramifications of our work for earthquake instabilities in the final section of this paper.

2. *Quartz Indents like a Glassy Polymer*

Experiments on the indentation hardness of quartz by EVANS (1984) revealed remarkable material properties of quartz. The hardness of quartz $H = C\,\tau_y$ is a function of the differential yield stress τ_y, the angle of the indenter with the test surface β, and the Young's modulus E

$$C = \frac{2}{3}\left[1 + \ln\left(\frac{1}{3}\frac{E}{\tau_y}\tan\beta\right)\right]. \tag{1}$$

C varies between 3 and 1 depending on whether E/τ_y is of the order of 10^2–10^3 or larger (e.g., most metals), or of the order of 10 (e.g., low-temperature, glassy polymers, glass), respectively (JOHNSON, 1970; TABOR, 1970, 1996). If the hardness and the elastic modulus are known at a given temperature the constraint factor C and the yield stress can be computed (EVANS and GOETZE, 1979; JOHNSON, 1970). The hardness data for quartz suggest that at temperatures below 600°C, the ratio of Young's modulus to the yield is very low, probably less than 25. Therefore, Evans concluded that the hardness behavior of quartz is more akin to that of a highly elastic material, such as a polymer, than to that of a metal.

On the other hand, olivine shows a totally different behavior more akin to metal. This notion is true for the fast laboratory time and space scale, where quartz behaves more like a glassy polymer. However, can this important material property also survive for longer time scales where viscous creep is certainly more significant? We have re-interpreted the original data of Brian Evans and co-workers and cast it into a finite-element thermo-mechanical analysis (ABAQUS/STANDARD, 2000) to pose the important question whether Evans ideas concerning the importance of elastic deformation for quartz also holds true for the geological time scale. If so, this observation would have important implications for the emergence of a short-time-scale seismic response out of a smooth long-term geological loading rate.

3. Thermal-mechanical Equations

We give here a summary of the equations based on thermal-mechanical coupling, which forms the basis of our work (e.g., REGENAUER-LIEB and YUEN, 2004). Based on the energy conservation principle, NEMAT-NASSER (1982) has shown that for an elastoplastic body the rate of deformation, decomposes additively as:

$$\dot{\varepsilon}_{ij}^{tot} = \dot{\varepsilon}_{ij}^{el} + \dot{\varepsilon}_{ij}^{pl} \tag{2}$$

for finite elastic (*el*) and plastic (*pl*) rate of deformation, provided that the strain increments are defined with respect to the same reference configuration and that the overall elastic strain remains small compared to the plastic strain. We extend this solid thermodynamic approach and consider thermal expansion, appearing as part of the elastic strain rate (*el*) and viscous creep (*cr*) components, i.e.,

$$\dot{\varepsilon}_{ij}^{tot} = \dot{\varepsilon}_{ij}^{el} + \dot{\varepsilon}_{ij}^{pl} + \dot{\varepsilon}_{ij}^{cr}. \tag{3}$$

For the elastic component, which is important for short time scales, we assume isotropic hypo-elasticity, and objective stress rates. Special care is exercised in the integration of rate constitutive equations to preserve objectivity. Hypo-elasticity is limited to small elastic strains (but large rotations) implying that elastic strain (but not rotations) remain small compared to their viscous or plastic counterparts. This assumption is valid for most geological materials, such as quartz and olivine; the two materials on which we will focus in this study.

Polymers (e.g., rubber) can exhibit significant elastic and plastic deformation of comparable magnitude and the computationally more expensive additive hyper-elastic approach is warranted which is in itself objective. In our mathematical treatment we note that, although quartz is stated to behave like a polymer in the laboratory experiments, hypo-elasticity is sufficiently accurate to describe the flux of the thermodynamic potentials (e.g., Helmholtz free energy). In this context we emphasize that Evans analogy between quartz and glassy polymers is only meant to describe the characteristics of the yield phenomenon underlying the indentation experiments at low temperatures. More precisely, the stress-strain characteristics of quartz is akin to polymers below the glass transition temperature. For such conditions the classical theory of elasticity assumes small elastic strain (hypo-elasticity) which is an adequate mathematical simplification for glassy polymers and for rocks. However, polymers above the glass transition temperatures are in a leathery or rubbery state. The latter transitions are clearly peculiarities in polymers that have not been observed in rocks.

However, the magnitude of visco-plastic strain can reach significantly much larger quantities than the elastic strain (≫1000%). For the viscous part we assume that the viscosity is made up of a combination of power law and Peierls Stress contribution, each with their lumped instantaneous nonlinear viscosity η and for the plastic deformation we assume von Mises plasticity, with a pressure (p) dependent and a temperature (T) yield stress τ (see the Appendix) combined within the visco-plastic flow rule thus

$$\dot{\varepsilon}_{ij} = \frac{1+\nu}{E}\overset{\triangledown}{\sigma}'_{ij} + \frac{\nu}{E}\dot{p} + \alpha \dot{T}_{\mathrm{equ}}\delta_{ij} + \frac{1}{2\eta}\sigma'_{ij} + \dot{\varepsilon}^{pl}\frac{\sigma'_{ij}}{2\tau}, \tag{4}$$

where the Young's modulus is E and the Poisson ratio is ν. The objective co-rotational stress rate is $\overset{\triangledown}{\sigma}_{ij}$. The prime symbol denotes the deviator of each tensor and the over-dot its substantial derivative with respect to time. The pressure p is defined as the trace of the stress tensor divided by 3 and the plastic strain rate by its second invariant. The third term in the equation (4) spells out the thermal-elastic strain through thermal expansion α where δ_{ij} is the Kronecker delta.

We solve the fully coupled mass conservation, momentum equation and energy equations within a Lagrangian framework. The classical equation for mass conservation of compressible media is

$$\frac{\partial \rho}{\partial t} + \rho \nabla \cdot \mathbf{u} = 0, \tag{5}$$

where ρ is the density and $\mathbf{u}$ the local material velocity.

The momentum equation for an infinitesimally small volume element dV is:

$$\nabla \cdot \sigma_{ij} + \mathbf{f} = 0, \tag{6}$$

where $\mathbf{f}$ is the body force and the first term describes the surface tractions. The set of basic thermal-mechanical equations is closed by the energy equation

$$\rho C_p \dot{T} = \chi \sigma'_{ij} : \dot{\varepsilon}_{ij} + \alpha \delta T_{\mathrm{equ}}\dot{p} - \rho C_p \kappa \nabla^2 T, \tag{7}$$

where again the over-dot denotes the Lagrangian (substantive) time-derivative, and c_p the specific heat. The first term on the right describes the shear-heating term, χ is the dimensionless shear-heating efficiency (1 for full efficiency) and this coefficient χ depends on the microstructure of the polycrystalline material (BERCOVICI and RICARD, 2003). The second term on the right is the isentropic work term, where α is the thermal expansivity which multiplied by the adiabatic temperature change δT_{equ} describes the recoverable elastic volume change.

Further details on the numerical solution technique (REGENAUER-LIEB and YUEN, 2004), our novel approach for calculating shear zones (REGENAUER-LIEB and

YUEN, 2003) and further details on the simple, but unified brittle-ductile model setup (REGENAUER-LIEB *et al.*, 2005) can be found elsewhere. We use high resolution calculations with a local resolution of 300 m × 300 m and a total initial spatial grid of 30 × 50 km.

The heart of our numerical approach lies in the energy equation that is solved in a fully coupled way with the displacement equation using a solver for highly nonlinear systems (ABAQUS/STANDARD, 2000). The novelty of our approach thus lies in abandoning empirical constitutive theories for faulting and abstaining from *ad hoc* weakening laws. We solve instead the energy equation including thermal-elastic feedback and shear-heating terms to provide a self-consistent framework for the nucleation, weakening and persistence of shear zones. We calculate here, for the first time, the dissipative structures developing in a quartz dominated crust in extension. Quarzite flow law is taken to be representative for a granitic composition where two types of perturbations are included. First, completely random heterogeneities are the presence of other minerals in the granite (e.g., feldspar crystals). These are mimicked by random nodal (5°C) thermal perturbations applied at the start of the model triggering a white noise of thermal stresses. Another set of sinusoidal perturbations is superposed with 5°C amplitude and a 10-km long wavelength in order to give the model some short wavelength structure. The motivations for these mild perturbations are aimed at mimicking some structure, which should be inherited from previous buckling or necking modes of the upper crustal units.

Another motivation for the short wavelength perturbation is to test for the influence of the model boundary which cannot be avoided. Since periodic boundary conditions are taken with respect to vertical displacement on the left and right side of the model. In a foregoing study (REGENAUER-LIEB *et al.*, 2005), we have varied the amplitude and wavelength of the short wavelength perturbations and find a perturbation independent pattern that emerges upon subjecting the model to pure shear horizontal extension. This pattern appears to be independent of the sinusoidal perturbations and we find a thermodynamically inspired criterion for the nucleation of listric faults, which is reported elsewhere (REGENAUER-LIEB *et al.*, 2005). This pattern is apparently due to decoupling above a maximum dissipation, subhorizontal layer marking the rheological transition from a brittle *p*-dependent to a ductile *T*-dependent localization phenomenon. We call this layer the brittle-ductile transition in a granitic crust.

We use here exactly the same setup and compare for the purpose of clarity the results to a virtual crust of the same dimensions and temperature profile but composed of the main mechanical constituent olivine. This crust is labeled "oceanic crust," bearing in mind that the temperature profile would be different for a real oceanic crust and the lower part of the crust would in fact be the mantle. We discuss the surprising difference in the mechanical behavior, in particular, the tendency for short time-scale instabilities.

4. Quartz versus Olivine

Indentation experiments on olivine (EVANS and GOETZE, 1979) and quartz (EVANS, 1984) suggest that quartz yields when it has stored a substantial amount of elastic energy. It is thus best compared to a polymer. On the other hand, olivine does not store elastic energy efficiently and fails in smooth manner. Its mechanical behavior is better compared to a metal. This could have strong implication for the genesis of and interpretation of earthquakes. It is surprising that these important implications have never been seriously investigated as a potential explanation for the apparent absence of earthquakes in the mantle part of the continental lithosphere. Instead theories where the continental mantle is assumed to be anomalously weak have been favored (JACKSON, 2002). This implies that the continental mantle is rather hot and/or anomalously wet. The absence and presence of earthquakes could have a simple mechanical reason, if one extrapolates the mechanical data by Brian Evans (EVANS, 1984).

However, there remains the open question, whether the laboratory results can be extrapolated to geological conditions. For this extrapolation the Peierls stress quantifies the stress required for the first dislocations to move (see Appendix). The data suggest that quartz has an anomalously high Peierls stress τ_y , which implies that the ratio of Youngs modulus over yield stress E/τ_y could be even lower than suggested by Evans. However, a high intracrystalline threshold for plasticity may have quite the opposite effect at lower temperatures. The material may fail by brittle mechanism or deform by a phase transition. It is well known that materials with a very high Peierls stress like silicon (higher than the ideal strength) can undergo a phase transition when deformed under low temperatures (GALANOV *et al.*, 2003). Thus a lattice instability may actually precede dislocation mobility at low temperatures and accommodate deformation (ROUNDY and COHEN, 2001). If water is available alpha quartz may for instance transform into amorphous silica (DI TORO *et al.*, 2004) thus allowing creep at considerably lower shear stress than extrapolated with the formulation in the Appendix. From the Appendix, we can see that the yield stress of alpha quartz is 1.1 kbar at 500 K. It is thus lower than the brittle failure stress but still possibly higher than the phase transition stress. In order to test Evans contention on the quartz-olivine difference for geological strain rates we use an order of magnitude lower stress for the onset of creep, thus allowing power-law creep, to dissipate elastic energy above an initial strain rate of 10^{-16} s^{-1} or an associated initial yield stress of 110 bar. Thus we assess the difference with the highest permissible E/τ_y for quartz but we leave the olivine rheology untouched, thus narrowing the differences between quartz and olivine.

We perform extension experiments with a homogeneous quartz rheology "granitic" crust (30-km thick to the basic symmetry plane, a steady-state initial geotherm with a 10-km thick radiogenic layer producing 40 mWm^{-2} radiogenic heat, and a bottom heat flow at 30 km depth of 30 mWm^{-2}) thus we obtain a surface heat flow of 70 mWm^{-2} and compare the extension results to the same thickness and temperature crust made of homogeneous olivine (200 ppm H/Si) labeled "oceanic" crust. The extension velocity is fixed at 1 cm/yr in both cases. The boundary conditions are perfectly symmetrical extension in x direction and there is a half symmetry in the z direction, i.e., the top is a free surface but the bottom is a zero vertical flow plane although free to glide in a horizontal plane. The motivation for using a fully symmetric setup is to test for the phenomenon of symmetry breaking on random thermal perturbations that have been inserted at the start of the model.

Recent work by B. Kaus (KAUS and PODLADCHIKOV, 2004) in his thesis has identified four different thermal feedback modes based on a single inclusion. These are listed as viscous, static-plastic, plastic and thermal runaway. Our solution has multiple inclusions and we cannot interpret the results in terms of a classical domain map. However, we link single inclusion and multiple inclusion studies by discussing the results in terms of a matrix problem in which we explore the potential for exciting multiple global eigenmodes in space and time (REGENAUER-LIEB and YUEN, 2000, 2004).

We have presented an indirect approach to the formation of shear zones, where a perfectly homogeneous olivine or quartz slab is slightly modified with initial thermal imperfections. During subsequent extension a smoothly varying deformation field gives way to one involving highly localized deformation. We interpret the results using the definition of the classical direct approach (RIKS, 1984). The localization phenomenon is there seen as a bifurcation phenomenon in which the velocity field aborts the continuous branch and takes a new discontinuous path. The problem may then be understood as a numerical solution to a nonlinear matrix problem where the tangential stiffness matrix K is a function of the nodal displacements D and we look for d the unknown increment of nodal displacement leading to bifurcation.

$$K_{(D)}d = F \tag{8}$$

This tangential stiffness is equal to a residual nodal force F. If the determinant of the stiffness matrix becomes zero a bifurcation is detected. The corresponding eigenvectors of $K_{(D)}$ are associated with zero eigenvalues. We may define this nodal displacement as the bifurcation eigenmode(s).

In our analysis we encounter two basically different bifurcation phenomena which can be attributed to two different families of elasto-visco-plastic eigenmodes of the system. One in which the shear zone nucleates on thermal perturbations in the ductile field, and the second which is fully associated with elasto-plastic (brittle, *pressure-dependent*) displacements.

The perturbations are exciting initial short wavelength shear zones which at 17 ka are present in both rheologies (Fig. 1). These coincide with the two left lateral and right lateral 45° shear characteristics (slip lines) of an ideal rigid-plastic body. After

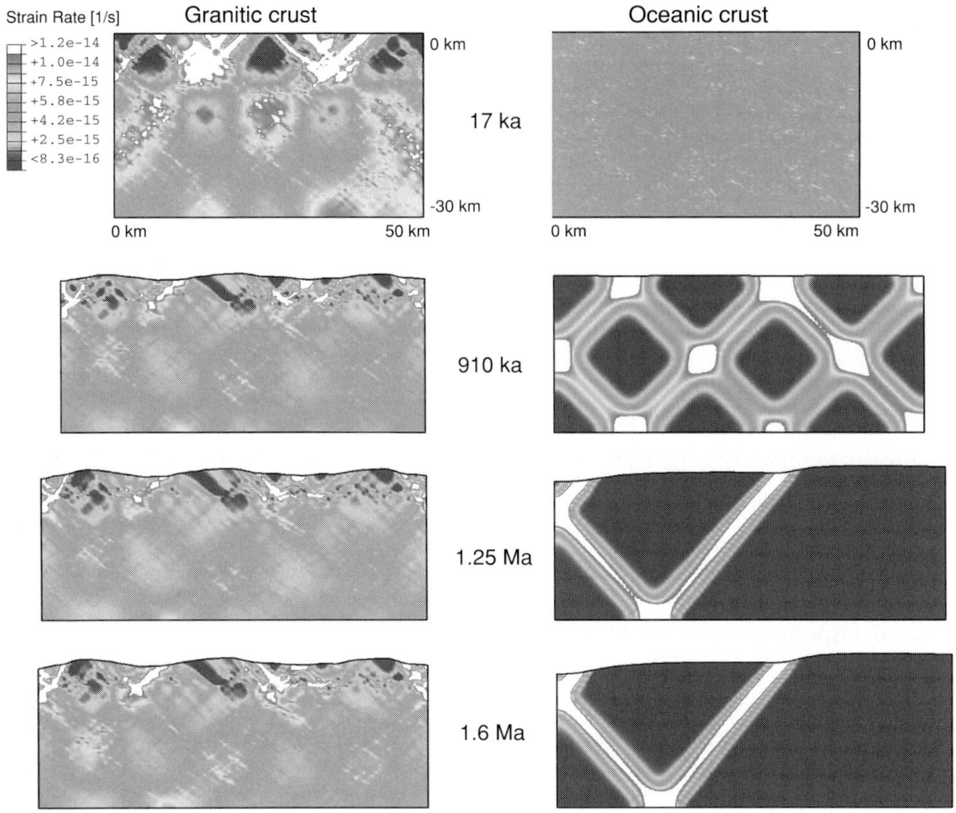

Figure 1

Synchronous time slices of identical (70 mW/m^2 surface heat flow) extending (1 cm/a) slabs consisting of quartz and olivine rheology, here called granitic and oceanic crust showing strain rate maps. The transition from pressure sensitive to T sensitive creep occurs when the T-dependent flow law produces a shear stress that is equal to the lithostatic stress. The left panels show the persistence of highly dynamic solutions in the strain rate map [up to 10^{-12} s^{-1}] of the 30 × 50 km crust while the right panels show relatively stable deformation [order of 10^{-14} s^{-1}] of the same olivine crust, respectively. The prominent subhorizontal decoupling zone at the brittle-ductile transitions in the left panels - see marker grid in the corresponding Figure 2 (ROSENBAUM *et al.*, 2005) - is completely missing in the olivine case.

17 ka there is, however, a significant departure from the ideal rigid-plastic solution in the quartz rheology, only. Quartz displays—owing to a mismatch of brittle and ductile localization mechanism as well as a maximum dissipation at a decoupling interface (Figs. 2 and 3)—shear localization at the surface that peters out with depth into a prominent listric fault system with extremely high strain (> 7000%). The color scale in the strain rate plot in Fig. 1 has been clipped (white) to show the tendency for the characteristics to turn around towards the subhorizontal decoupling layer. However, such listric faulting is better seen in the strain plots on a marker grid shown

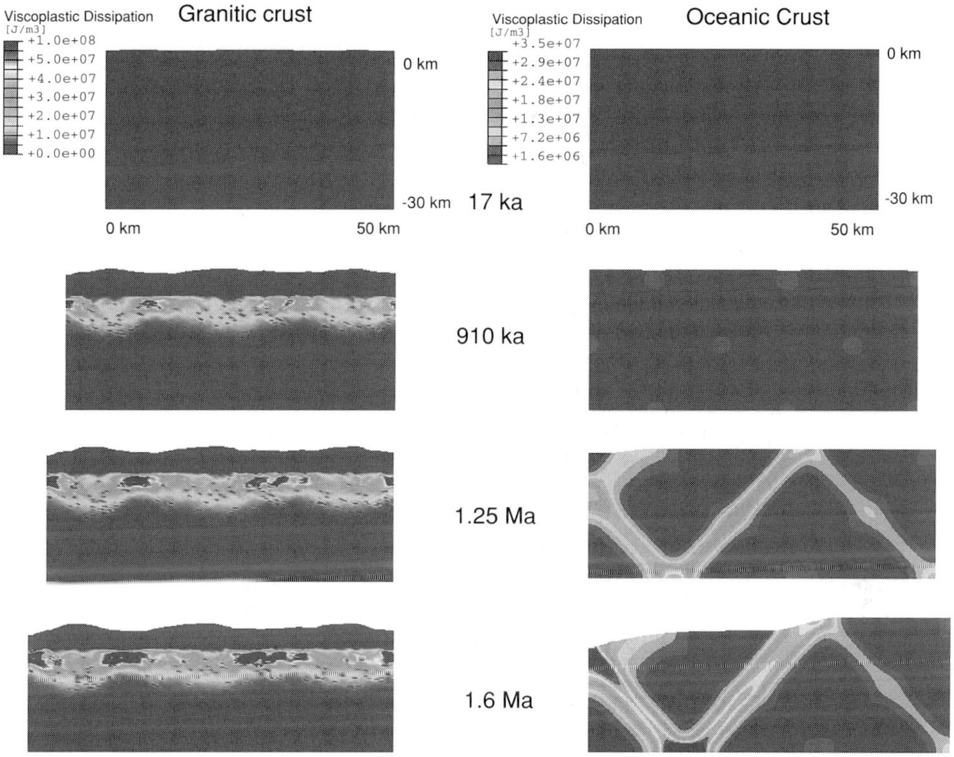

Figure 2

Time slices of the visco-plastic dissipation (shear heating) corresponding to the sequence shown in Figure 1. Comparing the left panel with Figure 1 one can see that even a weak shear heating (light blue patches @ 17 ka) has a strong effect on producing the weak detachment and localizing the faults. Shear heating on detachments intensifies with time and produces a low viscosity zone at the strongest part of the quartz plate (REGENAUER-LIEB *et al.*, 2005). For the olivine the low viscosity zone develops as well. It is fully developed as a crust cutting oblique fault at 1.6 Ma. However, in spite of the higher activation energy of olivine, shear heating is lower by a factor of 3 than in the equivalent red patches in the quartz detachments. This is due to the fact that for the olivine case the oblique low viscosity zone traverses the entire crust.

elsewhere (Fig. 2, in ROSENBAUM *et al.*, 2005). The brittle-ductile decoupling surface in the quartz slab leads to a clear separation of pressure (elasto-plastic) and temperature-dependent (elasto-visco-plastic) eigenmodes operating simultaneously in three different depth levels.

The olivine slab, on the contrary, can keep the ideal plastic characteristics and develops them further into long wavelength features through thermal-mechanical weakening on the largest elasto-visco-plastic (*temperature-dependent*) eigenmode of the system. This governs the entire model absorbing the shorter wavelength shear bands. The olivine slab appears to lack the brittle elasto-plastic (*pressure-dependent*) eigenmode.

ductile faults evident in other minerals. Quartz behaves totally different and retains its jerky slip behavior for prolonged deformation. An example has been discussed here in which a 30 × 50 km piece of a wet granitic crust is extended for about 2 Ma. For quartz the associated total displacement field clearly shows the unstable slipping events which have a characteristic time frame of one to several years. The same olivine slab on the other hand is very stable and has a characteristic thermal-mechanical instability time frame of 100 kyrs. The vast difference in thermal-mechanical behavior hinges on two main thermal-mechanical differences.

- Olivine has an activation enthalpy of around 500 kJ/mol while quartz has an activation enthalpy of only around 134 kJ/mol. This results in stronger weakening through thermal-mechanical instabilities for olivine than for quartz, which in turn implies extreme localization and selection of single master faults, which does not interact with other faults. Hence, olivine lacks seismicity induced by thermal-mechanical feedback. However, as the activation energy or the yield stress increases or the Young's modulus is reduced, the vigor of the instability decreases until a lower limit.
- Quartz has a considerably lower ratio of yield stress over Young's modulus. Therefore quartz is closer to the optimum for shear heating feedback through multiple interacting instabilities than olivine. In effect, micro-indentation experiments have shown that quartz deforms like a polymer while olivine deforms like a metal (EVANS, 1984). Hence, this can explain the jerky creep phenomenon observed in quartz.

Here, we have compared only two specific materials olivine and quartz. We have also explored key parameter ranges, varying elastic modulus, activation energy and yield stress over several orders of magnitude. This analysis prompts us to qualify quartz as a material that is marked by prominent short-time-scale feedback processes. We have discussed that the results indicate at least two different families of eigenmodes, which for the case of quartz appear to communicate in a flip-flop manner along the major subhorizontal detachment on top of the maximum dissipation layer in Figure 2. We find indications of a power-law time series (Fig. 4). The analysis of this phenomenon clearly calls for an extended analysis for fractal space-time evolutions. Calculations are very time-consuming and we therefore refrain from discussing our parameter runs here, which are far from comprehensive. A different model setup is required. We are now investigating this phenomenon more closely by characterizing only the behavior of the critical layer.

We conclude by also encouraging more laboratory experiments on the low stress, low temperature end of the Peierl's stress mechanism in quartz, which needs to be clearly looked at to understand the mechanism for ductile instabilities. An additional interesting phenomenon that may boost the tendency of quartz for creating earthquakes is phase transitions and associated changes in elastic modulus. Apart from results for dry silicon, no experimental data exist for the low temperature regime of alpha quartz since the pioneering experiments performed by Brian Evans in

1984. Upon reinterpreting the original data we find that the classical Vickers indentation experiment is not suited to deliver reliable estimates of the Peierls stress, based on the elasto-plastic contact theory of Johnson (JOHNSON, 1970) which has recently been improved by accounting for phase transitions (GALANOV *et al.*, 2003). New experiments need to be designed (GOLDSBY *et al.*, 2004). A potential avenue for directly deriving micro-mechanical data is for instance given by a quantitative comparison of atomic force microscopy (AFM) indents with fully coupled finite element (FEM) and molecular dynamics (MD) calculations.

Using a simple generic setup in extension we have conducted numerical experiments in order to compare the fundamental behavior of quartz versus olivine. The motivation for this study was based on the surprising difference of the two materials in laboratory indentation experiments. Since under laboratory conditions quartz behaves much like a polymer and olivine much like a metal, we expect fundamental differences in terms of creep instabilities at crustal scale and geological conditions. The focus of the paper was hence to assess whether quartz according to its low yield stress/Youngs modulus ratio, may contribute at crustal scale to fast ductile slip events. Unexpected short-term instabilities were obtained and followed to a single day at which stage we stopped the calculations because we did not consider inertial terms in our formulation. Our models were initially not geared at specific earthquake simulations, but they nevertheless suggest that seismic instabilities may be triggered by such creep events. We suggest that this important phenomenon is a possible source for slow earthquakes.

Our simplified crustal setup has produced quartz slip events with a typical time period of a year. Although this is still outside the realm of slow earthquakes (e.g., SACKS *et al.*, 1978), more realistic boundary conditions may lead to shorter time-scale instabilities. Slow earthquakes with time scales of days to weeks, such as the recent Sumatran event in December, 2004, may be more common than previously thought (www.eri.u-tokyo.ac.jp).

Fully coupled thermal-mechanical deformation can lead to counterintuitive results. An example is the paradoxical observation of lack of earthquakes in the continental mantle. Although crustal rocks are generally softer than the mantle, earthquakes seem to occur only just above the Moho and not below (JACKSON, 2002; MAGGI *et al.*, 2000). The paradox may be resolved by considering the stabilizing effect of olivine described in this paper. Another puzzling aspect is that olivine has the tendency to develop a weak and aseismic, ductile master fault cross-cutting the lithosphere (REGENAUER-LIEB and YUEN, 2000). We show here a comparison for the same models discussed before (Fig. 5). We show here the elastic stress of the 1 Ma case of Figure 5. This figure illustrates the counterintuitive phenomenon that the strongest material is becoming the weakest material through the effect of deformation.

For an interpretation of Figure 5 it is useful to note that the shear stress maps the elastic energy (Fig. 3) and also the dissipation (Fig. 2). Thermal-mechanical

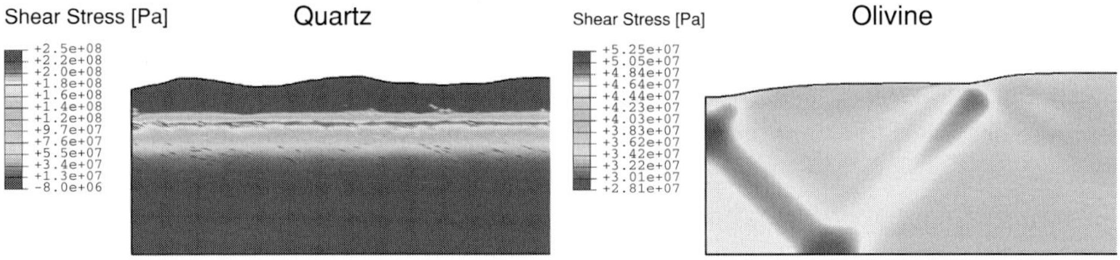

Figure 5
Second invariant of the deviatoric stress tensor after 1 Ma of extension. As expected from the jelly cake rheology, the granitic crust (Quartz rheology) has initially (not shown) substantially lower strength than the oceanic (olivine). However, after thermal-mechanical feedback the situation shown here is quite the opposite. Quartz is stronger (peak stress 2.5 kbar) than olivine (peak stress 525 bars). Quartz also has a significantly higher tendency for instabilities which potentially turn seismic.

weakening is hence highest in areas with high shear-stress/dissipation potential. Olivine can be seen to create a weak crustal cross-cutting fault, while quartz weakens on a subhorizontal layer. Through thermal-mechanical feedback the "strong" olivine crust (Kohlstedt *et al.*, 1995) becomes weaker than the same crust composed of "weak" quartz rheology (Fig. 5).

Acknowledgments

We would like to acknowledge reviews of Matt Davis and Klaus Gottschalk. This work was carried out at CSIRO Exploration and Mining Perth and continued at the Johannes Gutenberg-Universität Mainz. The use of computational facilities at the ETH Zürich and the Minnesota Supercomputer Institute at the University of Minneapolis are gratefully acknowledged. We appreciate discussions with Boris Kaus and Satoru Honda. We also received support from CSEDI, Math-GEO and ITR grants from the National Science Foundation and the predictive mineral discovery Cooperative Research Centre pmd*CRC.

Appendix

Temperature-sensitive Yield Stress

Following the work of Wang and Shimamoto (Wang, 1996; Wang and Shimamoto, 1994) there exists in the low dislocation density, low stress regime a threshold-like phenomenon, which is linked to a lower yield stress phenomenon. A visco-plastic rheology threshold appears associated with the activation of dislocations at low stress. This logic implies that, in addition to the well-known Peierls mechanism as a stress delimiter at high stress (Ashby and Verall, 1977; Kameyama *et al.*, 1999), there also exists a low stress delimiter (KOCKS, 1987; KOCKS *et al.*, 1975) attributed to the Peierls mechanism (REGENAUER-LIEB *et al.*, 2001). This mechanism assists and inhibits power-law creep. It is hence important for the capacity of low temperature material to store elastic energy before the onset of creep. Hence we also expect significant impact for lithosphere or crustal behavior.

The low stress-yield stress can be derived by two independent theoretical considerations. One approach requires knowledge of the activation enthalpy; a characteristic strain for the onset of creep and knowledge of the thermally influenced lattice vibration (KOCKS, 1987; KOCKS *et al.*, 1975). This approach has been applied for olivine rheology (e.g., BRANLUND *et al.*, 2001; REGENAUER-LIEB and YUEN, 1998, 2000) and has recently been discussed in full details (REGENAUER-LIEB *et al.*, 2004). The necessary data can be obtained from Vickers indentation experiments and extrapolated from the high stress experiments of the Peierls mechanism. The required material constants for low temperature have, however, only been derived with confidence for olivine indents.

We have attempted to derive the Peierls stress flow law from the quartz indentation data (EVANS, 1984) using the same approach as proposed for the olivine case (EVANS and GOETZE, 1979). We find that we cannot use the elastic contact formulae because of the polymer-like behavior of quartz. Nano-mechanical constants cannot be derived for quartz because the indentation formula combines a rigid-plastic assumption with an elastic approach. The low ratio of elastic modulus over yield stress invalidates the basic assumption of infinite rigidity, i.e., insignificant elastic strain upon yield. Elastic strain plays a considerably larger role in quartz than in olivine.

Additional aspects for failure of deriving the Peierls stress flow law may be deformational processes other than those accommodated by dislocations, such as solid-solid phase transformations, flash heating under the indenter and surface energy effects. These phenomena are currently being assessed (DI TORO *et al.*, 2004; DIETERICH and KILGORE, 1996; GALANOV *et al.*, 2003; GOLDSBY *et al.*, 2004; ROUNDY and COHEN, 2001), but it is premature to use nano-indents for the estimation of the Peierls stress. An experimental technique, which is very well suited for the low stress branch of the Peierls mechanism has been devised (WANG, 1996; WANG and SHIMAMOTO, 1994) but this experiment cannot be performed for the low temperatures of alpha quartz. A classical Bingham-type rheology was found for low stress—high temperature deformation. This implies no creep (just elastic deformation) before the onset of yield and then Newtonian creep (Harper-Dorn creep) finally assisting the onset of the classical power-law creep for higher stress. For simplicity we just use a lower threshold stress for power creep (equation A3) of 110 bar, without incorporating the Peierls stress mechanism.

The ideal Peierls stress can also be calculated theoretically from WANG (1996) and WANG and SHIMAMOTO (1994) and the experiments and the theory are compared in Table 1.

Table 1

Predicted and observed Peierls stress for Olivine and Quartz rheology (Wang, 1996; Wang and Shimamoto, 1994)

	Material	T [K]	ν	d/b	$\tau_{\text{Peierls }(T)}$	$\tau_{\text{Peierls }(T)}$	$\tau_{\text{Peierls }(T=0)}$
		Temperature Experiment	Poissons Ratio	Unit cell-burgers ratio	Peierls Stress Experiment	Peierls Stress Predicted	Peierls Stress Predicted
Mg2SiO4	Forsterite	1823	0.24	0.635	1.85E+07	6.43E+07	4.21E+08
$(MgFe)_2SiO_4$	Olivine	1673	0.25	0.714	2.07E+07	3.24E+07	2.12E+08
SiO_2	Quartz (beta)	1073	0.25	0.55	5.24E+07	8.25E+07	5.40E+08
SiO_2	Quartz (alpha)		0.17	0.55		1.06E+08	7.61E+08

$$\tau_{\text{Peierls }(T=0)} = \mu \frac{1}{1-\nu} \exp\left(-\frac{d}{b}\frac{2\pi}{1-\nu}\right) \tag{A1}$$

where the parameters are explained and given in Table 1 and the elastic shear modulus μ is of the order of 40 GPa for quartz and 60 GPa for olivine for the temperatures of the experiment.

The temperature sensitivity of the Peierls lower yield stress phenomenon can be approximated by a linear relation (Wang, 1996).

$$\tau_{\text{Peierls }(T)} = \frac{\tau_{\text{Peierls }(T=0)} T_m}{13T}, \tag{A2}$$

where T_m is the melting point temperature.

Power-law Creep

For calculating viscous strain rates we only use power-law creep.

$$\dot{\varepsilon}_{ij}^{cr} = A\sigma'_{ij} J_2^{n-1} \exp\left(-\frac{H^{\text{Power}}}{RT}\right). \tag{A3}$$

We use a flow law of wet quartzite where 0.4% wt water has been added in a sealed capsule (Kronenberg and Tullis, 1984). The material constant $A = 3.98$ 10^{-21} $Pa^{-n}s^{-1}$, the activation enthalpy is $H^{\text{Power}} = 134$ kJ/mol and $n = 2.6$. Material constants for the wet (200 ppm H/Si) olivine power-law rheology are listed in Regenauer-Lieb *et al.*, (2001).

References

ABAQUS/Standard, *User's Manual* vol. 1, version 6.1 (Hibbit, Karlsson and Sorenson Inc. (2000)).

Ashby, M.F. and Verall, R.A. (1977), *Micromechanisms of flow and fracture, and their relevance to the rheology of the upper mantle*, Phil. Trans. Roy. Soc. London *288*, 59–95.

BERCOVICI, D. and RICARD, Y. (2003), *Energetics of a two-phase model of lithospheric damage, shear localization and plate-boundary formation,* Geophys. J. Internat. *152*, 581–596.

BRANLUND, J., REGENAUER-LIEB, K., and YUEN, D. (2001), *Weak zone formation for initiating subduction from thermo-mechanical feedback of low-temperature plasticity,* Earth Planetary Sci. Lett. *190,* 237–250.

DI TORO, G., GOLDSBY, D.L., and TULLIS, T.E. (2004), *Friction falls towards zero in quartz rock as slip velocity approaches seismic rates,* Nature *427*(6973), 436–439.

DIETERICH, J.H. and KILGORE, B.D. (1996), *Imaging surface contacts: Power-law contact distributions and contact stresses in quartz, calcite, glass and acrylic plastic,* Tectonophysics *256* (1–4), 219–239.

EVANS, B. (1984), *The effect of temperature and impurity content on indentation hardness of quartz* J. Geophys. Res. *89*(B6), 4213–4222.

EVANS, B. and GOETZE, C. (1979), *The temperature variation of the hardness of olivine and its implications for the polycrystalline yield stress,* J. Geophys. Res. *84,* 5505–5524.

EVANS, B. and KOHLSTEDT, D.L. (1995), *Rheology of Rocks, Rock Physics and Phase Relations: A Handbook of Physical Constants* (AGU Reference Shelf 3), pp. 148–165.

GALANOV, B.A., DOMNICH, V., and GOGOTSI, Y. (2003), *Elastic-plastic contact mechanics of indentations accounting for phase transformations,* Experim. Mechan. *43*(3), 303–308.

GOLDSBY, D.L., RAR, A., PHARR, G.M., and TULLIS, T.E. (2004), *Nanoindentation creep of quartz, with implications for rate- and state-variable friction laws relevant to earthquake mechanics,* J. Mater. Res. *19* (1), 357–365.

ISHII, M., SHEARER, P.M., HOUSTON, H., and VIDALE, J.E. (2005), *Extent, duration and speed of the 2004 Sumatra-Andaman earthquake imaged by the Hi-Net array,* Nature *435*(7044), 933–936.

JACKSON, J. (2002), *Strength of the continental lithosphere: Time to abandon the jelly sandwich?* GSA Today, *12*(9), 4–10.

JOHNSON, K.L. (1970), *The correlation of indentation measurements.* J. Mechan. Phy. Sol. *18,* 115–126.

KAMEYAMA, C., YUEN, D.A., and KARATO, S. (1999), *Thermal-mechanical effects of low temperature plasticity (the Peierls mechanism) on the deformation of a viscoelastic shear zone,* Earth Planet. Sci. Lett. *168*, 159–162.

KAUS, B. and PODLADCHIKOV, Y. (2004), Chapter 4: *Initiation of shear localization in visco-elasto-plastic rocks,* Ph.D. Thesis, ETH, ZÜrich.

KOCKS, U.F. *Constitutive behaviour based on crystal plasticity, In Unified Equations for Creep and Plasticity.* A. K. Miller, (Ed.) (Elsevier Applied Science, London 1987) pp. 1–88.

KOCKS, U.F., ARGON, A.S., and ASHBY, M.F., *Thermodynamics and kinetics of slip* (Pergamon Press, Oxford 1975) 293 pp.

KOHLSTEDT, D.L., EVANS, B., and MACKWELL, S.J. (1995), *Strength of the lithosphere: Constraints imposed by laboratory measurements.* J. Geophys. Res. *100*(B9), 17587–17602.

KRONENBERG, A.K. and TULLIS, J. (1984), *Flow strength of quartz aggregates: grain size and pressure effects due to hydrolytic weakening,* J. Geophys. Res. *89,* 42981–4297.

KRUGER, F. and OHRNBERGER, M. (2005), *Tracking the rupture of the M–w = 9.3 Sumatra earthquake over 1,150 km at teleseismic distance,* Nature *435*(7044), 937–939.

MAGGI, A., JACKSON, J.A., MCKENZIE, D. and PRIESTLEY, K. (2000), *Earthquake focal depths, effective elastic thickness, and the strength of the continental lithosphere,* Geology *28*(6), 495–498.

NI, S., KANAMORI, H., and HELMBERGER, D. (2005), *Seismology - Energy radiation from the Sumatra earthquake,* Nature, *434*(7033), 582–582

NEMAT-NASSER, S. (1982), *On finite deformation elasto-plasticity,* Int. J. Sol. Struct. *18,* 857–872.

REGENAUER-LIEB, K., HOBBS, B., and ORD, A. (2005), *On the thermodynamics of listric faults,* Earth Planets and Space, 56, 1111–1120.

REGENAUER-LIEB, K. and YUEN, D. (1998), *Rapid conversion of elastic energy into shear heating during incipient necking of the lithosphere,* Geophys. Res. Lett. *25*(14), 2737–2740.

REGENAUER-LIEB, K., YUEN, D., and BRANLUND, J. (2001), *The initation of subduction: criticality by addition of water?* Science *294,* 578–580.

REGENAUER-LIEB, K. and YUEN, D.A. (2000), *Fast mechanisms for the formation of new plate boundaries,* Tectonophysics *322,* 53–67.

REGENAUER-LIEB, K. and YUEN, D.A. (2003), *Modeling shear zones in geological and planetary sciences: Solid- and fluid- thermal- mechanical approaches,* Earth Sci. Rev. *63,* 295–349.

REGENAUER-LIEB, K. and YUEN, D.A. (2004), *Positive feedback of interacting ductile faults from coupling of equation of state, rheology and thermal-mechanics,* Phys. Earth and Planet. Inter. *142*(1–2), 113–135.

ROSENBAUM, G., REGENAUER-LIEB, K., and WEINBERG, R.F. (2005), *Continental extension: From core complexes to rigid block faulting,* Geology *33*(7), 609–612.

RIKS, E. (1984), *Bifurcation and stability - A numerical approach,* Internat. Conf. *Innovative Methods in Nonlinear Problems* (Pineridge press, 1984) pp. 313–344.

ROUNDY, D. and COHEN, M.L. (2001), *Ideal strength of diamond*, Si, and Ge. Physical Rev. B *6421*(21), art. no-212103.

SACKS, I., SUYEHIRO, S., LINDE, A., and SNOKE, J. (1978), *Slow earthquakes and stress redistribution,* Nature *275*, 599–602.

STEIN, S. and OKAL, E.A. (2005), *Speed and size of the Sumatra earthquake,* Nature *434*(7033), 581–582

TABOR, D. (1970), *The Hardness of Solids,* Rev. Phys. Tech. *1* (145–179).

TABOR, D. (1996), *Indentation hardness: Fifty years on - A personal view,* Phil. Magazine, *74*(5), 1207–1212.

WANG, J.N. (1996), *Microphysical model of Harper-Dorn creep,* Acta Mater. *44*(3), 855–862.

WANG, J.N. and SHIMAMOTO, T. (1994), *On Newtonian Viscoplastic Dislocation Creep and the Effect of the Peierls Stress,* Materials Science and Engineering Structural Materials Properties Microstructure and Processing *188*(1–2), 175–184.

YUEN, D. and SCHUBERT, G. 1979, *The role of shear heating in the dynamics of large ice masses,* J. Glaciology *25,* 185–212.

(Received December 30, 2004, revised December 1, 2005, accepted December 30, 2005)
Published Online First: August 31, 2006

Pure appl. geophys. 163 (2006) 1933–1947
0033–4553/06/091933–15
DOI 10.1007/s00024-006-0105-3

Pure and Applied Geophysics

Recurrence Interval Statistics of Cellular Automaton Seismicity Models

DION WEATHERLEY

Abstract—The recurrence interval statistics for regional seismicity follows a universal distribution function, independent of the tectonic setting or average rate of activity (CORRAL, 2004). The universal function is a modified gamma distribution with power-law scaling of recurrence intervals shorter than the average rate of activity and exponential decay for larger intervals. We employ the method of CORRAL (2004) to examine the recurrence statistics of a range of cellular automaton earthquake models. The majority of models has an exponential distribution of recurrence intervals, the same as that of a Poisson process. One model, the Olami-Feder-Christensen automaton, has recurrence statistics consistent with regional seismicity for a certain range of the conservation parameter of that model. For conservation parameters in this range, the event size statistics are also consistent with regional seismicity. Models whose dynamics are dominated by characteristic earthquakes do not appear to display universality of recurrence statistics.

Key words: Earthquakes, recurrence statistics, cellular automata.

Introduction

Long-term seismic hazard assessment relies upon detailed knowledge of the recurrence statistics of regional earthquakes and the spacial and temporal variability of these statistics. Recurrence statistics encompass not only the average frequency of earthquakes versus magnitude, the Gutenberg-Richter law (GUTENBERG and RICHTER, 1954), but also the distribution of recurrence intervals. Traditional seismic hazard assessment studies have used *ad hoc* assumptions for the recurrence interval distribution (e.g., a Poisson distribution). This was largely due to a lack of instrumental records with which to determine the empirical recurrence interval distribution for regional seismicity.

CORRAL (2004) revisited the question of recurrence interval statistics in a recent study of the worldwide seismic catalogue. Corral demonstrated that earthquake recurrence intervals follow a universal distribution when recurrence interval

ESSCC, The University of Queensland, Brisbane, 4072, Australia. E-mail: dion@quakes.uq.edu.au probability density functions are rescaled by dividing by the average rate of activity. The universal distribution is a generalised gamma distribution:

$$f(\theta) = C\frac{1}{\theta^{1-\gamma}}\exp(-\theta^{\delta}/B), \quad (1)$$

where θ is the rescaled recurrence interval and B, C, γ and δ are fit parameters. For seismicity, $B = 1.58 \pm 0.15$, $C = 0.50 \pm 0.10$, $\gamma = 0.67 \pm 0.05$ and $\delta = 0.98 \pm 0.05$ (CORRAL, 2004).

Corral also demonstrated that aftershock sequences follow the same universal distribution and provided evidence for inverse-time scaling of very short recurrence intervals.

In the hope of gaining an increased understanding of the physics of earthquakes and the spatial and temporal variability of earthquakes, a variety of earthquake simulation models have been developed. These range from highly simplified cellular automaton models capable of simulating long sequences of earthquakes, to quasi-static and dynamic models able to accurately model earthquake rupture and stress change, yet only able to produce short sequences of earthquakes due to computational limitations. A lingering question hanging over seismicity studies using simulations is: Do the models accurately simulate the long-term statistics of earthquakes?

The apparent universality of earthquake recurrence interval statistics may offer a new approach for gauging the accuracy of earthquake simulation models. WEATHERLEY and ABE (2004) examined the recurrence interval statistics of two elastodynamic models for one-dimensional faults: The inertial Burridge-Knopoff model of CARLSON and LANGER (1989) and the Lattice Solid Model (MORA and PLACE, 1994). The recurrence statistics for neither model matched that of regional seismicity nor did the recurrence statistics of the models agree. However, there was evidence for universality of recurrence statistics for earthquake sequences obtained from simulations with differing model parameters. The inertial Burridge-Knopoff model produced recurrence distributions that scaled with inverse time, similar to that for very short recurrence intervals of natural seismicity. The Lattice Solid Model produced an exponential distribution of recurrence intervals, with no power-law scaling (i.e., $\gamma = 1$ in Equation 1).

The results of WEATHERLEY and ABE (2004) were somewhat encouraging, suggesting that examination of recurrence statistics may permit different models to be grouped into differing universality classes. In the following, we examine the recurrence statistics obtained from simulations of four different cellular automaton seismicity models. Previous research has demonstrated that these models can simulate earthquakes with a variety of different event size distributions, as the model parameters are varied.

We find that models with markedly different stress redistribution mechanisms can produce identical recurrence statistics for a broad range of model parameter values. The universal distribution for these models does not match that of regional seismicity, however. It is a modified exponential distribution with no power-law scaling of short recurrence intervals. Such recurrence distributions are typical of Poisson point processes, i.e., individual events occur independently of one-another.

An interesting, as yet unexplained, result of the research is that the Olami-Feder-Christensen (OFC) model (OLAMI *et al.*, 1992) does produce recurrence statistics matching regional seismicity, for a particular range of values of its tuning parameter. For this range, the OFC model has been shown to produce event size distributions with b values matching regional seismicity. The OFC model is arguably the simplest of the seismicity models examined, yet it is the only model that reproduces both the event size and recurrence distributions of regional seismicity. Further research is needed to determine the key differences between the OFC model and the others, giving rise to the realistic event statistics.

Overview of Models

The models selected for this study are all cellular automaton seismicity models. All the models consist of a rectangular grid of cells, each of which is assigned a constant, scalar failure threshold and a scalar variable representing shear stress. Various failure threshold distributions are employed in the models: Constant failure threholds, a random distribution of thresholds, and a statistical fractal distribution with variable fractal dimension. Initially the stress of each cell is assigned a random value selected from a uniform random distribution ranging from zero to the failure threshold of the cell.

The increase in stress due to tectonic loading is modelled by incrementing the stress of each cell by a small stress increment. The size of the increment is computed dynamically as the increment required to increase the stress of at least one cell to the failure threshold of that cell.

When the stress of a cell equals or exceeds its failure threshold, the cell fails (modelling the stick-slip transition) and redistributes stress to neighbouring cells. The stress of the failed cell is reduced to a lower value and some stress is dissipated from the model, simulating the loss of energy due to seismic wave propagation and heat. Numerous stress redistribution mechanisms have been proposed in the literature. In the following, four different stress redistribution mechanisms are employed. Stress redistribution may trigger failure of neighbouring cells, initiating a failure cascade. Failure cascades proceed until no more failures occur, at which time tectonic loading is repeated.

The first model is a reproduction of the Olami-Feder-Christensen automaton (OLAMI *et al.*, 1992). A variant of the BAK *et al.* (1987) sandpile automaton, the OFC model was designed to be a cellular automaton version of the BURRIDGE-KNOPOFF

(1967) block-slider model for earthquakes. In the OFC model, the stress of a failed cell is reduced to zero and stress is redistributed to the four nearest neighbouring cells. The amount of stress redistributed is controlled by a model parameter, $0 < \alpha \leq 0.25$. Let σ_i be the stress of a failed cell (i) and let j be the index of a neighbour of cell i. The stress redistribution equation for the OFC model is given by:

$$\sigma_j := \sigma_j + \alpha\sigma_i. \tag{2}$$

The parameter α is called the conservation parameter. A value of $\alpha < 0.25$ results in dissipation of an amount of stress equal to $(1 - 4\alpha)\sigma_i$. OLAMI *et al.* (1992) demonstrated that the conservation parameter is a tuning parameter for this model. The b value of the event size distribution is related to α. For $\alpha \approx 0.20$, the b value is close to the empirical values for seismicity (PACHECO *et al.*, 1992).

The second model termed the Nearest Neighbour Crack Model (NNCM) is a nearest neighbour automaton representing earthquake rupture within the plane of an earthquake fault. It is a generalisation of models by STEACY and MCCLOSKEY (1998, 1999). These models differentiate between cells that have previously failed in a given failure cascade and cells that are yet to fail. In the first model, all of the stress of failed cells is shared amongst the nearest neighbouring cells that have not failed in the current failure cascade. This stress redistribution mechanism results in crack-like propagation of failure cascades and the event size distribution contains an overabundance of large events compared with that predicted by extrapolation of the power-law distribution of smaller events, i.e., a characteristic earthquake distribution (WEATHERLEY *et al.*, 2000). JAUMÉ *et al.* (2000) demonstrated a systematic change in the frequency of moderate to large events preceding the largest events. The second model (STEACY and MCCLOSKEY, 1989) produces a Gutenberg-Richter distribution of events and there are no systematic changes in earthquake frequency (JAUMÉ *et al.*, 2000).

WEATHERLEY and MORA (2004) performed parameter space studies of a generalisation of these models. A model parameter termed the stress transfer ratio (κ) defines the ratio of stress redistributed to failed cells relative to that of unfailed neighbouring cells. A second model parameter, the dissipation factor (δ) governs the amount of stress dissipated from failed sites prior to stress redistribution. Characteristic earthquakes preceded by accelerating energy release are observed in the corner of parameter space corresponding to low values of the stress transfer ratio and low dissipation (WEATHERLEY and MORA, 2004). The stress redistribution equations for the NNCM model are, for unfailed neighbours:

$$\sigma_j := \sigma_j + \frac{\delta\sigma_i}{(4 - N_f) + \kappa N_f}, \tag{3}$$

and for failed neighbours:

$$\sigma_j := \sigma_j + \frac{\delta\kappa\sigma_i}{(4 - N_f) + \kappa N_f}, \tag{4}$$

where N_f is the number of failed neighbours of failed cell i.

The third model is the Rundle-Jackson-Brown (RJB) automaton (RUNDLE and JACKSON, 1977; BROWN *et al.*, 1991). This model was also motivated by the Burridge-Knopoff model. In this model, long-range redistribution of stress and a stress Green's function is introduced, in recognition that elasticity is long-range in the continuum limit. The neighbourhood of a failed cell is a rectangular region centred on the failed cell. The fraction of stress redistributed to a cell (j) within the neighbourhood of a failed cell (i), is given by a stress Green's function:

$$T_{ji} = \frac{1}{K_l + K_c} R_{ij}^{-p}, \tag{5}$$

where K_l is the elastic coupling of each cell to the loading plate, R_{ij} is the distance to cell j from cell i and p is the interaction exponent. In its original form, the RJB model employed an interaction exponent $p = 0$, i.e. each neighbouring cell received an equal fraction of the stress of a failed cell. More recently other values of the interaction exponent have been considered. In the following we examine RJB models with interaction exponents in the range ($0 \le p \le 3$).

The elastic coupling of a cell to its neighbourhood (K_c) is given by:

$$K_c = \sum_{j=1}^{N_{\mathrm{nb}}} R_{ij}^{-p}, \tag{6}$$

where N_{nb} is the number of cells in the neighbourhood of a failed cell.

Thermal noise is included in the RJB model by reducing the stress of failed cells to a residual value (σ_R) given by a random value uniformly distributed in the range $0 < \sigma_R < \eta$. Consequently, the stress redistribution equation for the RJB model is:

$$\sigma_j := \sigma_j + T_{ji}(\sigma_i - \sigma_R). \tag{7}$$

The RJB model is also a dissipative model. The amount of stress dissipated from a failed cell is $\frac{K_l}{K_l+K_c}(\sigma_i - \sigma_R)$.

The final model is a variant of the RJB model in which failed cells are unable to support stress until the conclusion of the failure cascade. Stress redistributed to previously failed cells is dissipated from the model. This model is called the Dissipative Healing Long-Range (DHLR) model in the following. WEATHERLEY *et al.* (2002) demonstrated that for models with long-range interactions ($p < 2$) and no thermal noise, the dynamics of the model is dominated by quasi-periodic cycles of stress accumulation and release. The event size distribution of such models is a characteristic earthquake distribution and large events are preceded by accelerating energy release following the BUFE-VARNES (1993) power-law time-to-failure relation.

Method

Synthetic earthquake catalogues were generated using each of the models described in the previous section. Each catalogue consisted of approximately 900,000 earthquakes. Occurrence times were measured in units of tectonic stress increment and for each model, catalogues were generated for a range of model parameter values. From the synthetic catalogues, probability density functions (PDFs) or recurrence intervals were constructed.

Following the procedure of CORRAL (2004), we employed a log-binning method to construct the recurrence interval PDFs. Bin sizes were defined so that the probability density grows exponentially as c^n (where $c > 1$ and n labels consecutive bins). The number of samples in each bin was normalised by the bin-size and the total number of samples, to obtain an estimate of the probability density over each bin. The PDFs from each simulation were rescaled by dividing by the average rate of activity.

Nonlinear regression was performed to obtain a fit of each PDF to the generalised gamma distribution (Equation 1). The automated nonlinear regression algorithm employed was the Levenberg-Marquardt algorithm (MARQUARDT, 1963; PRESS *et al.*, 2002). The fit parameters and χ^2 for each PDF from a given model were averaged to obtain an estimate of the mean and standard error. The results are tabulated in Tables 1 and 2.

Due to the nonlinearity of the fitting function, the automated nonlinear regression yielded a relatively poor fit to the exponential tail of the PDFs. A

Table 1

Parameter values obtained from fits of the generalised gamma distribution (Equation 1) to rescaled recurrence time PDFs from various cellular automaton fault models and regional seismicity (CORRAL, 2004). The Trial-And-Error Fit parameters were obtained via trial-and-error and the goodness-of-fits $\tilde{\chi}^2$ were computed as the mean of the χ^2 deviation of each distribution from the trial-and-error fit

	RJB	DHLR[1]	NNCM[1]	Trial-And-Error Fit	Seismicity
$\bar{C}$	1.012 ± 0.015	0.9931 ± 0.0044	1.007 ± 0.019	1.00	0.50 ± 0.10
$\bar{\gamma}$	1.001 ± 0.0027	0.9971 ± 0.0008	0.9987 ± 0.0032	1.0	0.67 ± 0.05
$\bar{\delta}$	0.9778 ± 0.0192	1.007 ± 0.0062	0.9973 ± 0.0258	0.90	0.98 ± 0.05
$\bar{B}$	1.044 ± 0.017	1.071 ± 0.0055	1.062 ± 0.0233	1.02	1.58 ± 0.15
$\bar{\chi}^2$	0.0149 ± 0.0027	0.0213 ± 0.0011	0.02762 ± 0.00430		$CV \approx 1.2$
$\tilde{\chi}^2$	0.081 ± 0.012	0.140 ± 0.007	0.139 ± 0.024	0.136 ± 0.006[2]	

[1]Simulations with no characteristic earthquakes.
[2]Mean of the χ^2 goodness of fit for each recurrence distribution from the RJB, DHLR and NNCM models.

Table 2

Results of trial-and-error fits of Equation (1) to recurrence time PDFs obtained from the OFC model with values of the conservation parameter, α as indicated. Simulations with $0.2 \leq \alpha \leq 0.225$ have event size distributions with b values matching that of regional seismicity (see Fig. 4) and the recurrence time distributions for this range also closely match those of regional seismicity

	$0.225 < \alpha \leq 0.25$	$0.2 \leq \alpha \leq 0.225$	$0.175 \leq \alpha < 0.2$	Seismicity
C	1.5	1.25	1.40	0.50 ± 0.10
γ	1.0	0.85	0.95	0.67 ± 0.05
δ	0.7	0.61	0.70	0.98 ± 0.05
B	0.68	0.68	0.70	1.58 ± 0.15

trial-and-error fit was also performed to model the tail better. The trial-and-error fit was found to yield only marginally higher χ^2 on average than the automated fits.

Results

Recurrence interval distributions for the RJB model all collapsed onto a single function, for a range of interaction functions, thermal noise and dissipation factors (Fig. 1. Nonlinear regression resulted in an exponential distribution with

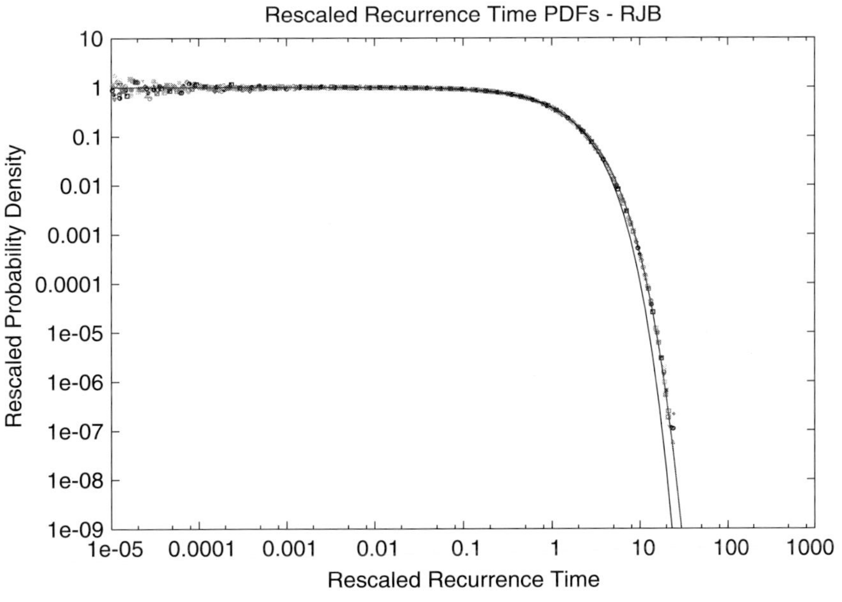

Figure 1
Recurrence interval statistics for the Rundle-Jackon-Brown automaton. A universal exponential distribution is obtained for a range of model sizes, interaction ranges and thermal noise levels. The solid lines are the results of nonlinear regression (black) and trial-and-error fits (blue) of the data to Equation (1). The fit parameters are tabulated in Table 1.

no power-law scaling of recurrence intervals shorter than the average rate of activity. An exponential distribution of recurrence intervals is indicative of a Poisson process.

The majority of DHLR and NNCM simulations produced the same exponential distribution as the RJB model (see Figs. 2 and 3), however some simulations produced different recurrence interval distributions. Inspection of event-size distributions confirmed that these outlying simulations were ones in which a significant population of characteristic earthquakes occurred. Recurrence interval distributions varied considerably for characteristic earthquake simulations with no clear evidence for universality.

Characteristic earthquake simulations, particularly those in which the entire model region fails repeatedly, are not considered indicative of regional seismicity although they may represent the long-term statistics of earthquakes occurring along an isolated fault (Wesnousky, 1994). When the characteristic earthquake simulations are separated out, the results of the nonlinear regression compare favourably with those of the RJB model (Table 1). As for the RJB model, the trial-and-error fit also better models the tail of the recurrence distributions than the automated fits.

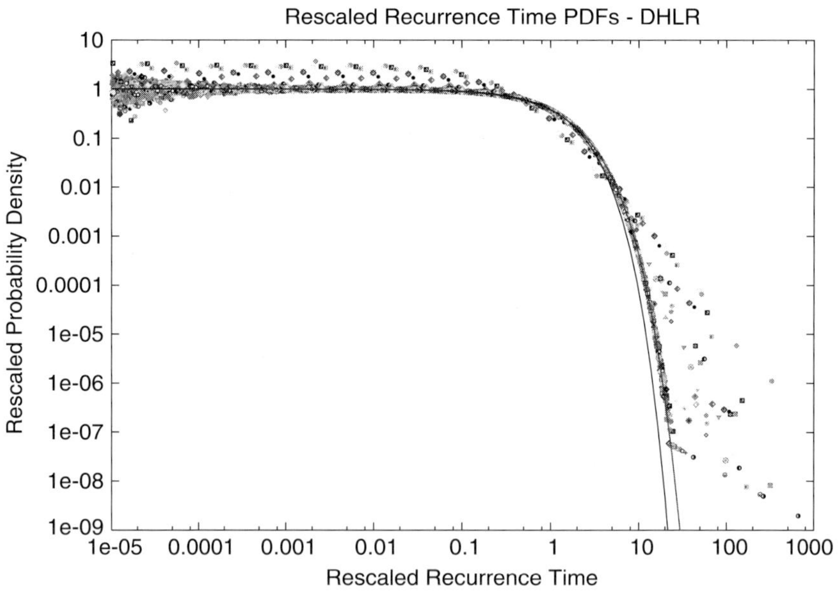

Figure 2
Recurrence interval statistics for the Dissipative Healing Long-Range automata. A universal exponential distribution is obtained for models with a power-law distribution of event sizes. Models with characteristic earthquakes are the outliers in the plot. There is no apparent universality of recurrence statistics for characteristic earthquake models.

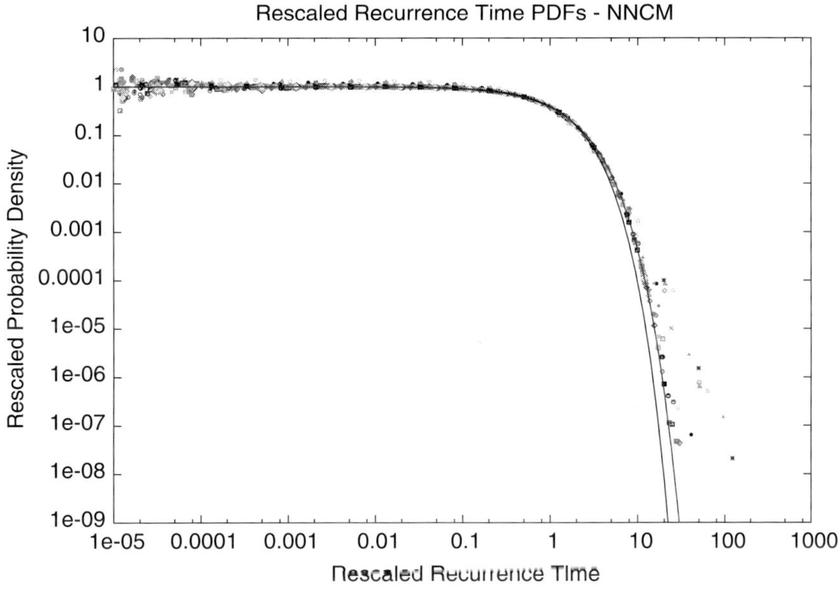

Figure 3
Recurrence interval statistics for the Nearest-Neighbour Crack Models. A universal exponential distribution is obtained for models with a power-law distribution of event sizes. Models with characteristic earthquakes do not have an exponentially decaying probability of large recurrence intervals.

The recurrence interval distributions for the OFC model displayed considerable variability, warranting close inspection. The OFC model has a single tuning parameter, the conservation parameter α. OLAMI *et al.* (1992) demonstrated that the b value of the event size distribution is related to α. For $\alpha \approx 0.20$, the b value is close to the empirical values for seismicity (PACHECO *et al.*, 1992). Cumulative event size distributions for $0.2 \leq \alpha \leq 0.225$ are shown in Figure 4.

Recurrence interval distributions for this regime (Fig. 6) display power-law scaling for short recurrence intervals similar to seismicity, albeit with a higher value of $\gamma \approx 0.85$ (compared with regional seismicity $\gamma \approx 0.67$ and a lower rate of exponential decay for large recurrence intervals. The scatter in the data for very short recurrence intervals is due to a lack of samples (<100) although there is evidence for inverse-time scaling for these intervals, in agreement with regional seismicity.

Higher values of the conservation parameter yielded an exponential recurrence interval distribution (Fig. 5) similar to the other models (see Table 2). Highly nonconservative models ($\alpha < 0.2$) also produced an exponential distribution (Fig. 7) with very similar values for the fit parameters (Table 2).

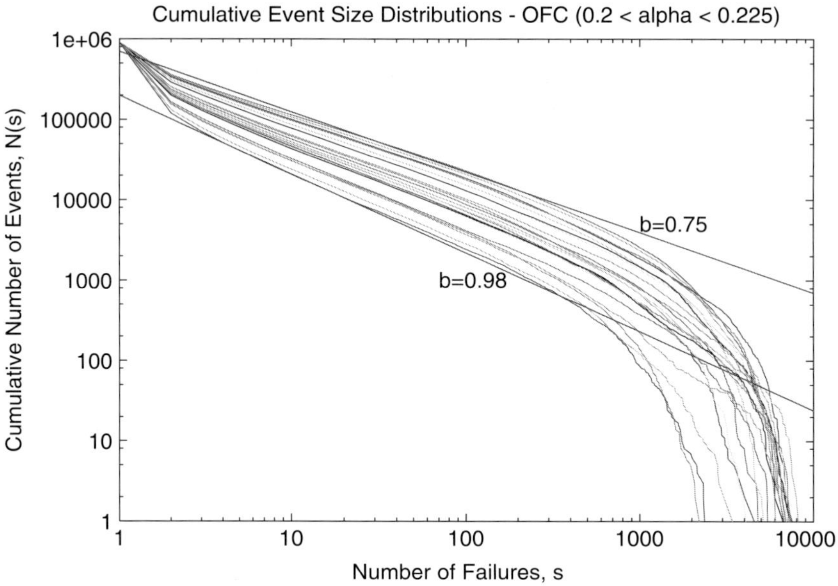

Figure 4

Cumulative event size distributions for OFC models with values of the conservation parameter in the range $0.2 < \alpha \leq 0.225$. For this range, the model produces event size distributions with b values comparable to regional seismicity.

Discussion and Conclusions

The recurrence statistics of regional seismicity follows a universal distribution in the form of a generalised gamma distribution (CORRAL, 2004) with power-law scaling for short recurrence intervals and exponential decay for long intervals. A precise explanation for the origin of the gamma distribution of recurrence intervals is not yet known, however the following argument provides some insight. The generalised gamma distribution consists of a power-law distribution of recurrence intervals shorter than some characteristic interval and an exponentially decaying distribution of longer intervals. It is well-known that a Poisson process, one in which events occur independently of one-another, has an exponential recurrence distribution. Suppose we select a time-series consisting only of a main-shock and its aftershock sequence, with a power-law decay in aftershock rate, according to Omori's law. During the aftershock sequence, the recurrence time distribution is also a power-law. Now suppose we consider a time-series consisting of a number of main-shocks, with associated aftershocks, occurring independently of one-another. If the duration of the aftershock sequences is significantly smaller than the average recurrence time of the main-shocks, then the recurrence time distribution for the entire time-series will consist of two populations: one with

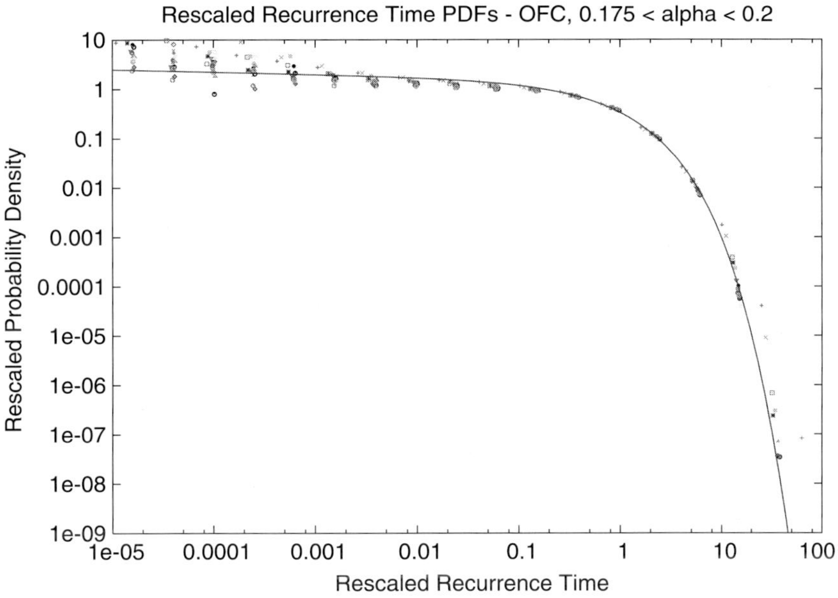

Figure 7

Recurrence interval statistics for OFC models with $0.175 < \alpha < 0.2$. These models also have a universal, exponential distribution with evidence for power-law scaling of very short recurrence intervals (similar to regional seismicity), however only limited data is available for these intervals.

least three types of universal distribution: An exponential distribution, a generalised gamma distribution with power-law exponent near -1 and a generalised gamma distribution with power-law exponent near -0.33. Even though there are multiple types of distribution, each is still a *universal* distribution: Multiple different models (or simulations with differing model parameters) produce the same distribution of recurrence intervals.

Analysis of recurrence interval distributions provides an effective method for classifying earthquake simulation models and comparison with regional seismicity. Thus far, only one model has been found to produce recurrence statistics matching regional seismicity. The majority of models produce an exponential distribution, the distribution of recurrence intervals for a Poisson process. Analysis of recurrence interval distributions of future seismicity models is recommended. One should exercise caution drawing any conclusions about long-term earthquake statistics from models that do not produce the recurrence interval distribution for regional seismicity.

Acknowledgments

The assistance of A. Corral and insightful comments of the reviewers is gratefully acknowledged. This research was funded by an ARC Discovery Grant and supported by the Australian Computational Earth Systems Simulator Major National Research Facility, The University of Queensland and SGI. The ACcESS MNRF is funded by the Australian Commonwealth Government and participating institutions (Univ. of Queensland, Monash U., Melbourne U., VPAC, RMIT) and the Victorian State Government. Computations were made using the ACcESS MNRF supercomputer - a 208 processor 1.1 TFlops SGI Altix 3700, which was funded by the Queensland State Government Smart State Research Facility Fund and SGI.

References

Bak, P., Tang, C., and Weisenfeld (1987), *Self-organised criticality: An explanation of 1/f Noise*, Phys. Rev. Lett. *59*(4), 381–384.

Brown, S.R., Scholz, C.H., and Rundle, J.B. (1991), *A simplified spring-block model of earthquakes*, Geophys. Res. Lett. *18*(2), 215–218.

Bufe, C.G. and Varnes, D.J. (1993), *Predictive modelling of the seismic cycle of the greater San Francisco Bay region*, J. Geophys. Res. *98*(B6), 9871–9883.

Burridge, R. and Knopoff, L. (1967), *Model and theoretical seismology*, Bull. Seismol. Soc. Am. *57*(3), 341–371.

Carlson, J.M. and Langer, J.S. (1989), *Mechanical model of an earthquake fault*, Phys. Rev. A *40*(11), 6470–6484.

Corral, A. (2004), *Long-term clustering, scaling, and universality in the temporal occurrence of earthquakes*, Phys. Rev. Lett. *92*(10), 108501.

Gutenberg, B. and Richter, C.F., *Seismicity of the Earth and Associated Phenomena* (Princeton University Press, Princeton, New Jersey (1954)).

Jaumé, S.C., Weatherley, D., and Mora, P. (2000), *Accelerating seismic energy release and evolution of event time and size statistics: Results from two heterogeneous cellular automaton models*, Pure Appl. Geophys. *157*, 2209–2226.

Marquardt, D. (1963), *An algorithm for least-squares estimation of nonlinear parameters*, SIAM J. Appl. Math. *11*, 431–441.

Mora, P. and Place, D. (1994), *Simulation of the frictional stick-slip instability*, Pure Appl. Geophys. *143*, 61–87.

Olami, Z., Feder, H.J.S., and Christensen, K. (1992), *Self-organized criticality in a continuous, nonconservative cellular automaton modelling earthquake*, Phys. Rev. Lett. *68*, 1244–1247.

Pacheco, J.F., Scholz, C.H., and Sykes, L.R. (1992), *Changes in frequency-size relationship from small to large earthquakes*, Nature *355*(6355), 71–73.

Press, W.H., Teukolsky, S.A., Vetterling, W.T., and Flannery, B.P., *Numerical Recipes in C: the art of Scientific Computing* (Cambridge University Press, Cambridge, New York (2002)) pp. 683–688.

Rundle, J.B. and Jackson, D.D. (1977), *Numerical simulation of earthquake sequences*, Bull. Seismol. Soc. Am. *67*(5), 1363–1377.

Steacy, S.J. and McCloskey, J. (1998), *What controls an earthquake's size? Results from a heterogeneous cellular automaton*, Geophys. J. Int. *133*, F11–F14.

Steacy, S.J. and McCloskey, J. (1999), *Heterogeneity and the earthquake magnitude-frequency distribution*, Geophys. Res. Lett. *26*(7), 899–902.

Weatherley, D. and Abe, S. (2004), *Earthquake statistics in a block slider model and a fully dynamic fault model*, Nonlinear Proc. Geophys. *11*, 553–560.

Weatherley, D., Jaume, S.C., and Mora, P. (2000), *Evolution of stress deficit and changing rates of seismicity in cellular automaton models of earthquake faults*, Pure Appl. Geophys. *157*, 2183–2207.

Weatherley, D. and Mora, P. (2004), *Accelerating precursory activity within a class of earthquake analogue automata*, Pure Appl. Geophys. *161*, 2005–2019.

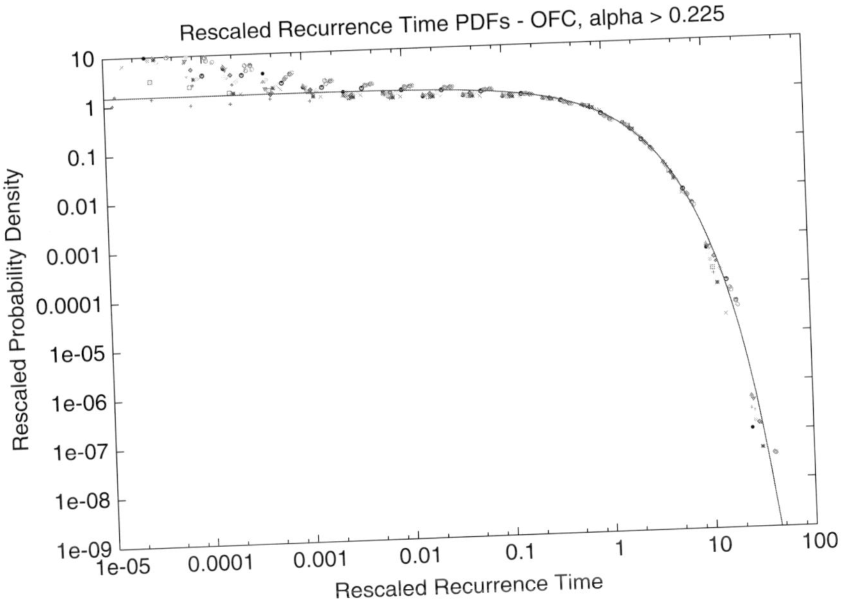

Figure 5
Recurrence interval statistics for OFC models with $\alpha > 0.225$. Such models have a universal, exponential distribution similar to the other models. The solid line is a trial-and-error fit of the data to Equation (1). The fit parameters are tabulated in Table 2.

power-law decay for short recurrence times and one with an exponential decay for long recurrence times. In other words, the recurrence distribution is a gamma distribution.

The recurrence interval distributions for synthetic earthquake catalogues generated by four different cellular automaton fault models were examined in this study. We find that the majority of synthetic catalogues have an exponential distribution of recurrence intervals, lacking power-law scaling for short recurrence intervals. The exponential distribution is universal: variation of model parameters such as model size, thermal noise, interaction range, stress redistribution mechanism and dissipation has no impact upon the recurrence intervals distribution.

Simulations producing a characteristic earthquake distribution do not display the same universal distribution. Such simulations result in a distribution that exponentially decays at a slower rate with greater probability density for short recurrence intervals. There is considerable variation in the recurrence interval distribution for simulations with differing characteristic earthquake distributions. Casual inspection failed to yield any obvious correlations between model parameters and the parameters of the recurrence distribution. Further research is required to determine how the event-size and recurrence interval distributions are related to model parameters for these characteristic earthquake simulations.

arthquakes: Power-
9, 2469–2490.
n, which is it?, Bull.
30, 2005)

In particular, Distinct Element Methods (DEMs) (CUNDALL and STRACK, 1979; MORA and PLACE, 1994, 1999) have successfully been employed to simulate the dynamics of fault shear regions (GUO and MORGAN, 2004; MORA and PLACE, 1998, 1999, 2002, MORA *et al.*, 2000; MORGAN, 1999, 2004; MORGAN and BOETTCHER, 1999; PLACE and MORA, 2000). To date, the majority of results have been produced from 2-D simulations, despite the fact that real fault zones are 3-D entities. With the continual performance improvement of computing hardware, comes the opportunity to study computationally demanding 3-D models of complex granular systems. Of specific interest is the influence of the out-of-plane particle dynamics on the macroscopic behavior of 3-D systems.

The Lattice Solid Model (LSM) (MORA and PLACE, 1994; PLACE and MORA, 1999) is a variant of the DEM which has been used to model the dynamics of fault gouge processes. LSM simulations of 2-D faults have yielded results which provide an explanation of gouge weakness and the heat flow paradox (MORA and PLACE, 1998, 1999) (HFP). Simulations with other 2-D DEMs, using round particles, also demonstrate low coefficients of friction due to rolling (MORGAN, 1999, 2004). Typical 2-D fault gouge models, using the LSM, have involved tens of thousands of particles. For comparable 3-D problems, particle numbers can readily increase into the millions. These large 3-D problems have remained intractable for serial implementations of the LSM. Parallel computing architectures, such as the SGI Altix 3700 super-cluster, provide the opportunity to solve much larger computational problems than traditional single processor systems. In order to take advantage of high performance systems, a Message Passing Interface version of the LSM has been implemented (ABE *et al.*, 2004). Recent benchmarks demonstrated an 80% parallel efficiency for the parallel LSM on 128 processors of the SGI Altix 3700 (LATHAM *et al.*, 2004). These results, for large 2-D wave propagation problems, indicate the potential for the LSM to simulate more computationally challenging 3-D fault gouge dynamics.

This paper describes results from simulations of 2-D and 3-D gouge regions using the parallel implementation of the LSM. Section 2 gives an overview of the LSM and the contact laws imposed on interacting particles, which is followed by a description of the fault gouge models in Section 3. A comparison of the *macroscopic friction* (also termed the *effective* or *fault friction*) for a 2-D model and varying *thicknesses* of 3-D model are presented in Section 4.

Power-law time-to-failure fits for the cumulative energy release and an evolution in size–frequency statistics in the lead up to large earthquake events are evidence that the crust behaves as a Critical Point (CP) system (BUFE and VARNES, 1993; JAUMÉ *et al.*, 1999; SYKES and JAUMÉ, 1990). If this is the case, then intermediate earthquake prediction is possible. However, the circumstances under which fault systems behave as CP systems is still an open question. 2-D LSM fault gouge simulations have been shown to exhibit accelerating energy release and stress correlation function evolution in simulated earthquake cycles (MORA *et al.*, 2000; MORA and PLACE, 2002). In

Table 1

Total number of particles, number of gouge particles, number of worker processes and run–times for the macroscopic friction simulations in Section 4

	2-D			3-D			
	Z0	Z2	Z4	Z8	Z16	Z24	Z32
Number of gouge particles	713	1169	2106	4123	7994	12003	16195
Total number of particles	1701	2881	5476	10690	20972	31441	41865
No. of worker processes	8	12	12	24	24	36	48
Runtime (hours) for 2000000 timesteps	2	4	7	8	14	16	17

Table 2

Model parameters

	2-D: Z0	3-D: Z2-Z32
Model size Width × Height	$34\ R_0 \times 64.6\ R_0$	$34\ R_0 \times 64.4\ R_0$
ZN model depth, $N = 2, 4, 8, \ldots, 32$		$R_0\left(\frac{3N-2}{\sqrt{3}}+2\right)$
Gouge particle radius R_j	$0.165\ R_0 \leq R_j \leq R_0$	$0.4\ R_0 \leq R_j \leq R_0$
Roughness particle radius R_j	$0.165\ R_0 \leq R_j \leq R_0$	$0.4\ R_0 \leq R_j \leq R_0$
Block Young's modulus	4.33 Gpa	5.66 Gpa
Block Poisson ratio	0.25	
Block particle radius R_0	500 μm	
Gouge region height	$\approx 10\ R_0$	
Normal stress	15 Mpa	
Density	2600 kg/m^3	
Shear velocity	$0.001\ R_0/s$, $0.0001\ R_0/s$	
$k^b = k^r = k^s$	735 MN/m	
μ_P	0.2, 0.4, 0.6	

dimensions and differing z dimension sizes. The 2-D model is referred to as the Z0 model and the 3-D models are referred to as the Z2, Z4, Z8, etc. models (2 regular z-layer packings, 4 regular z-layer packings, 8 regular z-layer packings, etc.). The same constant pressure is applied to all models by the uppermost wall (which moves in the y dimension to maintain constant pressure) and the bottom wall remains fixed in the y dimension. Table 2 summarises the parameters and properties of the gouge models.

4. Macroscopic Friction

The shear stress on the driving walls divided by the normal stress on these walls is termed the instantaneous macroscopic friction and denoted μ_I. In the macroscopic friction simulations, each of the driving walls was sheared at a speed of $0.001\ R_0/s \approx 0.001\ V_P$ (where V_P is the P-wave velocity). Figure 2 shows snapshots of

Table 3

Macroscopic friction and normalised dilatency statistics $\hat{h} = h - E[h]$. *(a) Macroscopic friction statistics*

		$\mu_P = 0.2$				$\mu_P = 0.4$				$\mu_P = 0.6$			
		$E[\mu_I]$	$\min[\mu_I]$	$\max[\mu_I]$	$\sqrt{V[\mu_I]}$	$E[\mu_I]$	$\min[\mu_I]$	$\max[\mu_I]$	$\sqrt{V[\mu_I]}$	$E[\mu_I]$	$\min[\mu_I]$	$\max[\mu_I]$	$\sqrt{V[\mu_I]}$
2-D	Z0	0.49	0.22	0.66	0.05	0.82	0.31	1.02	0.09	1.26	0.67	1.62	0.14
	Z2	0.55	0.39	0.69	0.04	0.89	0.54	1.12	0.10	1.43	0.65	1.95	0.18
	Z4	0.57	0.43	0.66	0.03	0.94	0.67	1.11	0.07	1.37	0.85	1.71	0.15
3-D	Z8	0.59	0.51	0.64	0.02	1.00	0.78	1.11	0.04	1.46	0.93	1.67	0.09
	Z16	0.59	0.51	0.64	0.01	0.96	0.75	1.04	0.03	1.42	0.93	1.60	0.08
	Z24	0.59	0.52	0.62	0.01	0.96	0.75	1.05	0.03	1.42	0.94	1.61	0.08
	Z32	0.59	0.52	0.62	0.01	0.98	0.75	1.05	0.03	1.44	0.94	1.59	0.07

(b) Normalised dilatency statistics

		$\mu_P = 0.2$				$\mu_P = 0.4$				$\mu_P = 0.6$			
		$\mathrm{range}[\hat{h}]$	$\min[\hat{h}]$	$\max[\hat{h}]$	$\sqrt{V[\hat{h}]}$	$\mathrm{range}[\hat{h}]$	$\min[\hat{h}]$	$\max[\hat{h}]$	$\sqrt{V[\hat{h}]}$	$\mathrm{range}[\hat{h}]$	$\min[\hat{h}]$	$\max[\hat{h}]$	$\sqrt{V[\hat{h}]}$
2-D	Z0	0.43	−0.29	0.14	0.05	0.89	−0.63	0.25	0.14	1.31	−1.00	0.31	0.27
	Z2	0.24	−0.14	0.10	0.04	0.77	−0.57	0.19	0.15	1.49	−1.04	0.45	0.37
	Z4	0.17	−0.10	0.07	0.03	0.77	−0.56	0.21	0.14	1.47	−0.94	0.53	0.29
3-D	Z8	0.15	−0.11	0.05	0.03	0.69	−0.59	0.10	0.13	1.42	−1.10	0.32	0.34
	Z16	0.14	−0.11	0.03	0.02	0.61	−0.49	0.12	0.12	1.09	−0.87	0.22	0.24
	Z24	0.15	−0.12	0.03	0.02	0.57	−0.48	0.09	0.10	1.22	−0.98	0.25	0.27
	Z32	0.14	−0.11	0.03	0.02	0.64	−0.54	0.10	0.13	1.17	−0.84	0.33	0.26

friction than the respective macroscopic friction values in the 2-D simulations. This has been observed in similar numerical experiments (HAZZARD and MAIR, 2003) and concurs with laboratory rod and bead experiments where the macroscopic friction was measured to have higher mean values in 3-D than 2-D (FRYE and MARONE, 2002; MAIR *et al.*, 2002). The minimum macroscopic friction value also differs significantly between the 2-D and 3-D models where the 3-D models show a smaller range of instantaneous macroscopic friction variation. While the minimum macroscopic friction value is greater in all the 3-D models than the 2-D model, the maximum macroscopic friction values for the larger 3-D models are comparable to the 2-D models.

The standard deviation values, in Table 3(a), of the Z2 and Z4 models are comparable with (and in some cases exceed) those of the 2-D model. It is expected that the variance in the macroscopic friction decreases as the number of particles in the model increases, and certainly this is the case for the Z8–Z32 models. In the thin z dimension Z2 and Z4 cases, the particles are hampered in the z direction of motion due to the close proximity of the frictionless confining walls. Hence, particle reorganisation in this dimension is restricted, promoting gouge dilation and subsequent larger oscillations in the instantaneous friction. This is supported by the larger dilation maximum values for the Z2 and Z4 models in Table 3(b).

The effect of the frictionless confining walls in the thin 3-D models becomes apparent when the gouge-particle trajectories are examined. Figure 4 plots gouge-particle displacement histograms for the Z0, Z2, Z4 and Z32 models in each of the x, y and z spatial dimensions. The 2-D model gouge deformation is, obviously, accommodated in the x and y dimensions only, and subsequently the Z0 model has the highest proportion of particles which move furthest in these directions. In the 3-D models, as the model size increases, there is a reduction in the amount of displacement of gouge particles in the x and y directions, and an increase in the displacement in the z direction. It is interesting to observe the dramatic difference between the Z0 and Z32 trajectory histograms in the x (shear) direction. Despite relatively modest (but significant) motion in the z direction, where >90% of Z32 gouge particles travel between 2 R_0 and 6 R_0 in the z direction, there is a significant change in the trajectory distance in the shear direction. In the Z0 model, >99% of particles travel between 20 R_0 and 40 R_0, while in the Z32 model only 50% of gouge particles travel between 20 R_0 and 40 R_0 with the remaining 50% travelling between 10 R_0 and 20 R_0 in the x direction.

While the 2-D vs 3-D values of Table 3(a) agree qualitatively with the experiment, laboratory mean macroscopic friction values for 3-D glass bead experiments (0.45) exceed the 2-D glass rod mean friction values (0.3) by more than 30% (FRYE and Marone, 2003; MAIR *et al.*, 2002). The 3-D mean values in Table 3(a) exceed the corresponding 2-D values by at most 22%. Again, this is likely to be an artifact of the absence of rotational dynamics within the models. Further numerical investigations,

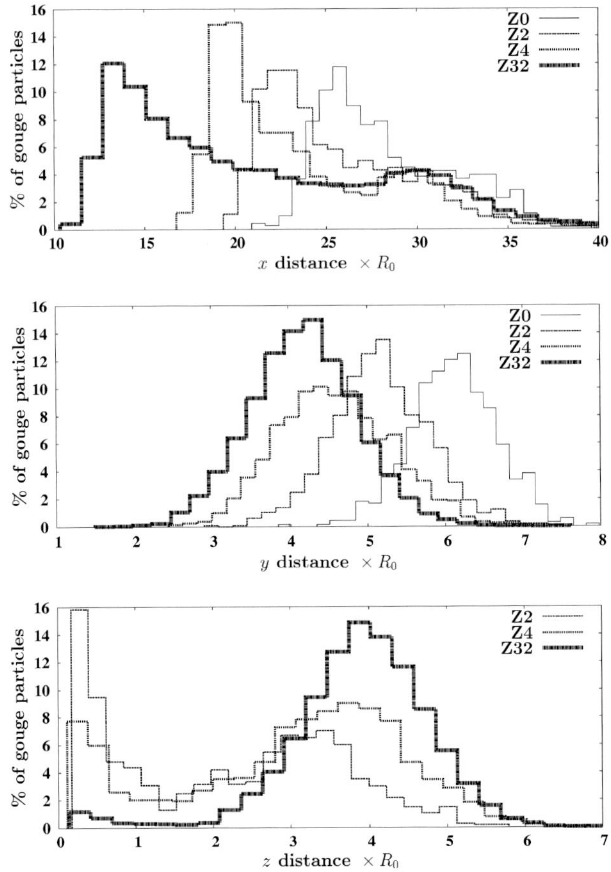

Figure 4

Gouge particle displacement histograms in the x (top), y (middle) and z (bottom) directions in the $\mu_P = 0.2$ models.

which incorporate rotational effects, need to be conducted in order to resolve the current quantitative differences between the laboratory experiment and the numerical simulation results.

5. *Accelerating Energy Release*

An increasing number of field observations suggest an evolutionary behavior of earthquakes. These observations support the Critical Point (CP) hypothesis, which suggests that the largest earthquakes can only occur after the system has reached a critical state (when stress correlations exist in the system at all length scales). A power-law time-to-failure function has been found to fit well with the observed

Benioff strain release in the lead-up to some large events (BUFE *et al.*, 1994; JAUMÉ and SYKES, 1999). Therefore, it is of interest to determine whether this same power-law time-to-failure function fits cycles in simulated earthquakes. 2-D LSM gouge models have shown accelerating energy release and stress correlation function evolution (MORA *et al.*, 2000; MORA and PLACE, 2002).

The 3-D models were also seen to exhibit the accelerating energy release behaviour. Figure 5 shows two earthquake cycles, one from a Z4 model and one from a Z8 model. In these models, the inter-particle friction was set as $\mu_P = 0.6$ and the shear velocity of the driving plates was slowed by a factor of 10 to 0.0001 V_P. Each small circle in Figure 5 represents the occurrence of a simulated earthquake. As well as the cumulative energy release for the cycle, this figure plots an associated fit of the power-law function

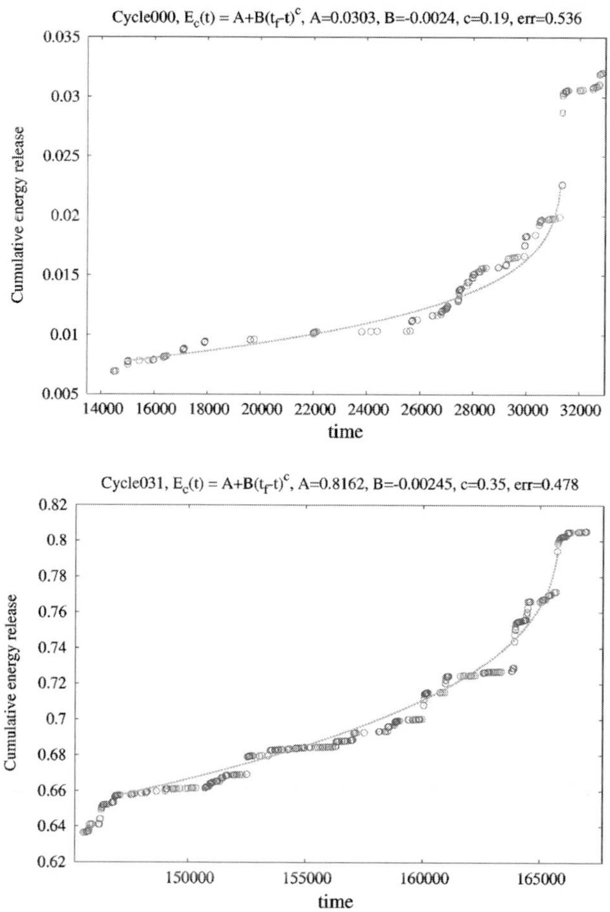

Figure 5
Examples of accelerating energy release (Z4 model top and Z8 model bottom) with power-law curve fit.

$$E_c(t) = A + B(t_f - t)^c, \tag{3}$$

where A, B, c and t_f are the fit parameters with t_f representing the failure time. The RMS error of the power-law time-to-failure fit divided by the RMS error of a linear fit provides a *quality measure* for the time-to-failure fits (BOWMAN *et al.*, 1998). The lower the quality measure value, the more superior the power-law fit when compared with the linear fit. The earthquake cycles of Fig. 5 show good fits of the power-law function, with quality–measure values of 0.536 for the Z4 cycle and 0.478 for the Z8 cycle.

These initial 3-D results provide further support for the CP hypothesis. Accelerating energy release is not present before all large earthquake events (simulated or real). It seems possible, however, that the analysis of 3-D numerical models will provide insight as to the conditions under which a fault (or fault network) behaves as a CP system.

6. Conclusions

Using a parallel/MPI implementation of the LSM, simulations of a simplified fault gouge produced higher mean macroscopic friction values in 3-D models than in 2-D models. These simulations support the hypothesis that friction is greater in 3-D than in 2-D models due to the extra dimension of particle interaction. For the thin 3-D models, an *end wall* effect was observed which restricted gouge particle motion in the z dimension and resulted in higher maximum macroscopic friction response than the thickest 3-D models. Particle trajectories, in the 3-D models, were shown to have a significant z displacement (Fig. 4), and this dramatically reduced particle displacement in the direction of shear. These results demonstrate the important influence of the third dimension on both the microscopic and macroscopic dynamics of shear zones.

The 3-D gouge simulations have also produced instances of accelerating energy release in the lead-up to the larger simulated earthquake events. This provides further support for the CP hypothesis.

The unrealisticly high values observed for the mean macroscopic friction in both the 2-D and 3-D models are due to the lack of rotational dynamics in the model. Future simulation studies will incorporate rotational dynamics both directly (by assigning angular properties to individual particles) and indirectly (through the use of bonded aggregates of particles in the gouge region). By generating 3-D models with bonded-aggregate grains, the effects of grain angularity and grain disintegration on the macroscopic behaviour of the 3-D system can be studied and their importance to earthquake processes evaluated.

Acknowledgements

The authors would like to thank Julia Morgan for her review comments, which improved the quality of the manuscript. Funding support for this work is gratefully acknowledged. Project work is supported by the Australian Computational Earth Systems Simulator Major National Research Facility, The University of Queensland and SGI. The ACcESS MNRF is funded by the Australian Commonwealth Government and participating institutions (Univ. of Queensland, Monash Univ., Melbourne Univ., VPAC, RMIT) and the Victorian State Government. Computations were made using the ACcESS MNRF supercomputer, a 208 processor 1.1 TFlops SGI Altix 3700, which was funded by the Queensland State Government Smart State Research Facility Fund and SGI.

References

Abe, S., Place, D., and Mora, P. (2004), *A parallel implementation of the lattice solid model for the simulation of rock mechanics and earthquake dynamics*, Pure Appl. Geophys. *161*, 2265–2277.

Bowman, D. D., Ouillon, G., Sammis, C. G., Sornette, D., and Sornette, A. (1998), *An observational test of the critical earthquake concept*, J. Geophys. Res. *94*, 15635–15637.

Bufe, C. G., Nishenko, S. P., and Varnes, D. J. (1994), *Seismicity trends and potential for large earthquakes in the Alaska-Aleutian region*, Pure Appl. Geophys. *142*, 83–89.

Bufe, C. G. and Varnes, D. (1993), *Predictive modeling of the seismic cycle in the Greater San Francisco Bay Region*, J. Geophys. Res. *98*, 9871–9983.

Cundall, P. A. and Strack, O. A. D. (1979), *A discrete numerical model for granular assemblies*, Geótechnique. *29*, 47–65.

Frye, K. M. and Marone, C. (2002), *The effect of particle dimensionality on granular friction in laboratory shear zones*, Geophys. Res. Lett. *29*. pg missing.

Guo, Y. and Morgan, J. (2004), *Influence of normal stress and grain shape on granular friction: Results of discrete element simulations*, J. Geoph. Res. *109*. doi:10.1029/2004JP003044.

Hazzard, J. F. and Mair, K. (2003), *The importance of the third dimension in granular shear*, Geophys. Res. Lett. *30*(13). 10.1029/2003GL017534.

Jaumé, S. C. and Sykes, L. R. (1999), *Evolving towards a critical point: a review of accelerating seismic moment/energy release prior to large and great earthquakes*, Pure Appl. Geophys. *155*, 279–305.

Latham, S., Abe, S., and Davies, M. (2004), *Scaling evaluation of the lattice solid model on the SGI Altix 3700*, In *HPCAsia2004, Proceedings of the 7th International Conference on High Performance Computing and Grid in the Asia Pacific Region*, pp. 226–233.

Mair, K., Frye, K. M., and Marone, C. (2002), *Influence of grain characteristics on the friction of granular shear zones*, J. Geophys. Res. *107*. pg missing.

Message Passing Interface Forum. (1997), *MPI-2: Extensions to the Message-Passing Interface*.

Mora, P. and Place, D. (1994), *Simulation of the frictional stick-slip instability*, Pure Appl. Geophys. *143*, 61–87.

Mora, P. and Place, D. (1998), *Numerical simulation of earthquake faults with gouge: towards a comprehensive explanation for the heat flow paradox*, J. Geophys. Res. *103*, 21067–21089.

Mora, P. and Place, D. (1999), *The weakness of earthquake faults*, Geophys. Res. Lett. *26*(1), 123–126.

Mora, P., Place, D., Abe, S., and Jaumé, S. *Lattice solid simulation of the physics of earthquakes: The model, results and directions*, In *GeoComplexity and the Physics of Earthquakes* (eds. Rundle, J., Turcotte, D., and Klein, W.), Geophysical Monograph Series, no. 120 (American Geophys. Union, Washington DC 2000), pp. 105–125.

in which Ω is a measure of seismic activity calculated from earthquake magnitude, K, k, and n ($n \neq 1$) are constants, $m = 1 - n$ is the scaling coefficient, and t_f is the failure time of the strong earthquake to be considered. In the processing of an earthquake catalogue, the acceleration behavior of seismic activity before strong earthquakes can be represented equally in three forms, namely accelerating moment/energy release, accelerating strain release, or simply the increase of the number of earthquakes (BUFE and VARNES, 1993). In this paper, following the conventions of present studies, we use accelerating strain release, but still abbreviate it as AMR. In this case, Benioff strain release is calculated from the magnitude of earthquakes using the empirical relation (VERE-JONES *et al.*, 2001)

$$S_i = E_i^{1/2} = 10^{2.4+0.75M_i} \tag{2}$$

which is similar to the classical definition of seismic energy and Benioff strain (KANAMORI and ANDERSON, 1975) with a changing coefficient considering new seismological data. In Equation (2), S_i is the Benioff strain release of the ith event, E_i the energy calculated from the magnitude, and M_i the magnitude of the earthquake. Accordingly $\sum\Omega$ in Equation (1) is the cumulative Benioff strain release, or the cumulative value of S_i.

Since recent years the study of AMR has caused wide-spread attention among seismologists partly because it provides a prospective approach to the prediction of strong and intermediate-strength earthquakes. The discussion regarding controversy over this phenomenon leads to the necessity for the study of its generality. As a matter of fact, currently, most of the studies addressing this phenomenon are case studies, providing both positive and negative results. In this study we systematically consider the earthquakes with $M_S \geq 6.0$ which occurred in China from 1978 to 2003. To investigate the behavior of seismic activity we use the earthquake catalogues starting from 1970. The monitoring capability of the Chinese seismograph network has been nearly homogeneous throughout China since that time.

2. Data Used and Method of the Analysis

The methodology used in this study is straightforward. Taking a strong or intermediate-strength earthquake as a "target earthquake." Seismic activity around this earthquake before its occurrence is taken for retrospective analysis. Cumulative Benioff strain curve is plotted for each earthquake. The aim of the analysis is simply to investigate whether there exists accelerating moment release before the "target earthquake" and how many "target earthquakes" under consideration possess AMR property.

We use the earthquake catalogue provided by the Center for Analysis and Prediction of the China Earthquake Administration (CEA, i.e., China Seismological

Bureau before 2002). The earthquake catalogue starts from 1970, with the overall completeness for earthquakes over $M_L3.0$, as shown by the frequency-magnitude relation statistics. In this investigation we follow the convention of present studies (e.g., ROBINSON, 2000), removing aftershocks from the catalogue using the scheme of KEILIS-BOROK *et al.* (1980). This process forms a declustered catalogue, eliminating the disturbance of the clustering properties of earthquakes.

According to BREHM and BRAILE (1998), "interfering events" are to be eliminated from the catalogue subject to analysis. Similar to BREHM and BRAILE (1998), an "interfering event" means an earthquake occurred before the "target earthquake" (falling within the temporal range subject to analysis), located near the "target earthquake" (falling within the spatial range subject to analysis), with its magnitude comparable to the magnitude of the "target earthquake" (the difference between these two magnitudes is within 1.2 magnitude unit). Usually the abrupt "jump" in the cumulative Benioff strain curve caused by an "interfering event" destroys the power-law fitting and the judgment of the existence of AMR.

In previous studies (e.g., BOWMAN *et al.*, 1998) the lower cutoff magnitude of catalogue is taken as 2 units below the magnitude of the "target earthquake." Following these previous works, and to obtain better results, in this study, if the magnitude of the "target earthquake" is M_L (converted from M_S), then the lower cutoff magnitude of the catalogue is taken as $M_L - 3$.

For the comparison between different "target earthquakes," regular searching areas around the "target earthquake" and a unified temporal range before the "target earthquake" are taken. The temporal range is taken as 8 years before the "target earthquake," because 8 years seems to be a characteristic time scale for the earthquakes in China—for time scales longer than 8 years the Benioff strain curve does not show a monotonous trend of variation. Circular searching areas centered at the epicenter of the "target earthquake" are used for the analysis. To analyze the stability of the accelerating property, we select three circular areas with different sizes. The minimum searching area is taken by gradually increasing the radius (with a step of 10 km) until there are at least 10 events falling into the circle. Then the radius is increased progressively with a step of 50 km for the "target earthquakes" with $M_S < 7.0$ or 100 km for the "target earthquakes" with $M_S \geq 7.0$. Previous studies tend to use the best fit of power-law type strain release to determine the size of the critical range and/or the scaling exponent (BOWMAN *et al.*, 1998; YANG *et al.*, 2000; YIN *et al.*, 2002). In this study we present a set of results rather than only the best-fitting curve. Whether the three curves exhibit the same behavior (accelerating or not) is used to judge the stability of the AMR property against the selection of the searching areas.

The judgment of accelerating moment release is both by visual inspection and by curve fitting. Using the nonlinear least-squares optimization programs in MatLab, Equation (1) is solved for three unknowns (K, k, and m), while t_f is taken as the time of the "target earthquake." JIANG and WU (2005) discussed the details

Table 1

Earthquakes over M_S 6.0 in China from 1978 to 2003 and related parameters

No.	Date	Time (GMT)	Latitude $\varphi_N/(^\circ)$	Longitude $\lambda_E/(^\circ)$	Magnitude M_S	Type of MR[1]	Second Radius/km	Focal Mechanism[2]
1	1978/ 04/ 04	00: 40	33.00	82.30	6.2	+	310	O
2	1978/ 05/ 18	12: 33	40.70	122.60	6.0	+	100	SS
3	1978/ 07/ 23	14: 42	22.60	121.00	7.6	+	250	T
4	1979/ 03/ 15	12: 52	23.12	101.25	6.8	+	170	SS
5	1979/ 03/ 29	02: 01	41.90	83.45	6.0	*	100	T
6	1979/ 03/ 29	07: 07	32.42	97.15	6.2	+	100	SS
7	1979/ 07/ 09	10: 57	31.45	119.25	6.0	+	220	O
8	1979/ 08/ 24	16: 59	41.23	108.10	6.0	+	170	N
9	1980/ 02/ 22	03: 02	30.60	88.60	6.6	+	340	N
10	1980/ 06/ 24	07: 35	32.20	87.70	6.2	+	250	SS
11	1981/ 01/ 23	21: 13	31.00	101.17	7.1	+	330	SS
12	1982/ 01/ 23	17: 37	31.70	82.30	6.6	*	320	N
13	1982/ 06/ 15	23: 24	31.83	99.85	6.0	*	130	SS
14	1983/ 12/ 16	13: 15	39.30	72.60	6.2	*	120	T
15	1985/ 08/ 23	12: 41	39.58	75.60	7.3	–	250	O
16	1986/ 03/ 22	12: 06	25.60	122.80	6.0	+	170	O
17	1986/ 06/ 20	17: 12	31.20	86.80	6.4	+	250	SS
18	1986/ 08/ 26	09: 43	37.70	101.57	6.2	*	110	T
19	1987/ 01/ 24	08: 09	41.45	79.28	6.2	*	130	T
20	1987/ 12/ 22	00: 16	41.33	89.80	6.2	*	150	T
21	1988/ 11/ 05	02: 14	32.90	90.70	6.9	+	240	SS
22	1988/ 11/ 06	13: 03	22.83	99.72	7.5	-	270	SS
23	1989/ 04/ 15	20: 34	29.93	99.25	6.7	+	310	N
24	1989/ 04/ 25	02: 13	29.93	99.40	6.7	+	310	N
25	1989/ 09/ 22	02: 25	31.55	102.38	6.6	+	180	T
26	1990/ 01/ 14	03: 03	37.90	92.00	6.6	*	270	O
27	1990/ 04/ 17	01: 59	39.45	74.55	6.4	+	100	O
28	1990/ 04/ 26	09: 37	36.12	100.13	6.9	*	270	O
29	1990/ 06/ 14	12: 47	48.10	85.50	7.3	+	590	O
30	1991/ 02/ 25	14: 30	40.33	79.00	6.5	+	110	T
31	1991/ 03/ 26	03: 58	21.82	121.67	6.7	–	160	SS
32	1992/ 07/ 30	08: 24	29.58	90.25	6.5	*	270	N
33	1993/ 01/ 18	12: 42	31.10	90.60	6.0	+	160	N
34	1993/ 01/ 26	20: 32	23.10	101.10	6.3	–	130	SS
35	1993/ 03/ 20	14: 52	29.40	87.20	6.6	+	300	N
36	1993/ 10/ 02	08: 42	38.30	88.68	6.6	+	200	O
37	1993/ 10/ 26	11: 38	38.47	98.62	6.0	+	130	Δ
38	1993/ 11/ 30	20: 37	39.32	75.63	6.0	–	100	SS
39	1993/ 12/ 10	08: 59	21.00	120.73	6.2	+	180	T
40	1994/ 06/ 29	18: 22	32.65	93.45	6.3	–	110	N
41	1994/ 09/ 16	06: 20	22.67	118.75	7.3	+	280	N
42	1994/ 12/ 31	02: 57	20.43	109.35	6.2	+	260	SS
43	1995/ 01/ 10	10: 09	20.48	109.35	6.2	+	260	SS
44	1995/ 07/ 11	21: 46	21.98	99.70	7.2	+	260	SS
45	1995/ 10/ 23	22: 46	25.83	102.32	6.5	–	160	SS
46	1995/ 12/ 18	04: 57	34.48	97.35	6.2	–	260	SS
47	1996/ 02/ 03	11: 14	27.30	100.22	7.1	+	270	N

Table 1

(Contd.)

No.	Date	Time (GMT)	Latitude $\varphi_N/(°)$	Longitude $\lambda_E/(°)$	Magnitude M_S	Type of MR[1]	Second Radius/km	Focal Mechanism[2]
48	1996/ 03/ 12	08: 43	48.40	88.23	6.2	–	220	SS
49	1996/ 03/ 19	17: 00	40.13	76.63	6.9	+	130	T
50	1996/ 05/ 03	03: 32	40.78	109.68	6.4	+	240	SS
51	1996/ 07/ 03	06: 44	30.25	88.60	6.0	*	100	N
52	1996/ 11/ 09	13: 56	31.83	123.10	6.2	+	210	SS
53	1996/ 11/ 19	10: 44	35.43	78.35	7.1	*	250	SS
54	1997/ 11/ 08	10: 02	35.20	87.30	7.4	–	470	SS
55	1997/ 12/ 30	13: 43	25.52	96.35	6.2	+	160	SS
56	1999/ 09/ 20	17: 47	23.93	120.82	7.6	–	250	T
57	2000/ 01/ 14	23: 37	25.58	101.12	6.3	+	140	SS
58	2000/ 06/ 07	21: 46	26.78	96.85	6.8	+	280	O
59	2000/ 09/ 12	00: 27	35.57	99.62	6.4	–	210	SS
60	2001/ 02/ 23	00: 09	29.42	101.10	6.0	+	150	N
61	2001/ 03/ 05	15: 50	34.20	86.50	6.2	+	360	SS
62	2001/ 11/ 14	09: 26	35.93	90.53	7.8	*	440	SS
63	2001/ 12/ 18	04: 02	23.92	122.77	7.5	+	250	O
64	2002/ 06/ 29	06: 54	34.10	94.30	6.0	+	210	SS
65	2003/ 02/ 24	02: 03	39.62	77.27	6.8	+	100	T

[1] "+" Accelerating-like moment release; "–" quiescence-like moment release; "*" Having no stable property of moment release.
[2] "SS" Strike-Slip; "N" Normal; "T" Thrust; "O" "Other" type according to the plunge of the P, T, or B axes; "Δ" No data of focal mechanism in the Harvard CMT database.

The selection of the minimum radius of the searching area is to some extent arbitrary since the criterion is that there are at least 10 events falling into the searching area. On the other hand, however, when putting all the "second radius" (corresponding to the curves in the middle in Figures 1 to 3 which have three fitting curves) against the magnitude of the "target earthquake," an approximate scaling relation can be seen, being close to the relations between the optimal "critical radius" and magnitude obtained by BOWMAN *et al.* (1998) and JAUMÉ and SYKES (1999), respectively. Figure 4 shows such a comparison. In the figure the magnitude of events was transformed to moment magnitude via the empirical relation of HANKS and KANAMORI (1979):

$$M_W = \frac{2}{3}\log M_0 - 6.1. \tag{3}$$

It may not be surprising that not all the earthquakes are accompanied by preshock AMR property. Among the 65 earthquakes considered, there are 40 events exhibiting the characteristics of AMR (similar to the case shown in Figure 1), 12

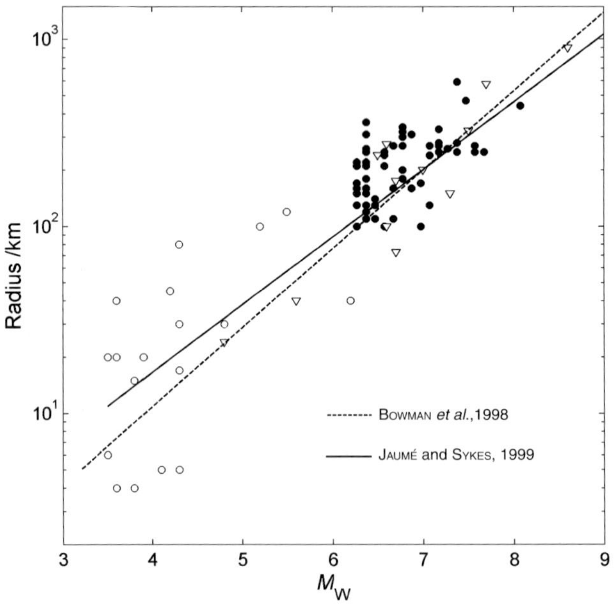

Figure 4

Size of the searching area (as represented by the "second radius") versus the magnitude of the "target earthquake." Solid dots are the 65 events listed in Table 1. Reverse triangles and the dashed line are from Bowman *et al.* (1998); open circles together with reverse triangles and the solid line are from Jaumé and Sykes (1999).

events showing quiescence-like moment release (similar to the case shown in Figure 2), and 13 events having no stable characteristics of moment release (similar to the case shown in Figure 3). Figure 5 shows the distribution of the events under consideration together with the types of moment release. It can be seen that the distribution of the types of moment release is somehow complicated. However, when considering the focal mechanisms of the "target earthquakes," some regularities can still be observed, although we are not sure at the present time whether such an apparent regularity makes sense in the physics of earthquakes.

Most (64 out of 65) of the earthquakes in the present study have the results of focal mechanisms in the Harvard CMT catalogue, as shown in Table 1. Specifying the type of focal mechanism follows the concept of Frohlich (1992) but with a simplified and more strict criterion: If the plunge of B axis is no less than 60°, then the focal mechanism is regarded as "strike-slip;" If the plunge of T axis is no less than 60°, then the focal mechanism is regarded as "thrust"; If the plunge of P axis is no less than 60°, then the focal mechanism is regarded as "normal." The only exception is the November 14, 2001 Kunlunshan Mountain Pass earthquake which was a perfect strike-slip event but in the Havard CMT catalogue the plunge of B axis is 59°. In our classification we consider this event as strike-slip. Table 2 shows the relation

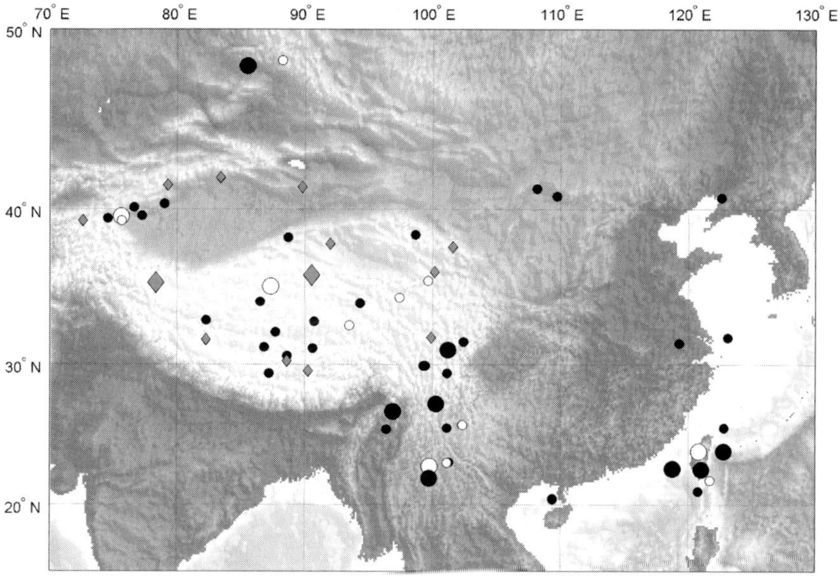

Figure 5
Distribution of earthquakes studied in this article with the types of moment release: Solid dots: accelerating-like; open circles: quiescence-like; diamonds: no stable characteristics of moment release. The two sizes of symbols denote the magnitude range of the earthquakes (for $M_S < 7.0$ and $M_S \geq 7.0$, respectively).

between the types of moment release and the focal mechanisms of the "target earthquakes." It may be seen that among the 28 strike-slip events, the ratio of AMR, quiescence-like, and "unstable" is 16:9:3. Considering the 25 dip-slip earthquakes including thrust and normal ones, however, this ratio is 15:2:8. Statistically it is hard to say that for the strike-slip earthquakes the stable moment release characteristics belong to AMR. For the dip-slip earthquakes, however, it can be said that to a considerable extent the stable moment release characteristics belong to AMR, but the cases of "unstable" moment release, i.e., whether the moment release is accelerating-like or quiescence-like depends sensitively on the radius of the searching area, occupy a significant portion. In general then the generality of AMR before strong and

Table 2

Focal mechanism types versus preshock moment release types

	Strike-Slip	Dip-Slip	"Other"	Sum
AMR-like ("+")	**16**	**15**	8	39
Quiescence-like ("–")	**9**	**2**	1	12
Unstable ("*")	3	8	2	13
Sum	28	25	11	64

intermediate-strength earthquakes in China from 1978 to 2003 is statistically marginal, given that the AMR is searched by a unified temporal scale (8 years) and a series of regular spatial areas (3 circles).

4. Conclusions and Discussion

To examine the generality of the AMR phenomena before strong and intermediate-strength earthquakes, we systematically considered the 65 earthquakes over M_S6.0 in China from 1978 to 2003. In the plotting of the cumulative Benioff strain release we used a unified temporal scale (8 years) and a series of regular spatial areas (3 circles centered at the strong or intermediate-strength earthquake under consideration). Whether or not AMR exists is justified by both visual inspection and by power-law curve fitting. It is observed that there are three types of moment release before the strong or intermediate-strength earthquakes under discussion: accelerating-like, quiescence-like, and "unstable" which means that the type of moment release is sensitive to the selection of searching area. Among the 65 events under consideration, there are 40 events exhibiting clear AMR characteristics, 12 events showing quiescence-like moment release, and 13 events having no stable characteristics of moment release. It is not surprising that not all the earthquakes are accompanied by AMR. The AMR-related earthquake cases are 62% of all the events. Therefore, approximately, two conclusions can be drawn from the analysis. Firstly, more than 3/5 of the earthquakes in China from 1978 to 2003 with magnitudes over 6.0 have AMR property before the earthquakes. Secondly, for 1/5 of the events, the preshock moment release seems to be sensitive to the searching area, i.e., changing of the searching area may change the type of moment release. The first characteristic gives hopes of the application of AMR in the intermediate-term medium-range earthquake prediction. The second characteristic implies that the simple searching area shape and unified temporal range may be problematic. One of the remedies to be considered is to include tectonic information within the analysis of seismic activity. However here it seems to be a dilemma: If the searching areas are case-specified, then it will be hard to evaluate the performance of the prediction. On the other hand, if a simple searching scheme is taken, then the generality of AMR will be significantly reduced.

Among the strike-slip events with stable moment release characteristics, the ratio of AMR and quiescence-like moment release is 16:9. Among the dip-slip earthquakes including thrust and normal ones, this ratio is 15:2. In the language of sign-test, the statistical significance of predominant AMR behavior is higher for dip-slip events than for strike-slip events. We are not clear as to what accounts for such a difference. But seen from other observations such as tidal triggering of earthquakes (TSURUOKA *et al.*, 1995), foreshock rates (REASENBERG, 1999), and other phenomena such as the scaling of apparent stress, clustering property of earthquakes, and stress triggering effect (Wu *et al.*, 2004) which are all apparently focal mechanism dependent, the focal mechanism dependence of AMR may also reflect some clues to the physics of earthquake preparation.

Acknowledgement

This work is supported by NSFC 40274013 and MOST Project 2004CB418406. Thanks are extended to Professor Yin Xiangchu for his invitation to participate in the ACES activities and Professors. Fu Zhengxiang, Zhang Guomin, Zhang Tianzhong, and Zhu Chuanzhen for helpful discussion, and to the anonymous referees and the editor for suggestions in revising the manuscript. Earthquake catalogue is provided by the Center for Analysis and Prediction (Now Institute of Earthquake Prediction) of the China Earthquake Administration.

References

Bowman, D. D. and King, G. C. P. (2001), *Accelerating seismicity and stress accumlation before large earthquakes,* Geophys. Res. Lett. *28*, 4039–4042.

Bowman, D. D., Ouillon, G., Sammis, C. G., Sornette, A., and Sornette, D. (1998), *An observational test of the critical earthquake concept,* J. Geophys. Res. *103,* 24359–24372.

Brehm, D. J. and Braile, L. W. (1998), *Intermediate-term earthquake prediction using precursory events in the New Madrid seismic zone,* Bull. Seismol. Soc. Amer. *88,* 564–580.

Bufe, C. G. and Varnes, D. J. (1993), *Predictive modeling of the seismic cycle of the Greater San Francisco Bay Region,* J. Geophys. Res. *98,* 9871–9883.

Bufe, C. G., Nishenko, S. P., and Varnes, D. J. (1994), *Seismicity trends and potential for large earthquake in the Alaska-Aleutian region,* Pure Appl. Geophys. *142,* 83–99.

Frohlich, C. (1992), *Triangle diagrams: Ternary graphs to display similarity and diversity of earthquake focal mechanisms,* Phys. Earth Planet. Interi. *75,* 193–198.

Hanks, T. C. and Kanamori, H. (1979), *A moment magnitude scale,* J. Geophys. Res. *84,* 2348–2350.

Jiang, C. S. and Wu, Z. L. (2005), *Test of the preshock accelerating moment release (AMR) in the case of the 26 December 2004 Indonesia M_W 9.0 earthquake,* Bull. Seismol. Soc. Amer. *95*(5), 2026–2035. doi: 10.1785/0120050018.

Jaumé, S. C. and Sykes, L. R. (1999), *Evolving towards a critical point: A review of accelerating seismic moment/energy release prior to large and great earthquake,* Pure Appl. Geophys. *155,* 279–306.

Kanamori, H. and Anderson, D. L. (1975), *Theoretical basis of some empirical relation in seismology,* Bull. Seismol. Soc. Amer. *65,* 1073–1096.

Keilis-Borok, V. I., Knopoff, L., Rotvain, I. M., and Sidorenko, T. M. (1980), *Bursts of seismicity as long-term precursors of strong earthquakes,* J. Geophys. Res. *85,* 803–811.

Knopoff, L., Levshina, T., Keilis-Borok, V. I., and Mattoni, C. (1996), *Increased long-range intermediate-magnitude earthquake activity prior to strong earthquakes in California,* J. Geophys. Res. *101,* 5779–5796.

Reasenberg, P. A. (1999), *Foreshock occurrence before large earthquakes,* J. Geophys. Res. *104,* 4755–4768.

Robinson, R. (2000), *A test of the precursory accelerating moment release model on some recent New Zealand earthquakes,* Geophys. J. Int. *140,* 568–576.

Sornette, D. and Sammis, C. G. (1995), *Critical exponents from renormalization group theory of earthquakes: Implications for earthquake prediction,* J. Phys. I. *5,* 607–619.

TSURUOKA, H., OHTAKE, M., and SATO, H. (1995), *Statistical test of the tidal triggering of earthquakes: contribution of the ocean tide loading effect,* Geophys. J. Int. *122,* 183–194.

VARNES, D. J. (1989), *Predicting earthquakes by analyzing accelerating precursory seismic activity,* Pure Appl. Geophys. *130,* 661–686.

VERE-JONES, D., ROBINSON, R., and YANG, W. Z. (2001), *Remarks on the accelerated moment release model: problems of model formulation and estimation,* Geophys. J. Int. *144,* 517–531.

WU, Z. L., WAN, Y. G., and ZHOU, G. W. (2004), *Focal mechanism dependence of a few seismic phenomena and its implications for the physics of earthquakes,* Pure Appl. Geophys. *161,* 1969–1978.

YANG, W. Z., VERE-JONES, D., MA, L., and LIU, J. (2000), *A method for locating the critical region of a future earthquake using the critical earthquake concept,* Earthquake *20*(4), 28–37, in Chinese with English abstract.

YIN, X. C., MORA, P., PENG, K., WANG, Y. C., and WEATHERLEY, D. (2002), *Load-unload response ratio and accelerating moment/energy release critical region scaling and earthquake prediction,* Pure Appl. Geophys. *159,* 2511–2523.

(Received November 1, 2004, revised September 10, 2005, accepted October 20, 2005)
Published Online First: September 2, 2006

Pure appl. geophys. 163 (2006) 1977–1990
0033–4553/06/091977–14
DOI 10.1007/s00024-006-0108-0

Pure and Applied Geophysics

Improvement in the Fault Boundary Conditions for a Staggered Grid Finite-difference Method

TAKASHI MIYATAKE and TAKESHI KIMURA

Abstract—The staggered grid finite-difference method is a powerful tool in seismology and is commonly used to study earthquake source dynamics. In the staggered grid finite-difference method stress and particle velocity components are calculated at different grid points, and a faulting problem is a mixed boundary problem, therefore different implementations of fault boundary conditions have been proposed. VIRIUEX and MADARIAGA (1982) chose the shear stress grid as the fault surface, however, this method has several problems: (1) Fault slip leakage outside the fault, and (2) the stress bump beyond the crack tip caused by S waves is not well resolved. MADARIAGA *et al.* (1998) solved the latter problem via thick fault implementation, but the former problem remains and causes a new issue; displacement discontinuity across the slip is not well modeled because of the artificial thickness of the fault. In the present study we improve the implementation of the fault boundary conditions in the staggered grid finite-difference method by using a fictitious surface to satisfy the fault boundary conditions. In our implementation, velocity (or displacement) grids are set on the fault plane, stress grids are shifted half grid spacing from the fault and stress on the fictitious surface in the rupture zone is given such that the interpolated stress on the fault is equal to the frictional stress. Within the area which does not rupture, stress on the fictitious surface is given a condition of no discontinuity of the velocity (or displacement). Fault normal displacement (or velocity) is given such that the normal stress on the fault is continuous across the fault. Artificial viscous damping is introduced on the fault to avoid vibration caused by onset of the slip. Our implementation has five advantages over previous versions: (1) No leakage of the slip prior to rupture and (2) a zero thickness fault, (3) stress on the fault is reliably calculated, (4) our implementation is suitable for the study of fault constitutive laws, as slip is defined as the difference between displacement on the plane of $z = +0$ and that of $z = -0$, and (5) cessation of slip is achieved correctly.

Key words: Finite-difference method, earthquake source dynamics.

Introduction

In studying earthquake source dynamics, the boundary integral method is commonly used to solve the dynamic equations of motion with fault boundary conditions in infinite elastic medium (e.g., ANDREWS, 1985; DAS and AKI, 1977; FUKUYAMA and MADARIAGA, 1995). This is an excellent method, however, it cannot

Earthquake Research Institute, University of Tokyo, Yayoi, 1-1-1, Bunkyo, Tokyo, 113-0032, Japan. E-mail: miyatake@eri.u-tokyo.ac.jp (T.M)

be applied to realistic problems in which the ground surface and the heterogeneity of the crust are taken into account, although the mirror image technique can be applied to a strike slip fault in elastic half space (AOCHI and FUKUYAMA, 2002). Finite-difference method (FDM) is a practical tool in seismology (e.g., HU and MCMECHAN, 1988; OLSEN *et al.*, 1977). Two types of FDM are used for solving the elasto-dynamic equations of motion; displacement-based FDM and staggered grid FDM. The staggered grid is used not only in problems of seismic wave propagation problems (e.g., GRAVES, 1996) but also in the study of source dynamics study, as the boundary condition of free surface or stress condition on the fault is presented as a Direclet's boundary condition and easily implemented. In the current study, we are concerned with discussing the staggered grid FDM, although displacement-based FDM also is used in source dynamics (ANDREWS, 1999; DAY, 1982; MIYATAKE, 1980).

MADARIAGA (1976), and VIRIUEX and MADARIAGA (1982) developed staggered grid finite-difference method implementations for study on earthquake source dynamics, however, VIRIUEX and MADARIAGA (1982) noted that the method is inaccurate in calculating the shear stress on a fault and does not clearly resolve the stress bump caused by S wave for a 2-D in-plane propagating crack problem with subshear speed. As demonstrated later in the text, we found that these problems resulted from an artificial viscous damping term employed in their method. We also emphasize that the leakage of the slip appears outside the crack, since particle velocity is calculated on the plane slightly off (1/2 grid size) from the fault plane. In 1998, MADARIAGA *et al.* proposed a new implementation; a thick fault model with more accurate stress calculation, however, leakage of the slip outside the rupture zone remains as problematic as for the earlier implementation of VIRIUEX and MADARIAGA (1982). This leakage of slip causes inaccuracy in estimating the frictional stress on the fault when slip-dependent or rate-dependent friction laws are applied; it also creates a problem related to the artificial thickness of the fault: discontinuity across the fault is poorly modeled.

In the present paper, we address the above issues by improving the implementation of the fault boundary conditions in the staggered grid FDM. First, we discuss the implementation of the fault boundary conditions in the early studies of VIRIUEX and MADARIAGA (1982) and MADARIAGA *et al.* (1998). Then we propose our improved implementation and again compare it with the previous studies.

Prevously Published Implementation of Boundary Condition in Staggered Grid FDM

For simplicity, we consider a 2-D in-plane shear crack problem adopted from previous studies by VIRIUEX and MADARIAGA (1982) and MADARIAGA *et al.* (1998).

The equations of motions to be solved are,

$$
\begin{aligned}
\rho \frac{\partial v_x}{\partial t} &= \frac{\partial \tau_{xx}}{\partial x} + \frac{\partial \tau_{xz}}{\partial z} + \eta \left(\frac{\partial^2 v_x}{\partial x^2} + \frac{\partial^2 v_x}{\partial z^2} \right) \\
\rho \frac{\partial v_z}{\partial t} &= \frac{\partial \tau_{zx}}{\partial x} + \frac{\partial \tau_{zz}}{\partial z} + \eta \left(\frac{\partial^2 v_z}{\partial x^2} + \frac{\partial^2 v_z}{\partial z^2} \right) \\
\frac{\partial \tau_{xx}}{\partial t} &= (\lambda + 2\mu) \frac{\partial v_x}{\partial x} + \lambda \frac{\partial v_z}{\partial z} \\
\frac{\partial \tau_{zz}}{\partial t} &= \lambda \frac{\partial v_x}{\partial x} + (\lambda + 2\mu) \frac{\partial v_z}{\partial z} \\
\frac{\partial \tau_{zx}}{\partial t} &= \mu \left(\frac{\partial v_x}{\partial z} + \frac{\partial v_z}{\partial x} \right),
\end{aligned}
$$

where η is a damping factor to dissipate spurious high-frequency oscillations. Although VIRIUEX and MADARIAGA (1982) adopted the second-order finite difference in spatial and time differentiation, we use the fourth-order finite difference in space except on the fault plane and near the absorbing boundaries. MADARIAGA *et al.* (1998) implemented a fourth-order finite difference in spatial differentiation in whole space.

In-plane Shear Crack Propagating with Fixed Speed

Consider an infinite, linear elastic medium subject to static prestress τ_{zx}. A crack is assumed to expand at a fixed speed of 0.8 times shear-wave velocity along the line of $z = 0$ from the origin and stop at $x = \pm 1$. In our calculation we used $\alpha = \sqrt{3}$, $\beta = 1$, $\rho = 1$, where α, β and ρ are P-wave velocity, S-wave velocity and density, respectively, and stress drop is assumed to be 1. To satisfy the stability condition, the grid size is 0.02 in space and 0.0017 in time.

We now briefly examine two different implementations of the fault boundary conditions of the staggered grid FDM.

(1) *VIRIUEX and MADARIAGA (1982)*

The grid configuration used by VIRIUEX and MADARIAGA (1982) is shown in Figure 1a. Shear stress is computed directly on the fault plane, while the particle velocity component v_x is calculated on a plane $h/2$ off from the fault plane, where h is a grid size. These settings cause a leakage of the slip outside the fault. Figure 2 shows the final slip distribution, slip-rate time functions, and stress distribution for $Dc = 0$. Slip is defined as the difference between $u_x(x, h/2)$ and $u_x(x, -h/2)$, which is written as $\Delta u(x, t) = u_x(x, h/2) - u_x(x, -h/2)$. The black and dark lines on Figure 2 indicate numerical and theoretical solutions for the case of $Dc = 0$. Leakage of slip appeared outside the fault. Figure 2c provides a comparison of a snapshot of the shear stress on the fault with the theoretical solution of KOSTROV (1964). As noted by VIRIUEX and MADARIAGA (1982), the stress bump caused by S-wave arrival and stress peak at the crack tip is not clearly resolved in the numerical solution. We found that this problem results from the effect of artificial damping (Fig 3). If the damping term is

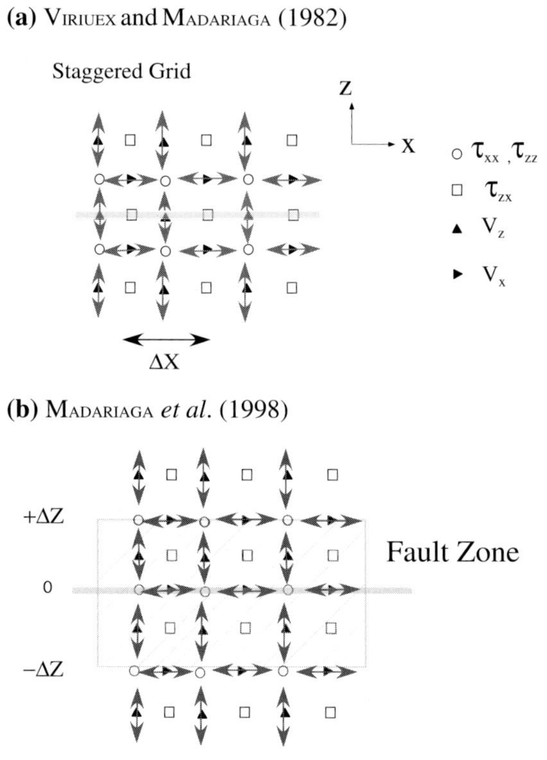

Figure 1
(a) Finite-difference grid for Viriuex and Madariaga (1982), and (b) Madariaga *et al.* (1998).

removed, the numerically calculated stress distribution fits the theoretical solution although spurious oscillations appear in the slip-rate time functions. It is natural that we add the damping term only on the fault that causes rapid stress changes. In this case, the numerically calculated stress distribution is in good agreement with the analytical solution of Kostrov (1964) in this case.

(2) *Madariaga et al. (1998): Thick fault model*

For Madariaga *et al.*'s (1998) implementation, slip is calculated as the difference between displacement at $z = +h$ and that at $z = -h$ (Fig. 1b). Leakage of slip still occurs outside the fault (Figs 4a,b). The stress distribution is in good agreement with the theoretical results (Fig. 4c) as shown in Madariaga *et al.* (1998). In the thick fault model, the displacement discontinuity across the fault plane is poorly simulated (Fig. 5). In both the thick fault model and the implementation by Viriuex and Madariaga (1982), slip is incorrectly defined, but the differences in displacement on the upper and lower planes of the thick fault, therefore it is not easy to control the slip. Fault slip must be stopped when the slip rate becomes zero or frictional stress increases. This is not easy to achieve for these two implementations. If we force the

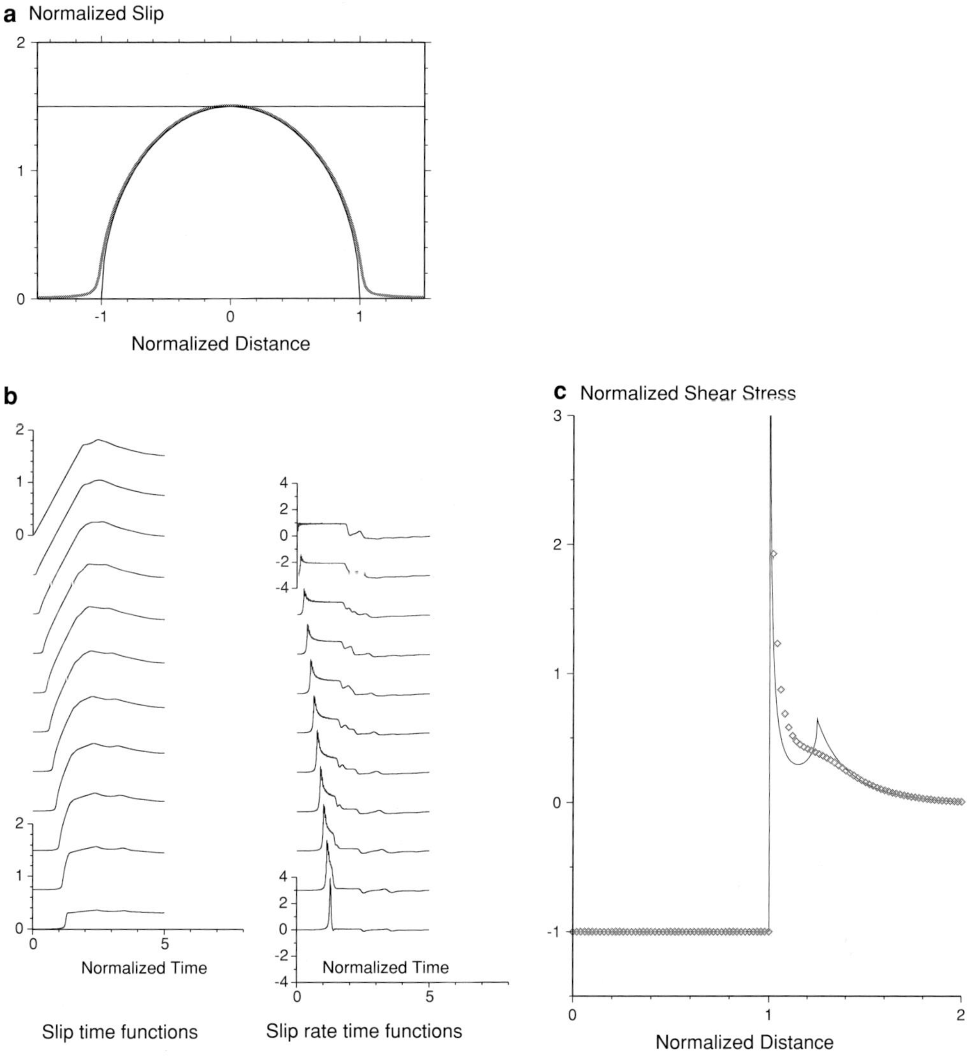

Figure 2

(a) Numerically calculated final slip distribution (thick line) for $Dc = 0$ using Viriuex and Madariaga's implementation and the theoretical static solution (thin line). (b) Numerically calculated slip and slip-rate time functions Viriuex and Madariaga's implementation. (c) Numerically calculated snapshot of shear stress using Viriuex and Madariaga's implementation and theoretical distribution (KOSTROV, 1994; TADA, 1995).

particle velocities on the upper and lower planes of the fault to be equal to zero, the grids act like a rigid body, which is not true for dip-slip and strike-slip faults in the heterogeneous medium. They must move in order to satisfy the equations of motions and zero slip rate.

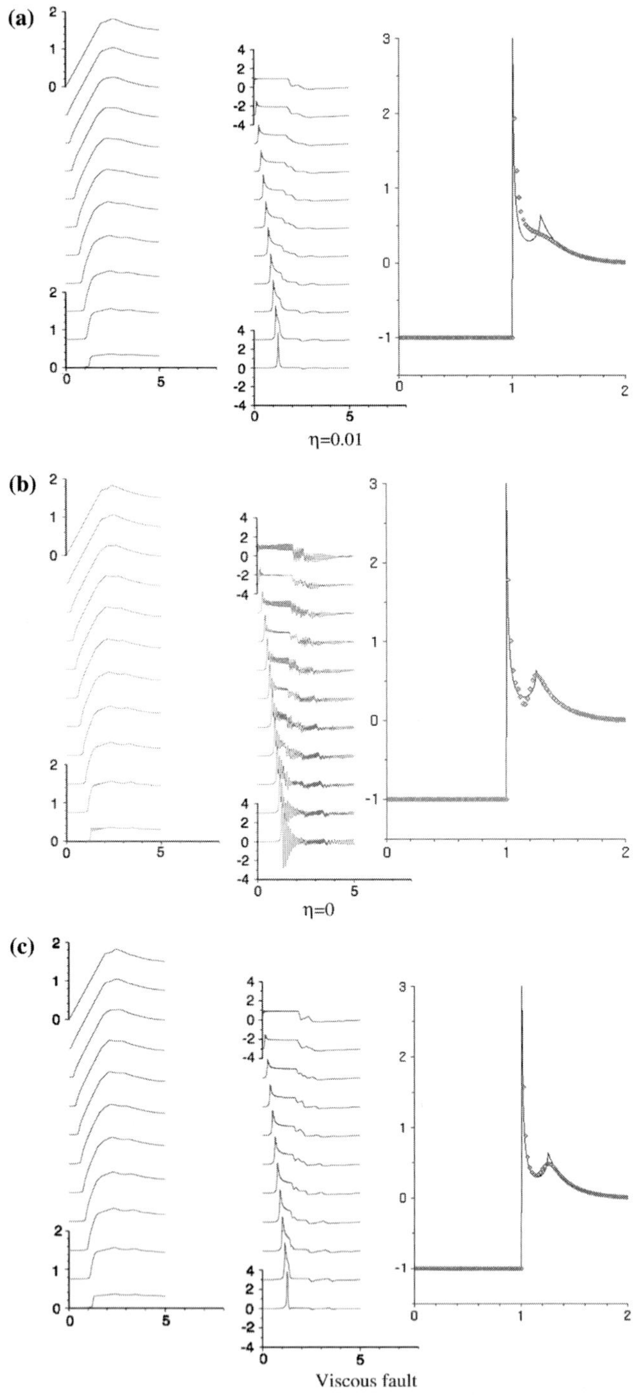
(a)
η=0.01
(b)
η=0
(c)
Viscous fault

◀

Figure 3
(a) The slip-time functions on the left-hand side, the slip-rate time functions on the center column, and the snapshot of the shear stress for the case of the damping factor, $\eta = 0.01$. (b) The slip-time functions on the left-hand side, the slip-rate time functions on the center column, and the snapshot of the shear stress for the case of the damping factor, $\eta = 0$. (c) The slip-time functions on the left-hand side, the slip-rate time functions on the center column, and the snapshot of the shear stress for the viscous fault model.

Modification of the Implementation of Boundary Condition

Having discussed the problems encountered by VIRIUEX and MADARIAGA (1982) and MADARIAGA *et al.* (1998) in this section, we modify the implementations of the boundary condition as a second-order FDM in spacc and time, as outlined in the following steps (Fig. 6).

(1) We assign the fault plane to two separate planes of particle velocity V_X; the planes of $z = +0$, and $z = -0$ correspond to the planes of $K = 1/2$ and $K = -1/2$ in finite-difference space (Fig. 6). If the fault is locked, then $v_x[i, 1/2] = v_x[1, -1/2]$ which corresponds to the condition of $v_x(x, +0) = v_x(x, -0)$, where $v_x[i, k]$ represents the x component of particle velocity at the staggered grid of $[i, k]$.

(2) A fictitious plane is added between the above two planes ($K = 0$ in Fig. 7).

(3) Particle velocity on the fictitious plane is calculated to satisfy the boundary conditions of continuity of the normal stress, $\tau_{zz}(x, +0) = \tau_{zz}(x, -0)$. As $\tau_{zz}(x, +0)$ is specified on the plane of $K = 1/2$, and $\tau_{zz}(x, -0)$ on $K = -1/2$, then

$$\tau_{zz}^{m+1/2}\left[i, \frac{1}{2}\right] = \tau_{zz}^{m-1/2}\left[i, \frac{1}{2}\right] + \lambda \frac{\Delta t}{\Delta x}\left[v_x^m\left[i + \frac{1}{2}, \frac{1}{2}\right] - v_x^m\left[i - \frac{1}{2}, \frac{1}{2}\right]\right] + (\lambda + 2\mu)\frac{\Delta t}{\Delta z}\left[v_z^m[i, 1] - v_z^m[i, 0]\right],$$

$$\tau_{zz}^{m+1/2}\left[i, -\frac{1}{2}\right] = \tau_{zz}^{m-1/2}\left[i, -\frac{1}{2}\right] + \lambda \frac{\Delta t}{\Delta x}\left[v_x^m\left[i + \frac{1}{2}, -\frac{1}{2}\right] - v_x^m\left[i - \frac{1}{2}, -\frac{1}{2}\right]\right] + (\lambda + 2\mu)\frac{\Delta t}{\Delta z}\left[v_z^m[i, 0] - v_z^m[i, -1]\right],$$

where $\tau_{mn}[i, k]$ represents the mn component of stress at the staggered grid of $[i, k]$. As, $\tau_{zz}^{m+1/2}[i, \frac{1}{2}] = \tau_{zz}^{m+1/2}[i, -\frac{1}{2}]$, then

$$v_z^m[i, 0] = \frac{\lambda}{2(\lambda + 2\mu)}\frac{\Delta z}{\Delta x}\left[v_x^m\left[i + \frac{1}{2}, \frac{1}{2}\right] - v_x^m\left[i - \frac{1}{2}, \frac{1}{2}\right] - v_x^m\left[i + \frac{1}{2}, -\frac{1}{2}\right] + v_x^m\left[i - \frac{1}{2}, -\frac{1}{2}\right]\right] + \frac{1}{2}\left[v_z^m[i, 1] + v_z^m[i, -1]\right].$$

(4) Shear stress on the fictitious plane is calculated by the fault constitutive law for the ruptured area.

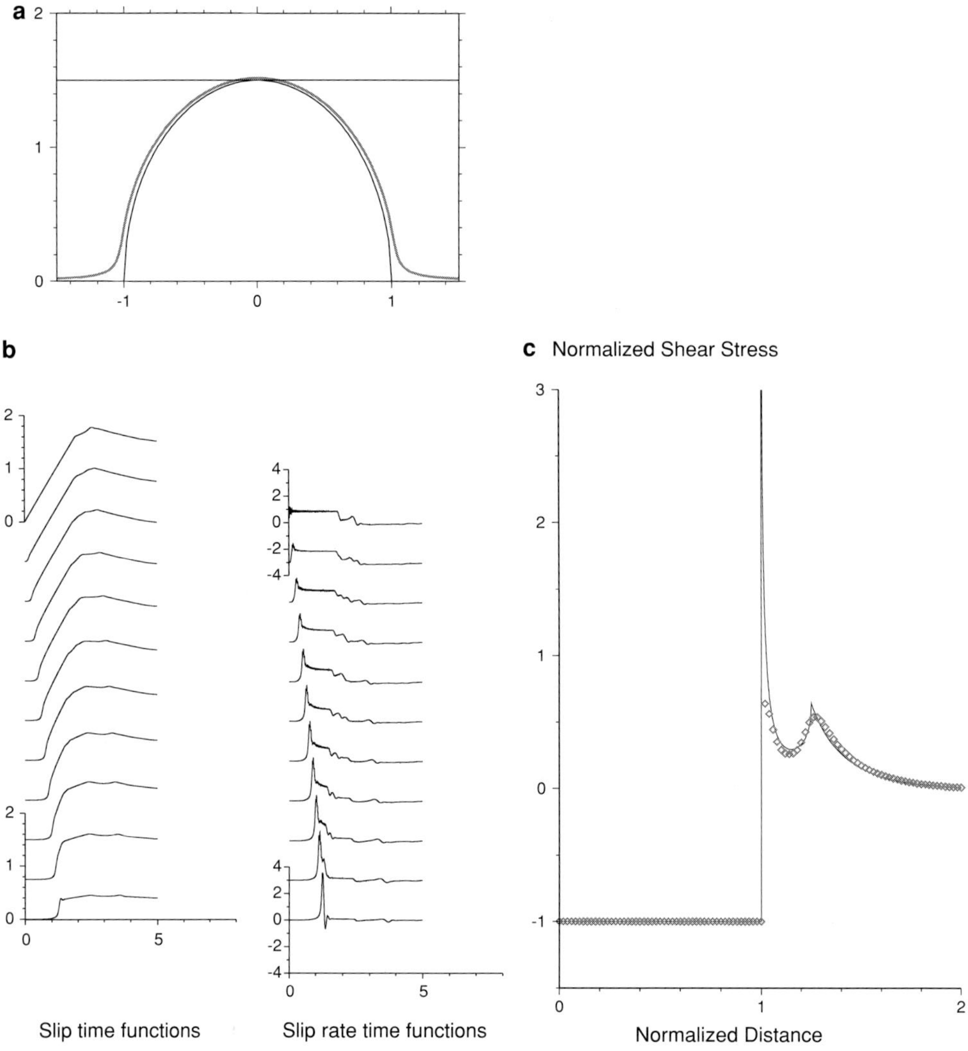

Figure 4
(a) Numerically calculated final slip distribution for the case of $Dc = 0$ using MADARIAGA *et al.*'s implementation. (b) Numerically calculated slip-time functions and slip-rate time functioins for the case of $Dc = 0$ using MADARIAGA *et al.*'s implementation. (c) Numerically calculated snapshot of shear stress for the case of $Dc = 0$ using MADARIAGA *et al.*'s implementation.

$$\tau_{zx}^{m+1/2}\left[i+\frac{1}{2},\frac{1}{2}\right] = \frac{\tau_{zx}^{m+1/2}\left[i+\frac{1}{2},1\right] + \tau_{zx}^{m+1/2}\left[i+\frac{1}{2},0\right]}{2} = -\Delta\sigma_f\left[i+\frac{1}{2}\right]$$
$$\tau_{zx}^{m+1/2}\left[i+\frac{1}{2},-\frac{1}{2}\right] = \frac{\tau_{zx}^{m+1/2}\left[i+\frac{1}{2},-1\right] + \tau_{zx}^{m+1/2}\left[i+\frac{1}{2},0\right]}{2} = -\Delta\sigma_f\left[i+\frac{1}{2}\right].$$

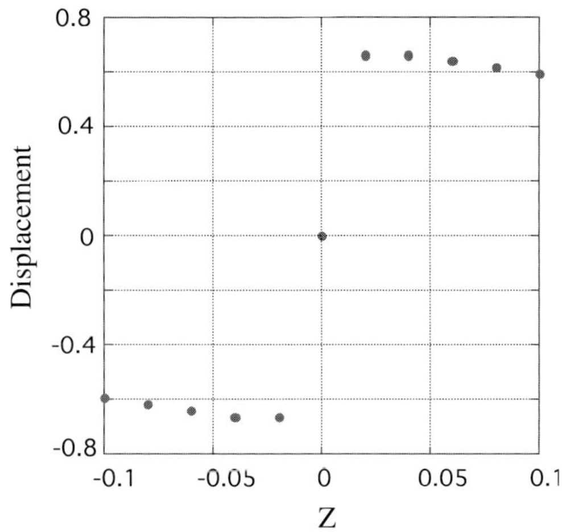

Figure 5
Displacement along the z axis at the center of the fault calculated by MADARIAGA *et al.*'s implementation.

Then,

$$\tau_{zx}^{m+1/2}\left[i+\frac{1}{2},0\right] = 2\Delta\sigma_f\left[i+\frac{1}{2}\right] - \frac{\tau_{zx}^{m+1/2}\left[i+\frac{1}{2},1\right]+\tau_{zx}^{m+1/2}\left[i+\frac{1}{2},\ 1\right]}{2}.$$

(5) Shear stress on the fictitious plane is calculated by the boundary condition of continuity of displacement (no slip) for the area that is locked.

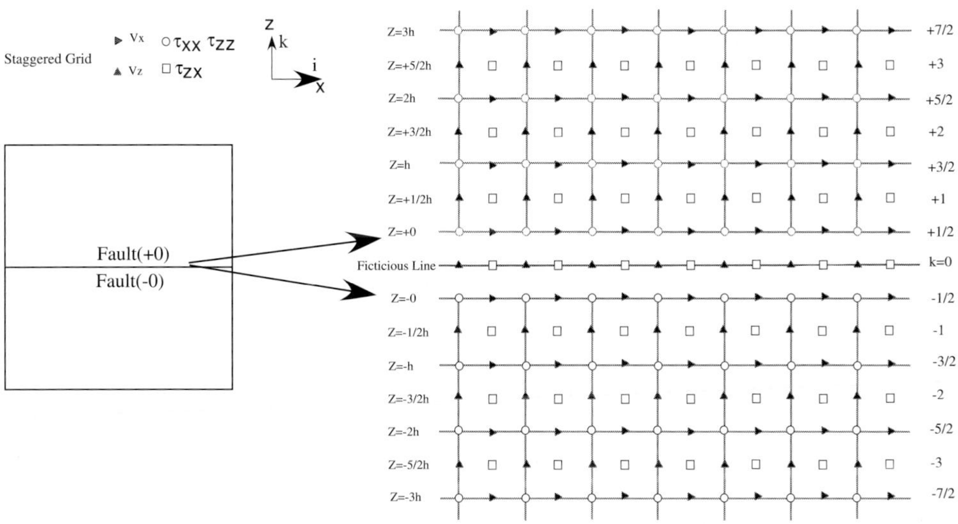

Figure 6
Finite-difference grids for this study.

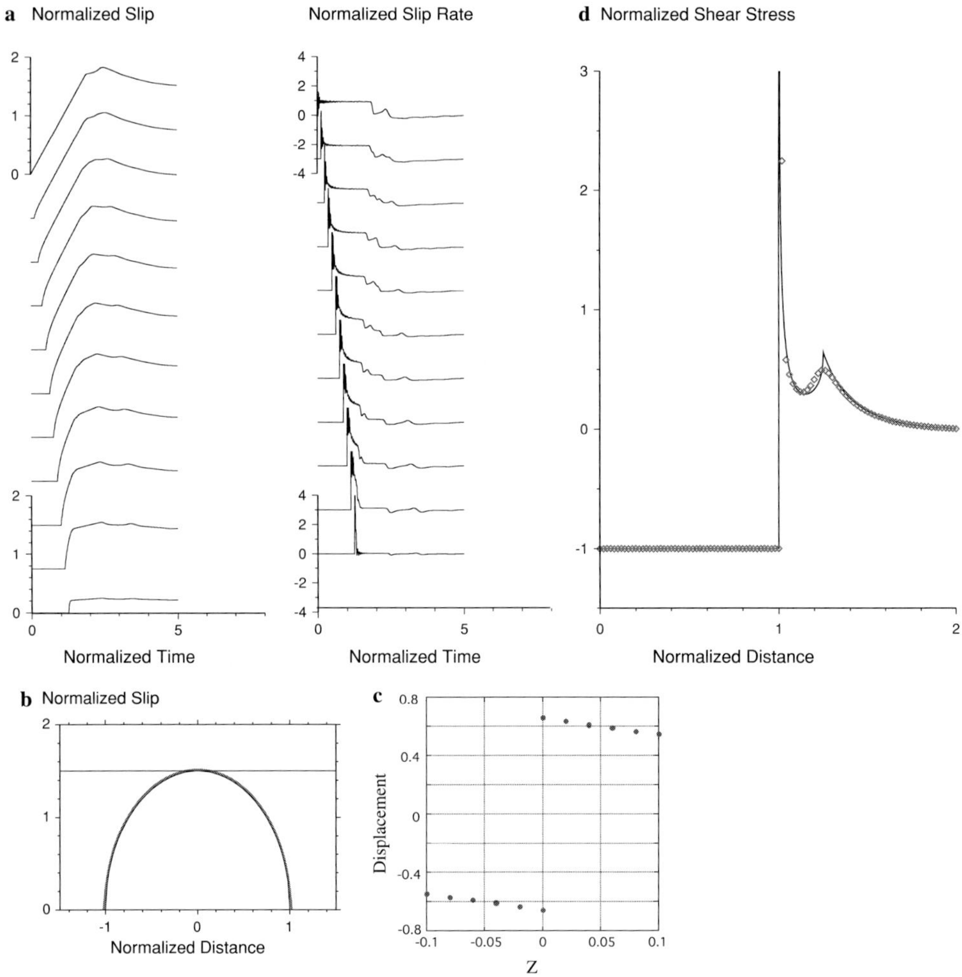

Figure 7
(a) Numerically calculated slip and slip-rate time functions for $Dc = 0$ using our implementation. (b) Numerically calculated final slip distribution on the fault for $Dc = 0$ using our implementation. (c) Numerically calculated the fault parallel displacement along z axis. (d) Numerically calculated snapshot of shear stress for the case of $Dc = 0$ using implementation.

$$
\begin{aligned}
v_x^{m+1}\left[i+\frac{1}{2},\frac{1}{2}\right] &= v_x^m\left[i+\frac{1}{2},\frac{1}{2}\right] \\
&\quad + \frac{\Delta t}{\rho \Delta x}\left[\tau_{xx}^{m+1/2}\left[i+1,\frac{1}{2}\right] - \tau_{xx}^{m+1/2}\left[i,\frac{1}{2}\right]\right] \\
&\quad + \frac{\Delta t}{\rho \Delta z}\left[\tau_{zx}^{m+1/2}\left[i+\frac{1}{2},1\right] - \tau_{zx}^{m+1/2}\left[i+\frac{1}{2},0\right]\right] + \frac{\eta}{2}\left[\nabla^2 v_x^m\left[i,\frac{1}{2}\right]\right]
\end{aligned}
$$

$$
\begin{aligned}
v_x^{m+1}\left[i+\frac{1}{2},-\frac{1}{2}\right] = {} & v_x^{m}\left[i+\frac{1}{2},-\frac{1}{2}\right] \\
& + \frac{\Delta t}{\rho \Delta x}\left[\tau_{xx}^{m+1/2}\left[i+1,-\frac{1}{2}\right] - \tau_{xx}^{m+1/2}\left[i,-\frac{1}{2}\right]\right] \\
& + \frac{\Delta t}{\rho \Delta z}\left[\tau_{zx}^{m+1/2}\left[i+\frac{1}{2},0\right] - \tau_{zx}^{m+1/2}\left[i+\frac{1}{2},-1\right]\right] + \frac{\eta}{2}\left[\nabla^2 v_x^m\left[i,-\frac{1}{2}\right]\right].
\end{aligned}
$$

As

$v_x^m\left[i+\frac{1}{2},\frac{1}{2}\right] = v_x^m\left[i+\frac{1}{2},-\frac{1}{2}\right]$ for the zone that is locked,

then

$$
\begin{aligned}
& \tau_{zx}^{m+1/2}\left[i+\frac{1}{2},0\right] \\
& = \frac{1}{2}\left[\tau_{xx}^{m+1/2}\left[i+1,\frac{1}{2}\right] - \tau_{xx}^{m+1/2}\left[i,\frac{1}{2}\right] - \tau_{xx}^{m+1/2}\left[i+1,-\frac{1}{2}\right] + \tau_{xx}^{m+1/2}\left[i,-\frac{1}{2}\right]\right] \\
& + \frac{1}{2}\left[\tau_{zx}^{m+1/2}\left[i+\frac{1}{2},1\right] + \tau_{zx}^{m+1/2}\left[i+\frac{1}{2},-1\right]\right] + \frac{\eta}{2}\left(\left[\nabla^2 v_x^m\left[i,\frac{1}{2}\right]\right] - \left[\nabla^2 v_x^m\left[i,-\frac{1}{2}\right]\right]\right).
\end{aligned}
$$

(6) Viscous damping term is added to the fault surface. As mentioned above, viscous damping on the fault removes spurious oscillations, but it does not blur stress ahead of the crack tip.

The equations of motion are implemented as a fourth-order finite difference in space and a second-order in time, although a second-order finite difference is used near the fault surface.

Numerical Test

We calculated the in-plane shear crack problems using our new implementation. The computational parameters are the same as those described in the previous section. We first show the results for $Dc = 0$, a sudden stress drop that generates spurious oscillations (Fig. 7). The slip-rate time functions and the slip time functions are shown in Figure 7a. No artificial pre-slip appeared prior to the onset of the slip. The final slip distribution is in good agreement with the theoretical solution (Fig. 7b) and the displacement discontinuity across the fault is well modeled (Fig. 7c). No artificial leakage of slip occurred outside the fault. The snapshot of the stress on the fault when the crack tip reaches the position of $x = 1$ is in good agreement with the theoretical solutions of KOSTROV (1964) and TADA (1995) (Fig. 7d). Figure 8 shows the slip rate and time functions for spontaneous rupture in which $Dc = 0.2$ and $S = 2$ ($S = \frac{\tau_p - \tau_0}{\tau_0 - \tau_f}$, where τ_p, τ_0 and τ_f are a peak stress, initial stress, and dynamic friction, respectively). The initial rupture length L_C is assumed to be $L_C = \frac{8\mu(\lambda+\mu)G}{\pi(\lambda+2\mu)(\tau_0-\tau_f)^2}$ (ANDREWS, 1976). The terminal velocity of rupture propagation is

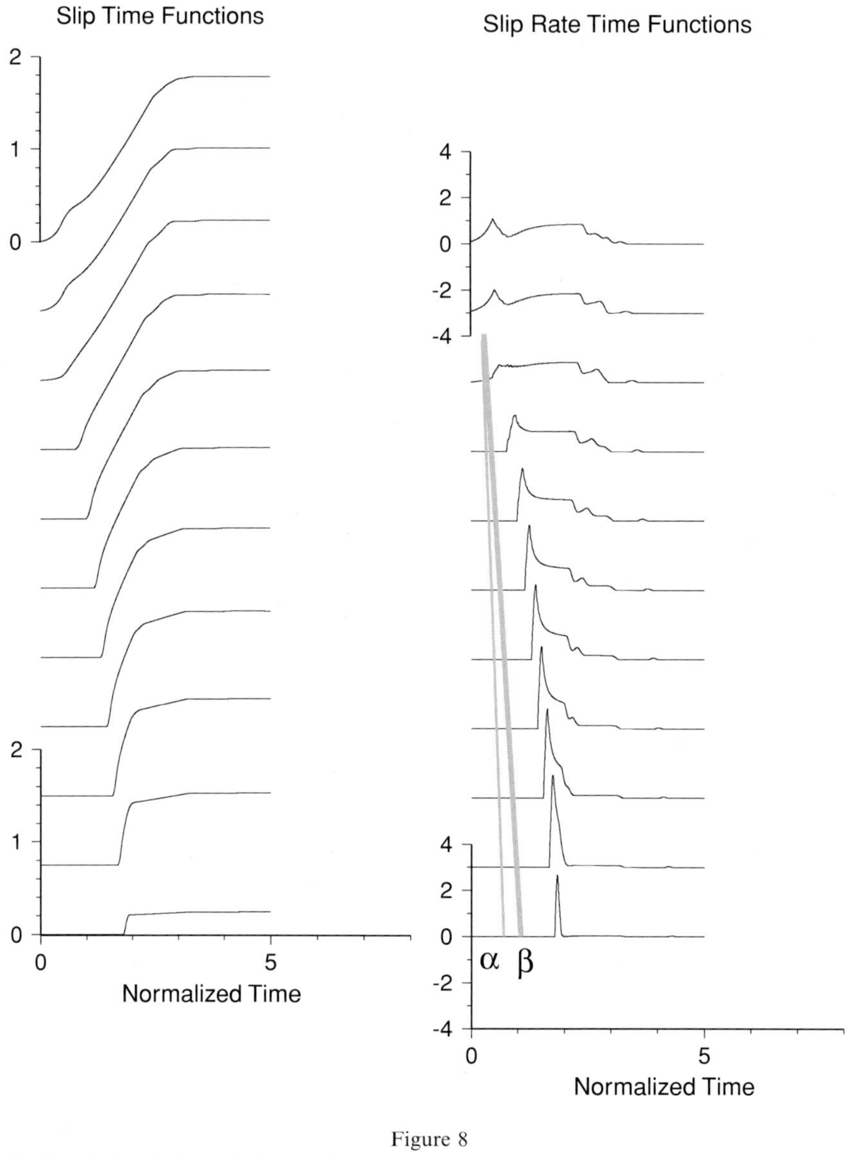

Figure 8
Numerically calculated slip and slip-rate time functions for $Dc = 0.2$. The condition of $\Delta\dot{u} \geq 0$ is applied on the fault.

approximately 0.8 β which is reasonable given that the rupture velocity is expected to be less than the Rayleigh wave velocity in the case of $Dc = 0$ and $S = 2$ (ANDREWS, 1976).

In our computation, we assumed that fault slip rate ($\Delta\dot{u}$) is greater than or equal to zero. This means that the fault slip stops at the point where the slip rate $\Delta\dot{u}(x, t)$

becomes negative in our numerical computations. Considering the frictional stress on the fault (KIKUCHI and FUKAO, 1976), this condition is reasonable for real earthquakes except for the very shallow part of the fault. We implemented this condition by switching the boundary condition of the fault plane for slipping from the condition of (4) for slipping to (5).

Conclusion

We improved the implementation of the fault boundary condition in a staggered grid finite difference method by introducing a fictitious surface to satisfy the fault boundary condition. In our implementation, we assign the fault plane to two separate planes of particle velocity V_X (plane of $z = +0$, and that of $z = -0$, and a fictitious surface of shear stress and V_Z are added between these two planes. The shear stress on the fictitious surface in the ruptured area is given such that the interpolated stress on the fault is equal to the frictional stress. Within the unruptured area, shear stress on the fictitious surface is given in the condition of no discontinuity of the particle velocity. The fault-normal component of particle velocity on the fictitious plane is given such that the normal stress on the fault is continuous across the fault. Artificial viscous damping is introduced on the fault to avoid the vibration caused by onset of the slip. Our implementation has three principal merits: (i) No leakage of the slip before rupture, (ii) zero thickness fault, and (iii) accurate calculation of shear stress on the fault (the stress bump caused by S-wave arrival and stress peak at the crack tip is clearly shown in the case of subsonic rupture propagation). We found that our modifications satisfy the above three checkpoints.

Acknowledgements

This study is partly supported by the special project for earthquake disaster mitigation in urban areas. The computations were carried out by the parallel computer of Earthquake Information Center in the Earthquake Research Institute, University of Tokyo.

REFERENCES

ANDREWS, J. (1976), *Rupture velocity of plane strain shear cracks*, J. Geophys. Res. *81*, 5679–5687.

ANDREWS, J. (1985), *Dynamic plane-strain shear rupture with a slip-weakening friction law calculated by a boundary integral method*, Bull. Seismol. Soc. Am. *85*, 1–21.

ANDREWS, J. (1999), *Test of two methods for faulting in finite-difference calculations*, Bull. Seismol. Soc. Am. *89*, 931–937.

AOCHI, H. and Fukuyama, E. (2002), *Three-dimensional nonplanar simulation of the 1992 Landers earthquake*, J. Geophys. Res. *107*, 10.1029/2000JB000061.

DAS, S. and AKI, K. (1977), *A numerical study of two-dimensional spontaneous rupture propagation*, Geophys. J. R. Astr. Soc. *50*, 643–668.

DAY, S. (1982), *Three-dimensional finite-difference simulation of fayult dynamics: rectangular faults with fixed rupture velocity*, Bull. Seismol. Soc. Am. *72*, 705–727.

FUKUYAMA, E. and MADARIAGA, R. (1995), *Integral equation method for plane crack with arbitrary shape in 3-D elastic medium*, Bull. Seismol. Soc. Am. *85*, 614–628.

GRAVES, R. W. (1996), *Simulating seismic wave propagation in 3-D elastic media using staggered-grid finite differences*, Bull. Seismol. Soc. Am. *86*, 1091–1106.

HU, L. and MCMECHAN, G. A. (1988) *Elastic finite-difference modeling and imaging for earthquake sources*, Geophys. J. *95*, 303–313.

KIKUCHI, M. and FUKAO, Y. (1976), *Seismic return motion*, Phys. Earth Planet. Inter. *12*, 343–349.

KOSTROV, B. (1964), *Self-similar problems of propagation of shear cracks*, J. Appl. Math. Mech. *28*, 1077–1087.

MADARIAGA, R. (1976), *Daynamics of an expanding circular fault*, Bull. Seismol. Soc. Am. *66*, 639–667.

MADARIAGA, R., OLSEN, K., and ARCHULETA, R. (1998), *Modeling dynamic rupture in a 3-D earthquake fault model*, (1982), Dynamic faulting studied by a finite-difference method, Bull. Seismol. Soc. Am. *88*, 1182–1197.

MIYATAKE, T. (1980), *Numerical simulations of earthquake source process by a three-dimensional crack model, Part 1. Rupture process*, J. Phys. Earth *28*, 565–598.

OLSEN, K., MADARIAGA, R., and ARCHULETA, R. (1977), *Three-dimensional dynamic simulation of the 1992 Landers earthquake*, Science *278*, 834–838.

TADA, T. (1995), *Boundary integral equations for the time-domain and time-independent analyses of 2-D non-planar cracks*, Thesis Ph. D. Earth and Planetary Physics at the University of Tokyo.

VIRIUEX, J. and MADARIAGA, R. (1982), *Dynamic faulting studied by a finite difference method*, Bull. Seismol. Soc. Am. *72*, 345–369.

(Received October 1, 2004, revised October 18, 2005, accepted November 22, 2005)

Pure appl. geophys. 163 (2006) 1991–2009
0033–4553/06/091991–19
DOI 10.1007/s00024-006-0109-z

Pure and Applied Geophysics

The Displacement and Strain Field of Three-dimensional Rheologic Model of Earthquake Preparation

ZHIPING SONG,[1] XIANGCHU YIN,[2] SHIRONG MEI,[2] YUCANG WANG,[3] CAN YIN,[3] HUIHUI ZHANG,[4] and LANGPING ZHANG[4]

Abstract—Based on the three-dimensional elastic inclusion model proposed by Dobrovolskii, we developed a rheological inclusion model to study earthquake preparation processes. By using the Corresponding Principle in the theory of rheologic mechanics, we derived the analytic expressions of viscoelastic displacement $U(r, t)$, $V(r, t)$ and $W(r, t)$, normal strains ε_{xx} (r, t), ε_{yy} (r, t) and ε_{zz} (r, t) and the bulk strain θ (r, t) at an arbitrary point (x, y, z) in three directions of X axis, Y axis and Z axis produced by a three-dimensional inclusion in the semi-infinite rheologic medium defined by the standard linear rheologic model. Subsequent to the spatial-temporal variation of bulk strain being computed on the ground produced by such a spherical rheologic inclusion, interesting results are obtained, suggesting that the bulk strain produced by a hard inclusion change with time according to three stages (α, β, γ) with different characteristics, similar to that of geodetic deformation observations, but different with the results of a soft inclusion. These theoretical results can be used to explain the characteristics of spatial-temporal evolution, patterns, quadrant-distribution of earthquake precursors, the changeability, spontaneity and complexity of short-term and imminent-term precursors. It offers a theoretical base to build physical models for earthquake precursors and to predict the earthquakes.

Key words: Three-dimensional elastic inclusion mode, rheological inclusion model, analytic expressions.

Introduction

The studies of the inclusion model started in the 1950s. ESHELBY (1957) first studied an inhomogeneous problem in which a region (or 'inclusion') with homogeneous elastic constants was embedded within a matrix with different

[1] Earthquake Administration of Shanghai Municipality, Shanghai 200062, China.
E-mail: zpsong@sohu.com
[2] Institue of Earthquake Prediction, China Earthquake Administration, Beijing 100036, China.
[3] Quakes, Earth Systems Science Computational Centre, The University of Queensland, St. Lucia, Brisbane, QLD 4072, Australia.
[4] State Key Laboratory of Nonlinear Mechanics, Institute of Mechanics, Chinese Academy of Sciences, Beijing 100080, China.

elastic constants and derived the elastic stresses in matrix and inclusion. Eshelby studied two kinds of inclusions, which were the hard inclusion in which the elastic modulus of inclusion is higher that of the matrix, and the soft inclusion in which the elastic modulus of inclusion is lower. Eshelby also studied the elliptic sphere inclusion model in the elastic medium and gave analytical solutions of the stress field of inclusion and matrix and applied to some problems of physics and mechanics.

BRADY (1974) proposed the soft inclusion model. The main hypothesis is that the intense crack concentrations developed within localized regions when a rock specimen is close to ultimate failure can be represented by inclusions of lower modulus. He proposed a scale-independent theory of rock failure, and studied the inclusion of crustal and deep earthquakes, and gave meticulous explanations for short-term premonitory phenomena of earthquakes. He pointed out that for the process of earthquake failures, it is appropriate to use the soft inclusion rather than the hard one.

DOBROVOLSKII (1991) further studied the hard inclusion model on the basis of elastic mechanics. He gave the elastic analytical solutions of the displace, the strain, the geo-electricity and the geomagnetism of random point (x, y, z) produced by an earthquake generating body of a three-dimension inclusion (spherical inclusion or rotating spheroid). He discussed the precursors' mechanism of seismicity, deformation, water-radon, geomagnetism, geo-electricity and so on and the relation between the distributive range of earthquake precursors and the magnitude of the earthquake. In fact, this is the first excellent model that remarks on features of long-term precursor fields. However it is still difficult to explain the mechanism of short-term and immediate precursors by using this model.

MEI (1994) set up the strong-body earthquake-generating model based on geophysics. By studying the tomographic image, it has been affirmed that the existence of a hard inclusion is an important condition for the high concentration of large amounts of strain energy. Numerous theoretical and experimental studies have consistently indicated that rock instability, sudden fracture and stress drop would be possible only if the stiffness of the source body is greater than the environmental stiffness.

They viewed the generating body as a three-dimension inclusion. Eshelby gave analytical solutions of the stress field of different inclusions. Brady proposed the soft inclusion model to explain short-term premonitory phenomena of earthquakes. Dobrovolskii studied the hard inclusion model to study long-term precursors. All of these models are elastic inclusion models. Rocks exhibit mainly elastic behaviors under low temperature, low pressure and short-time external forces, but they exhibit rheologic behavor under high temperature, high pressure and long-time external forces. Therefore it is appropriate that earthquake rupture process could be analyzed using the elastic theory, considering the fact that it occurs within seconds to hundreds of seconds. However, for the earthquake generating process lasting over hundreds or thousand of years, the rheologic property of rock must be taken into account (ZHANG *et al.*, 1987; YIN *et al.*, 1982; MEI *et al.*, 1993).

Our previous work extended Dobrovolskii's elastic theory of the hard inclusion model into the rheologic inclusion model (SONG, 1996). We analyzed preliminarily the spatial and temporal evolution process of bulk strain field for this model (SONG *et al.*, 2000). In this paper we give the displacement field and the strain field of the rheologic inclusion model, and endeavor to discuss the characteristics of spatial-temporal evolution, patterns and quadrant-distribution of earthquake precursors. Additional and detailed derivations can be found in two papers (SONG *et al.*, 2003, 2004).

Basic Theory

Assume there is an inclusion in the semi-infinite medium (Fig. 1a). S_0 is the surface of the Earth where the effects of the atmosphere pressure ignored. S_1 is a hypothetical surface of the Earth and its boundary condition may be given in terms of displacement, stress or other physical parameters. With the first-order approximation, the external forces for a shallow earthquake may include body forces, whereas the gravity and the tidal force may be neglected. Although the tidal disturbing effect is small, it may be significant due to its long and periodical nature, therefore in detailed analysis this effect should be considered.

In a rectangular coordinate system, the plane *XOY* is in the surface of the Earth. The *Z*-axis points to the center of the Earth. The center of the inclusion is at the point (0, 0, H). The boundary condition is that except for $\sigma_{xy}^{\infty} = \sigma_{yx}^{\infty} = -\tau$ (Fig. 1), all other

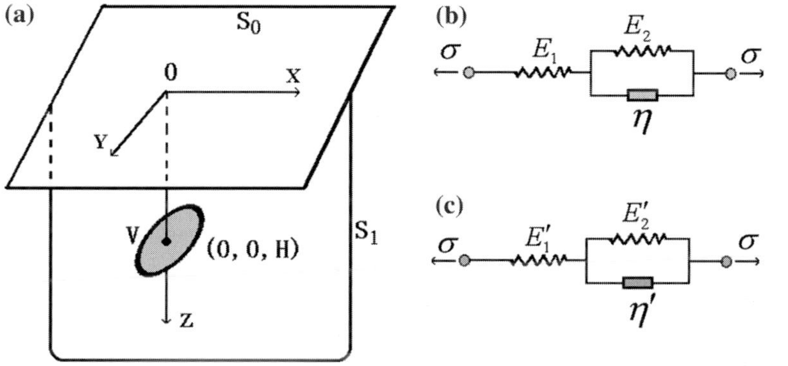

Figure 1
Three-dimensional inclusion and its rheologic medium model. (a) Three-dimensional inclusion model. (b) Rheologic medium model outside the inclusion. (c) Rheologic medium model inside the inclusion.

components are zero. That is, uniform shear stresses are applied at the infinitude and the bottom is a free surface. If the boundary condition is a function of time, it is expressed as $\tau H(t)$, where $H(t)$ is a unit jump function. Therefore, we choose the standard linear solid model (series connection between a spring and a Kelvin body or parallel connection between a spring and a Maxwell body) (YIN *et al.*, 1982) as the medium model inside and outside the three-dimensional inclusion in Figures.1b and 1c (SONG *et al.*, 2003).

On the basis of Dobrovolskii's hard inclusion elastic theory and the Corresponding Principle of rheologic mechanics (YIN, 1985), we derived the viscoelastic expression of displacement produced by the inclusion in the rheologic medium.

From the analytic expression $U_i^e(r)$ of displacement of an elastic inclusion, we can derive the solution $\bar{U}_i^e(r,s)$ of elastic displacement in the image-plane by using Laplace Transform. Then by substituting the elastic parameters μ and K with $\overline{\mu}(s)$ and $\bar{K}(s)$ of the image-plane, we can derive the solution $\bar{U}_i^v(r,s)$ of visco-elastic displacement in the image-plane. At last, using the Inverse Laplace Transform, the viscoelastic expression $U_i^v(r,t)$ can be obtained.

Displacement Field

The analytic solution of visco-elastic displacement $U(r,t)$ in the X axis, $V(r,t)$ in the Y axis and $W(r,t)$ in the Z-axis has been obtained in SONG *et al.* (2003).

For example, the analytic solution of visco-elastic displacement $U(r,t)$ in the X axis of a point (x,y,z) is as follows:

$$U(r,t) = \frac{Vy\tau}{8\pi}\left\{G_{u0}H(t) + (G_{u1} + G''_{u1} + tG'_{u1})\exp\left(-\frac{q_0}{q_1}t\right) + G_{u2}\exp\left(-\frac{1}{p'_1}t\right)\right.$$
$$\left. + G_{u3}\exp\left(-\frac{3K+2q_0}{3Kp_1+2q_1}t\right) + G_{u4}\exp\left(-\frac{6K+q_0}{6Kp_1+q_1}t\right)\right\} \quad (1)$$

Defining the time parameters T_1, T_2, T_3, T_4 as

$$T_1 = \frac{q_0}{q_1}, \quad T_2 = \frac{1}{p'_1}, \quad T_3 = \frac{3K+2q_0}{3Kp_1+2q_1}, \quad T_4 = \frac{6K+q_0}{6Kp_1+q_1}.$$

The Parameters G_{ui} in Equation (1) are

$$\begin{cases} G_{u0} = 2\left(\frac{q'_0}{q_0} - 1\right) \cdot \left[\frac{S_u(0)}{q_0(3K+q_0)} + P_{u2} \cdot \frac{3}{3K+2q_0} + P_{u4} \cdot \frac{3}{6K+q_0}\right] \\ G_{u1} = \frac{2(1-p_1T_1)}{q_0} \left\{ \begin{array}{l} \frac{S_u(T_1)}{(3K+2q_0)-(3Kp_1+2q_1)T_1} - \frac{3(1-p_1T_1)(q'_0-q'_1T_1)}{1-p'_1T_1} \bullet \\ \left[\frac{P_{u2}}{(3K+2q_0)-(3Kp_1+2q_1)T_1} + \frac{P_{u4}}{(6K+q_0)-(6Kp_1+q_1)T_1}\right] \end{array} \right\} \\ G_{u2} = -2(1-p_1T_2)^2 \\ \quad \times \frac{q'_0-q'_1T_2}{q_0-q_1T_2} \left\{ \begin{array}{l} \frac{S_u(T_2)}{(q_0-q_1T_2)[(3K+2q_0)-(3Kp_1+2q_1)T_2]} \\ + \frac{3P_{u2}}{(3K+2q_0)-(3Kp_1+2q_1)T_1} + \frac{3P_{u4}}{(6K+q_0)-(6Kp_1+q_1)T_1} \end{array} \right\} \\ G_{u3} = -\frac{2(1-p_1T_3)}{3K+2q_0}\left[\frac{1-p_1T_3}{1-p'_1T_3} \cdot \frac{q'_0-q'_1T_3}{q_0-q_1T_3} - 1\right]\left[\frac{S_u(T_3)}{q_0-q_1T_3} + 3P_{u2}\right] \\ G_{u4} = -\frac{6(1-p_1T_4)}{6K+q_0}\left[\frac{1-p_1T_4}{1-p'_1T_4} \cdot \frac{q'_0-q'_1T_4}{q_0-q_1T_4} - 1\right] \cdot P_{u4} \\ G'_{u1} = \frac{-2(q'_0-q'_1T_1)(1-p_1T_1)^2}{q_0q_1(1-p'_1T_1)[(3K+2q_0)-(3Kp_1+2q_1)T_1]} \cdot S_u(T_1) \\ G'_{u1} = 2\{P_{u1} \cdot S(6K+q_0, 6Kp_1+q_1) + P_{u3} \cdot S(6K+7q_0, 6Kp_1+7q_1)\} \end{cases} \tag{2}$$

where

$$S_u(T) = P_{u1} \cdot [6K + q_0 - (6Kp_1 + q_1)T] + P_{u3} \cdot [6K + 7q_0 - (6Kp_1 + 7q_1)T]. \tag{3}$$

The universal function $S(m, n)$ is

$$\begin{aligned} S(m,n) = {} & [AA(3K+2q_0, 3Kp_1+2q_1) + BB(3K+2q_0, 3Kp_1+2q_1)] \cdot (m - nT_1) \\ & + CC(3K+2q_0, 3Kp_1+2q_1) \cdot n. \end{aligned} \tag{4}$$

In Equation (4), $AA(m,n)$, $BB(m,n)$, $CC(m,n)$ and $DD(m,n)$ are

$$\begin{cases} AA(m,n) = -\frac{(1-p_1T_1)[q'_1(1-p_1T_1) - 2p_1(q'_0-q'_1T_1)]}{q_0q_1(1-p'_1T_1)} \cdot \frac{1}{m-nT_1} \\ BB(m,n) = -\frac{(1-p_1T_1)^2(q'_0-q'_1T_1)}{q_0(1-p'_1T_1)} \cdot \frac{DD(m,n)}{m-nT_1} \\ CC(m,n) = -\frac{(1-p_1T_1)^2(q'_0-q'_1T_1)}{q_0q_1(1-p'_1T_1)} \cdot \frac{1}{m-nT_1} \\ DD(m,n) = \frac{1}{q_0} - \frac{p'_1}{q_1(1-p'_1T_1)} - \frac{n}{q_1(m-nT_1)} \end{cases} \tag{5}$$

The parameters related to the coordinate system are

$$\begin{cases} P_{u1} = -\dfrac{6Hz}{r_2^5} + x^2\left(-\dfrac{3}{r_1^5} + \dfrac{30Hz}{r_2^7}\right) \\ P_{u2} = \dfrac{1}{r_2^3} - \dfrac{1}{r_1^3} \\ P_{u3} = -\dfrac{3x^2}{r_2^5} \\ P_{u4} = -\dfrac{4}{r_2(r_2+H+z)^2} + \dfrac{4(3r_2+H+z)}{r_2^3(r_2+H+z)^3}\cdot x^2 \end{cases} \tag{6}$$

When $t = 0$, the displacement $U(r,0)$ in the X axis is

$$U(r,0) = \frac{Vy\tau}{8\pi}\{G_{u0} + G_{u1} + G''_{u1} + G_{u2} + G_{u3} + G_{u4}\}. \tag{7}$$

The analytic solution of visco-elactic displacement $V(r,t)$ in the Y axis of the point (x, y, z) is as follows:

$$\begin{aligned} V(r,t) = \frac{Vx\tau}{8\pi}\bigg\{ & G_{v0}H(t) + (G_{v1} + G''_{v1} + tG'_{v1})\exp\left(-\frac{q_0}{q_1}t\right) + G_{v2}\exp\left(-\frac{1}{p'_1}t\right) \\ & + G_{v3}\exp\left(-\frac{3K+2q_0}{3Kp_1+2q_1}t\right) + G_{v4}\exp\left(-\frac{6K+q_0}{6Kp_1+q_1}t\right)\bigg\} \end{aligned} \tag{8}$$

where the parameters G_{vi} $(i = 0,1,2,3,4)$ are G_{ui} in the Equations (2)~(6) by substituting the parameters P_{ui} with P_{vi}.

When $t = 0$, the displacement $V(r,0)$ in the Y axis is

$$V(r,0) = \frac{Vx\tau}{8\pi}\{G_{v0} + G_{v1} + G''_{v1} + G_{v2} + G_{v3} + G_{v4}\}\cdot \tag{9}$$

The analytic solution of viscoelastic displacement $W(r,t)$ in the Z direction of the points (x,y,z) was obtained.

$$\begin{aligned} W(r,t) = \frac{V\tau xy}{8\pi}\bigg\{ & G_{w0}H(t) + (G_{w1} + G''_{w1} + tG'_{w1})\exp\left(-\frac{q_0}{q_1}t\right) + G_{w2}\exp\left(-\frac{1}{p'_1}t\right) \\ & + G_{w3}\exp\left(-\frac{3K+2q_0}{3Kp_1+2q_1}t\right) + G_{w4}\exp\left(\frac{6K+q_0}{6Kp_1+q_1}t\right)\bigg\} \end{aligned} \tag{10}$$

where parameters $G_{wi}(i = 0, 1, 2, 3, 4)$ are as follows

$$\begin{cases} G_{w0} = 2\left(\frac{q_0'}{q_0} - 1\right) \cdot \left[\frac{S_u(0)}{q_0(3K + q_0)} + \frac{3P_{w3}}{6K + q_0}\right] \\ G_{w1} = \frac{2(1 - p_1 T_1)}{q_0} \left\{ \frac{S_w(T_1)}{(3K + 2q_0) - (3Kp_1 + 2q_1)T_1} \right. \\ \qquad \left. - \frac{(1 - p_1 T_1)}{1 - p_1' T_1} \cdot \frac{3(q_0' - q_1' T_1)P_{w3}}{(6K + q_0) - (6Kp_1 + q_1)T_1} \right\} \\ G_{w2} = -2(1 - p_1 T_2)^2 \frac{q_0' - q_1' T_2}{q_0 - q_1 T_2} \\ \qquad \times \left\{ \frac{S_w(T_2)}{(q_0 - q_1 T_2)[(3K + 2q_0) - (3Kp_1 + 2q_1)T_2]} + \frac{3P_{w3}}{(6K + q_0) - (6Kp_1 + q_1)T_2} \right\} \\ G_{w3} = -\frac{2(1 - p_1 T_3)}{(3K + 2q_0)(q_0 - q_1 T_3)} \left[\frac{1 - p_1 T_3}{1 - p_1' T_3} \cdot \frac{q_0' - q_1' T_3}{q_0 - q_1 T_3} - 1\right] \cdot S_w(T_3) \\ G_{w4} = -\frac{6(1 - p_1 T_4)}{6K + q_0} \left[\frac{1 - p_1 T_4}{1 - p_1' T_4} \cdot \frac{q_0' - q_1' T_4}{q_0 - q_1 T_4} - 1\right] \cdot P_{w3} \\ G_{w1}' = -\frac{2(q_0' - q_1' T_1)(1 - p_1 T_1)^2}{q_0 q_1 (1 - p_1' T_1)[(3K + 2q_0) - (3Kp_1 + 2q_1)T_1]} \cdot S_w(T_1) \\ G_{w1}' = 2\{P_{w1} \cdot S(6K + q_0, 6Kp_1 + q_1) + P_{w2} \cdot S(6K + 7q_0, 6Kp_1 + 7q_1)\} \end{cases} \tag{11}$$

where

$$\begin{cases} P_{w1} = -\frac{3(H - z)}{r_1^5} + \frac{30Hz(H + z)}{r_2^7} \\ P_{w2} = \frac{3(H - z)}{r_2^5} \\ P_{w3} = -\frac{4(2r_2 + H + z)}{r_2^3 (r_2 + H + z)^2} \end{cases} \tag{12}$$

$$S_w(T) = P_{w1} \cdot [6K + q_0 - (6Kp_1 + q_1)T] + P_{w2} \cdot [6K + 7q_0 - (6Kp_1 + 7q_1)T]. \tag{13}$$

For the universal function $S(n, m)$ see Equation (4). When $t = 0$, the displacement $W(r, 0)$ in the Z axis is given by

$$W(r, 0) = \frac{V\tau xy}{8\pi} \{G_{w0} + G_{w1} + G_{w1}'' + G_{w2} + G_{w3} + G_{w4}\}. \tag{14}$$

Strain Field

The normal strain $\varepsilon_{xx}(r,t)$ in X axis, $\varepsilon_{yy}(r,t)$ in Y axis, $\varepsilon_{zz}(r,t)$ in Z axis and then the bulk strain $\theta(r,t)$ has been obtained by SONG (1996).

The normal strain variation with time for the rheologic incluson model in X axis is (SONG *et al.*, 2004)

$$\varepsilon_{xx}(r,t) = \frac{3V\tau xy}{8\pi}\left\{M_{x0}H(t) + (M_{x1} + M''_{x1} + tM'_{x1})\exp\left(-\frac{q_0}{q_1}t\right) + M_{x2}\exp\left(-\frac{1}{p'_1}t\right)\right.$$
$$\left. + M_{x3}\exp\left(-\frac{3K+2q_0}{3Kp_1+2q_1}t\right) + M_{x4}\exp\left(-\frac{6K+q_0}{6Kp_1+q_1}t\right)\right\} \tag{15}$$

The parameters $M_{xi}(i = 1,4)$ are as follows:

$$\begin{cases}
M_{x0} = 2\left(\dfrac{q'_0}{q_0} - 1\right)\cdot\left[\dfrac{S_{\varepsilon xx}(0)}{q_0(3K+q_0)} + \cdot\dfrac{3P_{x5}}{6K+q_0}\right] \\
M_{x1} = \dfrac{2(1-p_1T_1)}{q_0} \\
\qquad \times\left\{\dfrac{S_{\varepsilon xx}(T_1)}{(3K+2q_0)-(3Kp_1+2q_1)T_1} - \dfrac{(1-p_1T_1)}{1-p'_1T_1}\cdot\dfrac{3(q'_0-q'_1T_1)P_{x5}}{(6K+q_0)-(6Kp_1+q_1)T_1}\right\} \\
M_{x2} = -2(1-p_1T_2)^2\dfrac{q'_0-q'_1T_2}{q_0-q_1T_2} \\
\qquad \times\left\{\dfrac{S_{\varepsilon xx}(T_2)}{(q_0-q_1T_2)[(3K+2q_0)-(3Kp_1+2q_1)T_2]} + \dfrac{3P_{x5}}{(6K+q_0)-(6Kp_1+q_1)T_2}\right\} \\
M_{x3} = -\dfrac{2(1-p_1T_3)}{(3K+2q_0)(q_0-q_1T_3)}\left[\dfrac{1-p_1T_3}{1-p'_1T_3}\cdot\dfrac{q'_0-q'_1T_3}{q_0-q_1T_3} - 1\right]\cdot S_{\varepsilon xx}(T_3) \\
M_{x4} = -\dfrac{6(1-p_1T_4)}{6K+q_0}\left[\dfrac{1-p_1T_4}{1-p'_1T_4}\cdot\dfrac{q'_0-q'_1T_4}{q_0-q_1T_4} - 1\right]\cdot P_{x5} \\
M'_{x1} = -\dfrac{2(q'_0-q'_1T_1)(1-p_1T_1)^2}{q_0q_1(1-p'_1T_1)[(3K+2q_0)-(3Kp_1+2q_1)T_1]}\cdot S_{\varepsilon xx}(T_1) \\
M''_{x1} = 2\cdot\left\{\begin{array}{l} P_{x1}\cdot S(6K+q_0, 6Kp_1+q_1) + P_{x2}\cdot S(6K+7q_0, 6Kp_1+7q_1) \\ +P_{x3}\cdot S(12K-q_0, 12Kp_1-q_1) + P_{x4}\cdot S(12K+17q_0, 12Kp_1+17q_1) \end{array}\right\}.
\end{cases} \tag{16}$$

In the Equation (16), $S_{\varepsilon xx}(T)$ is given by

$$S_{\varepsilon xx}(T) = P_{x1}\cdot[6K+q_0-(6Kp_1+q_1)T] + P_{x2}\cdot[6K+7q_0-(6Kp_1+7q_1)T]$$
$$+ P_{x3}\cdot[12K-q_0-(12Kp_1-q_1)T] + P_{x4}\cdot[12K+17q_0-(12Kp_1+17q_1)T]. \tag{17}$$

For the universal function $S(n, m)$ in Equation (16), see Equations (4) and (5). The parameters P_{xi} (i = 1,4) in Equations (16) and (17) are given by

$$\begin{cases} P_{x1} = \dfrac{30Hz}{r_2^7} + x^2\left(\dfrac{5}{r_1^7} - \dfrac{70Hz}{r_2^9}\right) \\ P_{x2} = \dfrac{5}{r_2^7} \cdot x^2 \\ P_{x3} = -\dfrac{1}{r_1^5} \\ P_{x4} = -\dfrac{1}{r_2^5} \\ P_{x5} = \dfrac{4(3r_2 + H + z)}{r_2^3(r_2 + H + z)^3} - \dfrac{4x^2}{r_2^5(r_2 + H + z)^2} - \dfrac{8(2r_2 + H + z)}{r_2^4(r_2 + H + z)^4} \cdot x^2. \end{cases} \tag{18}$$

In Equation (15), when $t = 0$, the normal strain $\varepsilon \hat{x}x(r, 0)$ in X axis is

$$\varepsilon_{xx}(r, 0) = \frac{3V\tau xy}{8\pi}\left\{M_{x0} + M_{x1} + M''_{x1} + M_{x2} + M_{x3} + M_{x4}\right\}. \tag{19}$$

The analytic expression of the viscoelastic normal strain $\varepsilon_{yy}(r, t)$ at a point (x, y, z) in Y axis is

$$\begin{aligned} \varepsilon_{yy}(r, t) = \frac{3V\tau xy}{8\pi}\Bigg\{ & M_{y0}H(t) + (M_{y1} + M''_{y1} + t.M'_{y1})\exp\left(-\frac{q_0}{q_1}t\right) + M_{y2}\exp\left(-\frac{1}{p'_1}t\right) \\ & + M_{y3}\exp\left(-\frac{3K + 2q_0}{3Kp_1 + 2q_1}t\right) + M_{y4}\exp\left(-\frac{6K + q_0}{6Kp_1 + q_1}t\right)\Bigg\} \end{aligned} \tag{20}$$

where the parameters M_{yi} are the parameters M_{xi} in Equations (16) and (17) by substituting P_{xi} with P_{yi}. Of these, the parameters P_{yi} which related to the coordinate system are given by

$$\begin{cases} P_{y1} = -\dfrac{30Hz}{r_2^7} + y^2\left(\dfrac{5}{r_1^7} - \dfrac{70Hz}{r_2^9}\right) \\ P_{y2} = \dfrac{5}{r_2^7} \cdot y^2 \\ P_{y3} = -\dfrac{1}{r_1^5} \\ P_{y4} = -\dfrac{1}{r_2^5} \\ P_{y5} = \dfrac{4(3r_2 + H + z)}{r_2^3(r_2 + H + z)^3} - \dfrac{4y^2}{r_2^5(r_2 + H + z)^2} - \dfrac{8(2r_2 + H + z)}{r_2^4(r_2 + H + z)^4} \cdot y^2. \end{cases} \tag{21}$$

In Equation (20), when $t = 0$, the normal strain in Y axis $\varepsilon_{yy}(r,0)$ is given by

$$\varepsilon_{yy}(r,0) = \frac{3V\tau xy}{8\pi}\left\{M_{y0} + M_{y1} + M''_{y1} + M_{y2} + M_{y3} + M_{y4}\right\}. \tag{22}$$

The analytic expression of the viscoelastic normal strain $\varepsilon_{zz}(r,t)$ of a point (x,y,z) in Z axis is

$$\begin{aligned}\varepsilon_{zz}(r,t) = &\frac{3V\tau xy}{8\pi}\left\{M_{z0}H(t) + (M_{z1} + M''_{z1} + tM'_{z1})\exp\left(-\frac{q_0}{q_1}t\right) + M_{z2}\exp\left(-\frac{1}{p'_1}t\right)\right.\\ &\left. + M_{z3}\exp\left(-\frac{3K+2q_0}{3Kp_1+2q_1}t\right) + M_{z4}\exp\left(-\frac{6K+q_0}{6Kp_1+q_1}t\right)\right\}\end{aligned} \tag{23}$$

where the parameters M_{zi} are as follows:

$$\begin{cases} M_{z0} = 2\left(\dfrac{q'_0}{q_0} - 1\right)\cdot\left[\dfrac{S_{\varepsilon zz}(0)}{q_0(3K+q_0)} + \cdot\dfrac{3P_{z3}}{6K+q_0}\right] \\ M_{z1} = \dfrac{2(1-p_1T_1)}{q_0} \\ \qquad \times\left\{\dfrac{S_{\varepsilon zz}(T_1)}{(3K+2q_0)-(3Kp_1+2q_1)T_1} - \dfrac{(1-p_1T_1)}{1-p'_1T_1}\cdot\dfrac{3(q'_0-q'_1T_1)P_{z3}}{(6K+q_0)-(6Kp_1+q_1)T_1}\right\} \\ M_{z2} = -2(1-p_1T_2)^2\dfrac{q'_0-q'_1T_2}{q_0-q_1T_2} \\ \qquad \times\left\{\dfrac{S_{\varepsilon zz}(T_2)}{(q_0-q_1T_2)[(3K+2q_0)-(3Kp_1+2q_1)T_2]} + \dfrac{3P_{z3}}{(6K+q_0)-(6Kp_1+q_1)T_2}\right\} \\ M_{z3} = -\dfrac{2(1-p_1T_3)}{(3K+2q_0)(q_0-q_1T_3)}\left[\dfrac{1-p_1T_3}{1-p'_1T_3}\cdot\dfrac{q'_0-q'_1T_3}{q_0-q_1T_3} - 1\right]\cdot S_{\varepsilon zz}(T_3) \\ M_{z4} = -\dfrac{6(1-p_1T_4)}{6K+q_0}\left[\dfrac{1-p_1T_4}{1-p'_1T_4}\cdot\dfrac{q'_0-q'_1T_4}{q_0-q_1T_4} - 1\right]\cdot P_{z3} \\ M'_{z1} = -\dfrac{2(q'_0-q'_1T_1)(1-p_1T_1)^2}{q_0q_1(1-p'_1T_1)[(3K+2q_0)-(3Kp_1+2q_1)T_1]}\cdot S_{\varepsilon zz}(T_1) \\ M''_{z1} = 2\cdot\{P_{z1}\cdot S(6K+q_0, 6Kp_1+q_1) + P_{z2}\cdot S(6K+7q_0, 6Kp_1+7q_1)\} \end{cases} \tag{24}$$

In Equation (23), $S_{\varepsilon zz}(T)$ is given by

$$S_{\varepsilon zz}(T) = P_{z1}\cdot[6K+q_0-(6Kp_1+q_1)T] + P_{z2}\cdot[6K+7q_0-(6Kp_1+7q_1)T]. \tag{25}$$

For the universal function $S(n, m)$ above see Equations (4) and (5). In Equations (24) and (25), the parameters P_{zi} which related to the coordinate system are as follows:

$$\begin{cases} P_{z1} = -\dfrac{1}{r_1^5} + \dfrac{5(H-z)^2}{r_1^7} + \dfrac{10H(2z+H)}{r_2^7} - \dfrac{70Hz(H+z)^2}{r_2^9} \\ P_{z2} = -\dfrac{1}{r_2^5} - \dfrac{5(H^2-z^2)}{r_2^7} \\ P_{z3} = -\dfrac{4}{r_2^5} \end{cases} \tag{26}$$

When $t = 0$, the normal strain in Z axis $\varepsilon_{zz}(r, 0)$ is

$$\varepsilon_{zz}(r, 0) = \frac{3V\tau xy}{8\pi}\left\{M_{z0} + M_{z1} + M''_{z1} + M_{z2} + M_{z3} + M_{z4}\right\}. \tag{27}$$

The viscoelastic analytic expression of the bulk strain $\theta(r, t)$ at a point (x, y, z) is (SONG *et al.*, 2004)

$$\begin{aligned} \theta(r,t) = \frac{3V\tau xy}{4\pi}\Bigg\{ & M_{\theta 0}H(t) + M_{\theta 1}.\exp\left(-\frac{q_0}{q_1}t\right) + M_{\theta 2}\exp\left(-\frac{1}{p'_1}t\right) \\ & + M_{\theta 3}\exp\left(-\frac{3K+2q_0}{3Kp_1+2q_1}t\right) + M_{\theta 4}\exp\left(-\frac{6K+q_0}{6Kp_1+q_1}t\right)\Bigg\} \end{aligned} \tag{28}$$

where the parameters $M_{\theta 0}$, $M_{\theta 1}$, $M_{\theta 2}$, $M_{\theta 3}$ and $M_{\theta 4}$ are as follows

$$\begin{cases} M_{\theta 0} = 6\left(\dfrac{q'_0}{q_0} - 1\right) \cdot \dfrac{S_\theta(0)}{3K+2q_0} \\ M_{\theta 1} = -6 \cdot \dfrac{(q'_0 - q'_1T_1)}{q_0} \cdot \dfrac{(1-p_1T_1)^2}{1-p'_1T_1} \cdot \dfrac{S_\theta(T_1)}{(3K+2q_0)-(3Kp_1+2q_1)T_1} \\ M_{\theta 2} = -6 \cdot \dfrac{q'_0 - q'_1T_2}{q_0 - q_1T_2} \cdot \dfrac{(1-p_1T_2)^2}{(3K+2q_0)-(3Kp_1+2q_1)T_2} \cdot S_\theta(T_2) \\ M_{\theta 3} = -6 \cdot \dfrac{(1-p_1T_3)}{(3K+2q_0)}\left[\dfrac{1-p_1T_3}{1-p'_1T_3} \cdot \dfrac{q'_0 - q'_1T_3}{q_0 - q_1T_3} - 1\right] \cdot S_\theta(T_3) \\ M_{\theta 4} = -6 \cdot \dfrac{(1-p_1T_4)}{6K+q_0} \cdot \dfrac{(6K+7q_0)-(6Kp_1+7q_1)T_4}{(3K+2q_0)-(3Kp_1+2q_1)T_4} \cdot \left[\dfrac{1-p_1T_4}{1-p'_1T_4} \cdot \dfrac{q'_0 - q'_1T_4}{q_0 - q_1T_4} - 1\right] \cdot P_{\theta 2} \end{cases} \tag{29}$$

where the universal function of bulk strain $S_\theta(T)$ is given by

$$S_\theta(T) = P_{\theta 1} + P_{\theta 2} \cdot \frac{(6K + 7q_0) - (6Kp_1 + 7q_1)T}{(6K + q_0) - (6Kp_1 + q_1)T}, \tag{30}$$

where the parameters $P_{\theta 1}$ and $P_{\theta 2}$ which are related to the coordinate system are

$$\begin{cases} P_{\theta 1} = \dfrac{1}{r_1^5} - \dfrac{10H(H+Z)}{r_2^7} \\ P_{\theta 2} = \dfrac{1}{r_2^5} \end{cases}. \tag{31}$$

When $t = 0$, the bulk strain $\theta(r, 0)$ is

$$\theta(r, 0) = \frac{3V\tau xy}{4\pi}\{M_{\theta 0} + M_{\theta 1} + M_{\theta 2} + M_{\theta 3} + M_{\theta 4}\}. \tag{32}$$

Characteristics of Bulk Strain on the Ground Surface

In the earthquake precursors, there are many earthquake observation parameters concerning bulk strain, such as geo-electricity, water level below the ground surface, radon content and so on (MEI *et al.*, 1993). We study mainly the bulk strain produced by the spherical inclusion on the ground surface.

The Property of Bulk Strain Function and the Estimation of Medium Parameters

In the bulk strain expression (32), the difference of parameters $q_0, q_1, p_1, q'_0, q'_1, p'_1$ for the medium rheologic model (SONG *et al.*, 2003) will produce the difference of the parameters $T_i (i = 1, 2, 3, 4)$, which determine the time history of bulk strain from initial state to asymptotic value. The parameters $M_{\theta 0}, M_{\theta 1}, M_{\theta 2}, M_{\theta 3}, M_{\theta 4}$ are related to the coordinate and the medium rheologic constants. These parameters define the space field of bulk strain.

When $t < 0$, for the boundary condition, $\tau H(t)$ is zero where $H(t)$ is the unit step function and τ is the shear stress. When $t > 0$, the $\tau H(t)$ is τ which is given as 10^8 Pa (DOBROVOLSKII,1991) in this paper. In the expression of the bulk strain $\theta(x,y,z,t)$, if z is zero, and $r_1 = r_2 = r = (x^2 + y^2 + H^2)^{1/2}$, where H is the depth of the earthquake focus, $\theta(x,y,0,t)$ is the expression of the bulk strain concerning the t on the random point of the ground surface. We may calculate the perturbation of bulk strain $\Delta\theta$ as $\Delta\theta(x, y, 0, t) = \theta(x, y, 0, t) - \theta(x, y, 0, 0)$. Then we may analyze the characteristics of the time history and the space distribution of bulk strain.

As to how to choose the elastic and rheologic constants of the medium (DOBROVOLSKII, 1991) given $\mu = 2 \times 10^4$ MPa for the shear modulus of the Earth's crust, the Poisson's ratio $\nu = 0.25$. ZHANG (1987) given $E = 10^5$ MPa for the Young's modulus of the Earth's crust, viscosity constant $\eta = 10^{22}$ Pa s. WANG (1979) and MEI *et al.* (1992) took viscosity constant $10^{17} \sim 10^{23}$ Pa s and the Young's modulus of the Earth's crust $10^4 \sim 10^5$ MPa. KASAHARA (1979) took the viscosity constant $\eta = 2 \times 10^{21}$ Pa s. MATSU'URA *et al.* (1999) and FUKAHATA *et al.* (2004) given the viscosity of 5×10^{18} Pa s.

Having referenced the above results, in this paper we take the Poisson's ratio of the rock medium of 0.25 for both in and outside the inclusion, and the viscosity constants and the Young's modulus of the medium outside the inclusion of 10^{21} Pa s and 10^5 Mpa, respectively. The parameters of the hard inclusion and the soft one arc shown in the Table 1.

Based on the relation of parameters of the rheologic medium model (SONG *et al.*, 2003) as follows, we obtain the parameters $q_0, q_1, p_1, q'_0, q'_1, p'_1$ and the bulk strain by the expression (32).

$$\begin{cases} p_1 = \dfrac{\eta}{E_1 + E_2} \\ q_0 = \dfrac{E_1 \cdot E_2}{E_1 + E_2} \\ q_1 = \dfrac{E_1 \eta}{E_1 + E_2} \end{cases} \qquad \begin{cases} p'_1 = \dfrac{\eta'}{E'_1 + E'_2} \\ q'_0 = \dfrac{E'_1 \cdot E'_2}{E'_1 + E'_2} \\ q'_1 = \dfrac{E'_1 \eta'}{E'_1 + E'_2} \end{cases}$$

where the situation of considerably smaller η corresponds mainly to the elastic medium, whereas the very great η corresponds mainly to the viscoelastic and the case of moderate η represents the elastic-viscoelastic rheologic medium. Therefore, the parameter selection for the rheologic medium model is very important.

The η value selection may influence the time constants $T_i (i = 1, 2, 3, 4)$ of viscosity. For the linear model, with a different physical mechanism or a different η or T_i, the scale of time may be varied, but its conclusion may still be accepted in principle.

Table 1

Parameters of rheologic medium model in and outside the inclusion

Parameter	Wall rock	Hard inclusion	Soft inclusion
E_1 or E'_1(MPa)	$E_1 = 1 \times 10^5$	$E'_1 = 1.5 \times 10^5$	$E'_1 = 0.5 \times 10^5$
E_2 or E_2(MPa)	$E_2 = 1 \times 10^5$	$E'_2 = 1.5 \times 10^5$	$E'_2 = 0.5 \times 10^5$
η or η'(Pa s)	$\eta = 1 \times 10^{21}$	$\eta' = 2 \times 10^{21}$	$\eta' = 0.5 \times 10^{21}$

Assuming both the radius and depth of inclusion of 25 km, we may analyze the space-time evolution characters for bulk strain on the ground ($z = 0$) produced by the hard inclusion or the soft one.

Quadrant of Bulk Strain Produced by a Hard Inclusion and a Soft One

The characteristics of the bulk strain spatial field of the hard inclusion DOBROVOLSKII (1991) and the soft one BRADY (1974) are similar in that both have quadrant distribution and the distribution inside the region with the size two times the extension of seismic source is contrary to the one outside this region. The dissimilarity is that the bulk strain high-value regions produced by the hard inclusion correspond rightly to the low-value regions of the soft inclusion.

Pattern features of the bulk strain versus time. Pattern features of the bulk strain versus time produced by a hard inclusion are as follows: (1) The bulk strain pattern varies with the distance to the epicenter. The bulk strain in the source region goes through a process of "increasing → maximum → decreasing → tend towards a stable value" and seems like an upward protruding (Fig. 2). For the points around the region with its distance to seismic foci about two times the extension of seismic source, their bulk strains go down initially, then ascend and show the characteristics of "N".

The bulk strains at the points outside the region with the size two times the extension of seismic source show the characteristics of downward protrution. The

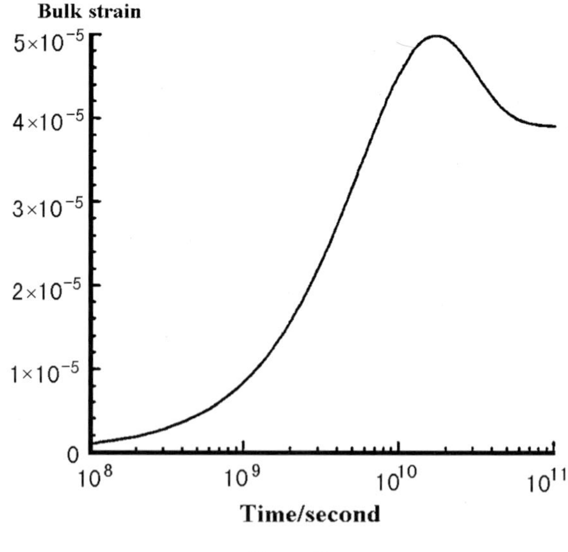

Figure 2
Bulk strain curve versus time in the near-source region of the hard inclusion.

patterns at points distant from the epicenter are basically similar to the seismic source, yet their variation values are comparatively smaller. (2) The variation value of bulk strain decreases as the distance increases.

It is meaningful that the pattern of bulk strain produced in the soft inclusion is different from that in the hard inclusion. The bulk strain values produced in the soft inclusion all increase gradually with the different epicenter distances, then tend towards a stable value, however there is no decreasing process after the maximum value.

Three-stage Characteristics of the Bulk Strain Produced in a Hard Inclusion

The bulk strain in the hard inclusion possesses three-stage characteristics of "increasing → maximum → decreasing to a stable value." That is, during the stage α, the bulk strain value increases gradually with the passage of time and the bulk strains in the near-source region and the far-source region all show outward extension. During the stage β, the bulk strain value and its range all vary slightly with time and reach an extreme state. Other relevant physical parameters such as strain, stress, etc. should be extreme too, then earthquake precursors during the stage β should be most significant and occur at a similar time in the near-source and far-source regions. During the stage γ, the bulk strain value weakens as time continues, and the corresponding spatial characteristics are that the bulk strain of the far-source region shows contraction to the epicenter and the bulk strain of the near-source area migrates outwards.

Characteristics of Bulk Strain Produced in a Soft Inclusion

The study shows that the bulk strains in the near-source region and the far-source area produced in the soft inclusion all diffuse outwards and their diffusing directions are conjugate. If a soft inclusion is regarded as an earthquake preparation model, the spatial-temporal evolution of precursors should diffuse outwards in the whole seismic preparation process.

Application of Theoretical Result to Earthquake Precursors

There exist three stages (namely α, β and γ) in the process of bulk strain versus time in the near-source region caused by the hard inclusion (Fig. 2). Having studied abundant observation data of crust deformation, MESHERIKOV (1968) and FUJII (1974) found that the strain curves have characteristics of three stages in the earthquake preparation process, namely, α, β and γ (Fig. 3). Though our theoretical curves cannot give the relative proportion of the three stages, the similarity of basic patterns of the theoretical and empirical curves supports the scientific reliability of

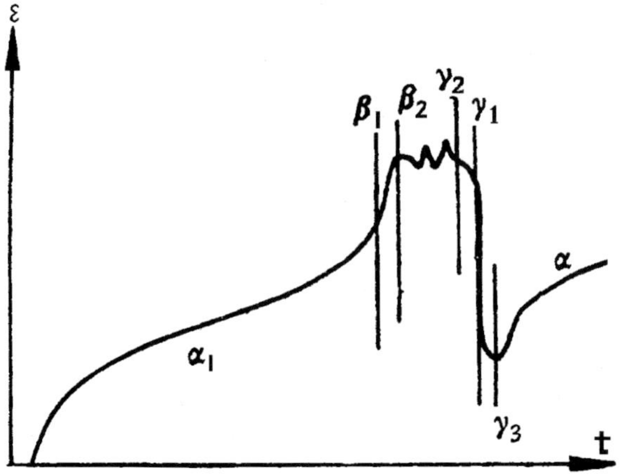

Figure 3
The three stages (α, β and γ) of strain (MESHERIKOV, 1968; FUJII, 1974).

the theoretical curves. This also supports the rationality of the strong-body earthquake preparation model.

The bulk strain curve pattern of the hard inclusion in the near-source region is "small → big → maximum → decreasing to stability" and shows an upward protruding variation. At points around the region with its distance to foci about two times the extension of seismic source, the pattern of bulk strain curves is "descending → ascending" and shows the characteristic of "N" type. The bulk strain types at points near the region of three times the extension of seismic source show downward protruding variation. The bulk strain at points distant from the inclusion has little variation. These are similar to the phenomena in the earthquake observation in that the bulk strain abnormal types vary with the increase of the distance (MEI *et al.*, 1993). In other words, this phenomenon does not occur occasionally, but possesses a physical basis.

LI (1981) found that the water-radon anomalies exhibit the development pattern outward half a year before the Tangshan earthquake. ZHENG (1979) had studied filtered water-level data and found that the water-level descending anomalies develop outwards in the directions of north-east and north-west a half year before the Tangshan earthquake. There also exist the characteristics about a half year before the Longling earthquake and the Songpan shock (MEI *et al.*, 1993). Song's studies (SONG *et al.*, 2000) reveal that the three stages (α, β and γ) are very evident. Particularly, there all exist the subphase γ_1 of precursors contracting from the periphery to the epicenter region and the subphase γ_2 of precursors diffusing from the epicenter area to the periphery during the γ-stage (corresponding to earthquake short-term and imminent-term stage). It can be seen that the existence of α-, β- and γ-stage and the spatial-temporal migration of precursors may be controlled by the earthquake preparation body. It may be a normal phenomenon and should be an important basis to predict earthquakes.

Quadrant distribution is one of the common characteristics of the spatial distribution of bulk strain on the ground with differences in some details associated with earthquake preparation. The spatial range and the values of the anomaly in bulk strain all increase with time and decrease after reaching the maximum value in the model of hard inclusion. For the soft model there is only trend variation, however. Experimental results (XU *et al.*, 1997) show compatible quadrant distribution. In earthquake precursors there exist some consistent quadrant distribution of anomalies in the deformation and georesistivity before the cases of Datong earthquake, Zhangbei earthquake and Baotou earthquake, etc. (SONG *et al.*, 2001). This feature would be useful for predicting the location of a strong earthquake.

By earthquake precursors, we refer to the phenomena which may occur before larger earthquakes. MEI *et al.* (1993 and 1994) elaborated the physical processes of earthquake generation. This paper will not describe it in details.

Conclusions

To date we have described the derivation of the bulk strain and ground deformation and the preliminary application based on the theory of a rheologic inclusion model in the seismogenic process. The results are as follows:

Based on the Correspondence Principle, we obtained the analytic expression of the viscoelastic displacement $U(r,t)$, $V(r,t)$ and $W(r,t)$, the viscoelastic normal strains $\varepsilon_{xx}(r,t)$, $\varepsilon_{yy}(r,t)$ and $\varepsilon_{zz}(r,t)$ and the bulk strain $\theta(r,t)$ produced a three-dimension inclusion in the semi-infinite rheologic medium defined by a standard linear rheologic model.

Applying the theoretical expression of the three-dimension rheologic inclusion model, we throughly computed the spatial-temporal evolution process of bulk strain on the ground produced by a spherical rheologic inclusion DOBROVOLSKII (1991) in the semi-infinite rheologic medium. The results show that spatial-temporal evolution of bulk strain produced by a hard inclusion has three stages of different characteristics, which are similar to most of the geodetic deformation curves, but not for the case of soft inclusion. The α-stage characteristic is a long stage in which the precursors in both the near-field and the far-field develop from the focal region to the periphery. The β-stage shows a very rapid propagation of the precursors, in which they appear everywhere. During the γ-stage, the precursors in the far-field converge from the periphery, and the precursors in the near-field develop outwards.

The theoretical results have been applied to explain tentatively the stage variation of the spatial-temporal evolution, the pattern feature and quadrant-distribution characteristics of earthquake precursors. It is conducive to better understand the process of seismogeny and occurrence, and to construct the physical model of the development of earthquakes.

Questions remain to be studied, for example, the viscoelastic expression of geo-electricity, geomagnetism, wave-velocity and so on.

Acknowledgements

Gratefull acknowledgment the support from the Chinese Joint Seismological Foundation (101105) and Committee of Science and Technology of Shanghai Municipality, NSFC under grant: 10232050, MOST (Ministry of Science and Technology, China, Grant No. 2004 CB418406) and Computer Network Information Center, CAS under grant INF105-SCE-2-02.

References

Brady, B. T. (1974), *Theory of earthquake,* Pure Appl. Geophys. *112*(4), 701–725.

Dobrovolskii, I. P., *Theory on the Preparation of Tectonic Earthquakes* (in Russian), Ivz. An SSSR (Moscow, 1991).

Eshelbym, J. D. (1957), *The determination of the elastic field of an ellipsoidal inclusion and related problem,* Proceed. Roy. Soc. *A241*, 376–396.

Fujii, Y. (1974), *Relationship between duration time of anomalous vertical deformation of the crust and magnitude before the occurrence of an earthquake* (in Japanese with English abstract), Earthquake *27*(3), 197–213.

Fukahata, Y., Nishitani, A., and Matsu'ura, M. (2004), *Geodetic data inversion using ABIC to estimate slip history during one earthquake cycle with viscoelastic slip-response functions,* Geophys. J. Int. *156*, 140–153.

Li, X. (1981), *Phenomena of diffusion and contraction of water-radon precursors* (in Chinese with English abstract), Earthquake *1*(5), 340–357.

Kasahara, K. (1979), *Migration of crust deformation*, Tectonophysics *52*, 329–341.

Matsu'ura, M., Nishitani, A., and Fukahata, Y. (1999), *Slip History during one earthquake cycle at the Nankai Subduction zone, inferred from the inversion analysis of leveling data with a viscoelastic slip response function,* 1st ACES Workshop Proceedings (http://quakes.earth.uq.edu.au/ACES/).

Mei, S. R. and Zhu, Y. Q., *Earthquake-generating Model of the Tangshan Earthquake* (in Chinese) (Seismological Press, Beijing, 1992).

Mei, S., Feng, D., Zhang, G., *et al.*, *Introduction to Earthquake Prediction in China* (in Chinese) (Beijing: Seismological Press, 1993).

Mei, S. (1994), *A study on the physical model of earthquake precursor field and the mechanism of precursors' time-space distribution ——Origin and evidences of the strong body earthquake-generating model*, Acta Seismologica. Sinica. *7*(3), 273–282.

Mesherikov, J. A. (1968), *Recent crust movements in seismic region: Geodetic and geomorphic data,* Tectonophysics *6*, 29–39.

Song, Z. *On the Spatial Temporal Evolution of the Load/Unload Response Ratio* (*LURR*) and Other Earthquake Precursors and their Implications (in Chinese with English abstract) (Ph. D. Thesis), Institute of Geophysics, China Seismological Bureau, (Beijing, 1996).

Song, Z., Yin, X., and Mei, S. (2000), *Theoretical analysis of the spatio-temporal evolution of the bulk strain field based on a rheologic inclusion model*, Acta Seismologica. Sinica. *13*(5), 525–535.

SONG, Z., XU, P., ZHANG, H., *et al.* (2001), *Quadrant characteristics of earthquake precursors and its theoretical analysis*, Acta Seismologica. Sinica. *14*(1), 58–65.

SONG, Z., YIN, X., and MEI, S. (2003), *Three-dimensional rheologic model of earthquake preparation (1)* ——displacement field, Acta Seismologica. Sinica. 16(6), 588–597.

SONG, Z., YIN, X., and MEI, S. (2004), *Three-dimensional rheologic model of earthquake preparation (2)* ——strain field and its applications, Acta Seismologica. Sinica. *17*(2), 131–143.

WANG, R., *Base of solid mechanics* (in Chinese) (Geological Press, Beijing, 1979).

XU, Z., WANG, B., ZHAO, J., *et al.* (1997), *An experimental study of fractural features of concrete samples containing hard inclusion*, Acta Seismologica. Sinica. *10*(1), 97–105.

YIN, X. and ZHEN, T. (1982), *Rheologic model of earthquake preparation process*, Scientia Sinica (series B) *10*, 922–930.

YIN, X., *Solid Mechanics* (In Chinese) (Seismological Press, Beijing, 1985).

ZHANG, G. (1987), *Application of rock rheologic model to earthquake-generating process and precursors* (*in Chinese with English abstract*), Acta Seismological. Sinica. *9*(4), 384–391.

ZHENG, Z. (1979), *On generation process of the Tangshan earthquake basing on water-level filtered data* (in Chinese with English abstract), Acta Geophysica. Sinica. *22*(3), 267–280.

(Received December 29, 2004, revised September 19, 2005, accepted December 26, 2005)
Published Online First: August 31, 2006

Pure appl. geophys. 163 (2006) 2011–2030
0033–4553/06/092011–20
DOI 10.1007/s00024-006-0110-6

Pure and Applied Geophysics

Deformations in Transform Faults with Rotating Crustal Blocks

E. Pasternak,[1] A. V. Dyskin,[2] and Y. Estrin[2,3]

Abstract—A continuum model of deformation of a transform fault is considered. The fault interior is modelled as a part of a plate fragmented into a large number of angular blocks. The blocks are not joined together by any binder, but are rather held together due to external compression such that they can, in principle, rotate independently. Significant numbers of blocks involved in the deformation process permit a continuum description of the deformation, which in view of the possibility of independent rotations, necessitates the application of the Cosserat theory. The crucial point in the described model is the interconnection between the rotations and the normal stresses associated with the angular shape of the blocks: the rotating blocks 'elbow' one another. Furthermore, elbowing produces compressive stresses independent of the direction of rotations. Consequently, the constitutive equations become non-linear involving *absolute values* of the components of the curvature tensor. The paper analyses a simple shear of a fault under constraining compression acting in the direction normal to the fault. An infinite layer subjected to opposite displacements and zero rotations at the edges is considered. It is shown that block rotations can lead to complex deformation patterns. There exists a displacement threshold proportional to the pressure: for imposed displacements below the threshold, the deformation pattern coincides with the conventional one as predicted for a classical elastic isotropic layer with uniform displacement gradient in the absence of block rotations. When the imposed displacement exceeds the threshold value, boundary zones of non-uniform rotations, displacement gradients and dilatation emerge. It is interesting to note that these features, which could be mistaken for indicators of non-elastic or localisation processes, occur in a situation where only elastic processes are acting at the scale of blocks and no friction or other energy dissipation processes take place.

Key words: Block rotation, homogenisation, Cosserat continuum, non-uniform deformations.

Introduction

Rotational degrees of freedom play an important part in modelling various phenomena, such as constitutive behaviour and strain localisation in granular materials. In the latter case, localisation is associated with a significant increase in rotations of the constituents of an assembly (e.g., Kishino, 1999; Tamura, 1999).

[1] School of Mathematics and Statistics, The University of Western Australia, 35 Stirling Highway, Crawley, WA 6009, Australia. E-mail: elena@maths.uwa.edu.au
[2] School of Civil and Resource Engineering, The University of Western Australia, 35 Stirling Highway, Crawley, WA 6009, Australia.
[3] Institut für Werkstoffkunde und Werkstofftechnik, Technische Universität Clausthal, Agricolastr. 6, D-38678, Clausthal-Zellerfeld, Germany.

Pronounced localisation of rotations has been observed in an interlocked plate-like assembly of cubes compressed in a lateral direction and loaded at the centre by concentrated load (ESTRIN *et al.*, 2003). Rotational degrees of freedom play an important role not only at the scale of engineering structures, but also at the scale of the solid Earth. Evidence is mounting that rotation of crustal blocks is one of the processes by which the edges of continents are modified (e.g., BURKART and SELF, 1985; JONES *et al.*, 1990; LYNENDYK, 1991; MOLNAR *et al.*, 1994; SCHELLE and GRÜNTHAL, 1996; WESNOUSKY, 2005). JONES *et al.* (1990) summarised the results of paleomagnetic measurements indicating the presence of large rotations of crustal blocks in the western part of the San Andreas Fault. This indicates that block rotations can constitute an important part of the deformation mechanism of transform faults. However, the fault mechanics based on conventional continuum theories leaves the independent block rotations out of the picture.

In many cases, the zones consisting of the crustal blocks can be modelled, at appropriately large scale, as interiors of large layers. Following JONES *et al.* (1990) we will call this layer a fault regardless of whether the relative plate motion producing this zone gives rise to a single fault or a series of faults parallel to the plate boundaries. Also, it is assumed that crust between two plates is brittle enough to crack and that it is the motion of the plates that produces block rotations. In this case, the fault interior can be visualised as a fragmented material.

Fragmented solids in general have two remarkable features. Firstly, the fragments (blocks) are, in a way, not connected (in the sense that there are no connectors nor is there a binder between them) or weakly connected such that that the integrity of the body can only be preserved by compression applied at the periphery, at least in one direction. This results in enhanced resistance to fracture propagation (increased fracture toughness, ESTRIN *et al.*, 2002). Secondly, the shape of the fragments becomes important, since the rotation of fragments and in some cases their shearing produce increased local contact pressure ('elbowing') which macroscopically is tantamount to coupling between rotation gradients and normal stresses on one hand and normal strains and moment stresses on the other hand. This makes the fragmented bodies essentially dilatant materials. Additionally, these effects are strongly pressure-dependent and non-linear, making modelling of the mechanical behaviour of fragmented bodies quite a challenging task.

We consider deformation of a material fragmented into a large number of rectangular blocks. The blocks are not joined together by any binder, but are rather held together due to external compression such that they can, in principle, rotate independently. Significant numbers of blocks involved in the process of deformation permit a continuum description of the deformation, which in view of the possibility of independent rotations, necessitates the application of the Cosserat theory. The crucial point in the model is the interconnection between the rotations and the normal stresses as illustrated by Figure 1. Consequently, the constitutive equations include terms that relate normal stresses with curvatures (rotation gradients), leading

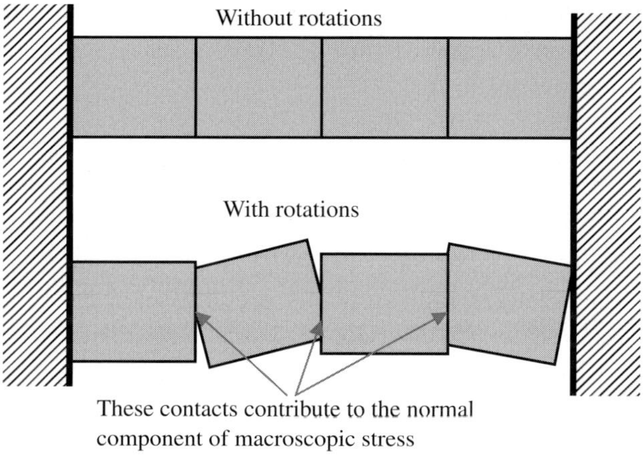

Figure 1
Schematics of the relation between the block rotations and the normal macroscopic stress.

to the so-called anisotropic Cosserat theory. Furthermore, because the rotating blocks 'elbow' one another, they produce compressive stresses independent of the direction of rotations. Consequently, the constitutive equations become non-linear involving *absolute values* of the components of the curvature tensor. It is also important to note that as the block size decreases the theory converges to the classical elasticity, loosing the 'elbowing' effect in the process.

The main goal of the paper is to demonstrate the effect of rotations on the overall deformation of a part of the Earth's crust that can be represented by the aforementioned model. As an example, the paper analyses a 2-D case of simple shear of a fault under constraining compression acting in the direction normal to the fault. The fault's interior consists of a part of a plate fragmented into numerous blocks. The fault is modelled as an infinite layer of the material of the kind described. The layer is subjected to opposite displacements at the edges, with the rotations being zero at the edges.

Continuum Mechanics of Fragmented Solids. Simple Shear of Fault in the Presence of Compression

Deformation of fragmented solids and, for that matter, the deformation of transform faults is controlled by the effect of elbowing of the rotating blocks. We will first introduce this effect using the simplest possible two-dimensional model of an infinite layer with decoupled cross sections under simple shear, which corresponds to the case of weak interaction between blocks in the adjacent cross sections. This models the deformations of transform faults of the type produced by spreading (e.g., KASAHARA, 1981; EARLE, 2004). In that case, the tensile stresses acting along the fault lead to virtually independent behaviour of the fault cross sections. We will then formulate the constitutive equations for the general case and consider an infinite layer with coupled cross sections.

Deformation of Transform Fault with Weakly Interacting Cross Sections

To demonstrate the effect the block rotations can have on the deformation of a transform fault we consider the case depicted in Figure 2a. In this case, the tensile stresses produced by the plate movement ease the compression holding the blocks of the fault cross sections together, which allows one to consider, in the first

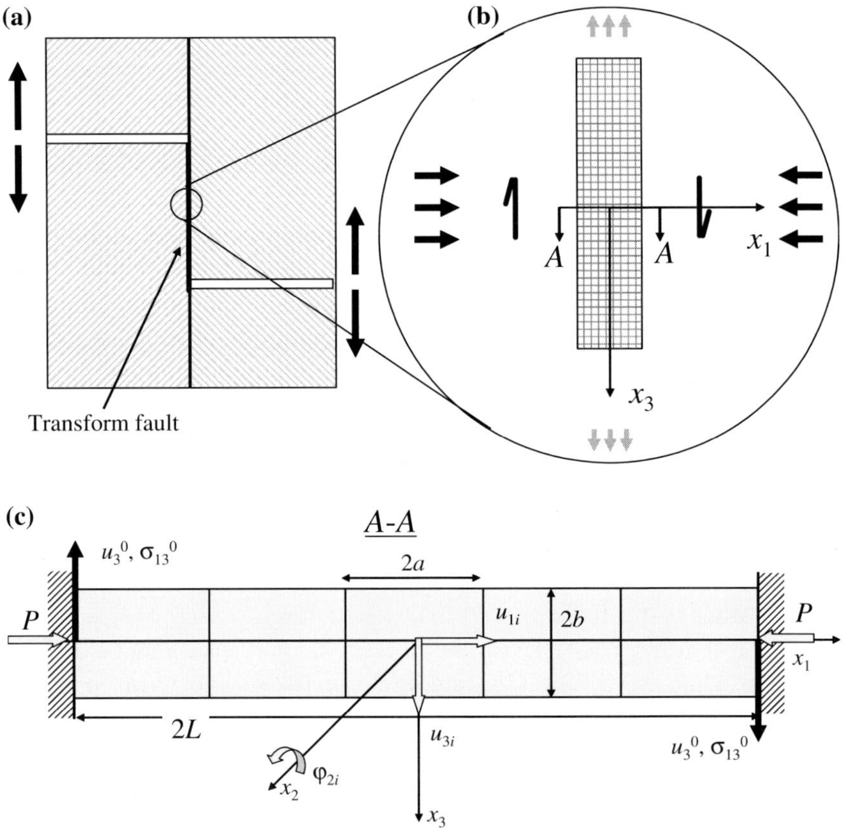

Figure 2
A model of transform fault with rotating angular blocks: (a) transform fault; (b) a section of the transform fault. The stretching produces tension that allows consideration of each section independently; (c) statics and kinematics of blocks in section AA, $2a$ and $2b$ denoting block length and height respectively, $2L$ is length of the section, u_3^0 is the applied shear displacement at the boundary, σ_{13}^0 is the corresponding shear stress at the boundary, P is an applied compression at the boundary, u_{1i}, u_{3i}, and φ_{2i} are displacements and rotation of the i-th block in the coordinate directions x_1, x_3 and x_2, respectively.

approximation, a single cross section (cross section *A-A* in Figs. 2b and 2c). Figure 2c shows displacements and rotation associated with each block. Hereafter the relative displacements of adjacent blocks are assumed small as compared to the block dimensions. Also, the block rotations are assumed small.

We suppose that the fragments (blocks) are identical, being of rectangular shape with the length $2a$, height $2b$ and width $2c$, and form a homogeneous elastic layer of length $2L$, occupying a strip $-L < x_1 < L$ in a Cartesian coordinate frame $(x_1x_2x_3)$ with specified displacement, rotation and a uniform compression at the boundaries given, respectively, by

$$u_3(\pm L, x_2, x_3) = \pm u_3^0, \quad \varphi_2(\pm L, x_2, x_3) = 0, \quad \sigma_{11} = -P/(4bc). \tag{1}$$

The block kinematics is expressed in terms of relative displacements Δu_{1i}, Δu_{3i} and relative rotation $\Delta\varphi_{2i}$ in the coordinate directions x_1, x_3 and x_2, respectively for the i-th block:

$$\Delta u_{1i} = u_{1i+1} - u_{1i} - \alpha_\varphi b \left|\varphi_{2i+1} - \varphi_{2i}\right|, \quad 0 < \alpha_\varphi \leq 1 \tag{2}$$

$$\Delta u_{3i} = u_{3i+1} - u_{3i} + u\left(\psi_{2i+1} + \varphi_{2i}\right), \tag{3}$$

$$\Delta\varphi_{2i} = \varphi_{2i+1} - \varphi_{2i}, \tag{4}$$

where α_φ is an 'elbowing coefficient' accounting for the fact that in reality blocks in the fragmented layer are not necessarily of rectangular shape.[1] Generally, when the blocks are not rectangular, their sides can be inclined to the layer boundary and the term $\alpha_\varphi b$ plays the role of an effective length instead of the side b of the rectangular block.

We take into account that the assembly of blocks can only withstand compression with no resistance to tension. We assume that the resistance is concentrated only at block contacts, Δu_{1i} being the blocks' relative displacement. It consists of the part associated with relative displacement of the centres of the blocks in horizontal direction, $u_{1i+1} - u_{1i}$, and the part associated with pure relative rotation $\varphi_{2i+1} - \varphi_{2i}$ (without any relative displacement of the centres of the blocks).

The effect of relative rotations on the relative block displacement is independent of their sense. That is why the absolute value of the second term appears in (2). Furthermore, the second term always works in such a way as to put the contacts under compression. That is why it is subtracted from the first term. The first term in (2) can be both positive and negative, depending upon the relative position of the block. The presence of the absolute value in (2) produces strong non-linearity in the constitutive behaviour.

[1] For a rectangular block $\alpha_\varphi = 1$, while for perfectly round fragments (blocks in the shape of circular cylinder) $\alpha_\varphi = 0$. The latter case is unrealistic in this context, since fragments caused by brittle fracturing are never circular. For $\alpha_\varphi = 0$ this theory yields the case of fault gouge considered in PASTERNAK *et al.* (2004).

The relative displacement Δu_{3i} is comprised of the relative displacement of the centres of the blocks in vertical direction $u_{3i+1} - u_{3i}$ and rotations of the blocks in the same direction, $\varphi_{2i+1} + \varphi_{2i}$, (again without any relative displacement of the centres of the blocks). The meaning of the relative rotation $\Delta\varphi_{2i}$ is straightforward.

The block kinematics relationship (2) differs from the one proposed by MÜHLHAUS (1993, 1997) and SULEM and MÜHLHAUS (1997) for blocky and masonry structures. The difference mainly results from the difference in assumed contact interactions. While MÜHLHAUS (1993), MÜHLHAUS *et al.* (1997) and SULEM and MÜHLHAUS (1997) considered a cement-between-blocks type of interaction which can withstand both compression and extension of the cement-interlayer, the contact interaction we consider responds to compression only, ruling out extension because of the absence of an inter-layer.

The discrete energy density associated with i-th block can be written as follows:

$$W_i = 1/2\{\Theta(-\Delta u_{1i})k_n\Delta u_{1i}^2 + k_s\Delta u_{3i}^2 + k_\varphi\Delta\varphi_{2i}^2\}, \tag{5}$$

where k_n, k_s, k_φ are generalised normal, shear and rotational stiffnesses of the contacts and Θ is the Heaviside step function.

The next step towards devising a continuum description is to homogenise the discrete energy (5). Following MÜHLHAUS and OKA (1996) we initially replace the discrete coordinates with the continuous ones, $ia \to x_1$, and then apply the homogenisation procedure by differential expansions, namely replacing the finite differences with corresponding Taylor expansions, keeping the leading terms of the series up to the second order (see for example PASTERNAK and MÜHLHAUS, 2005 for details). The result reads:

$$W(x_1) = a(4bc)^{-1}\left\{\Theta[-(\gamma_{11} - \alpha_\varphi b|\kappa_{12}|)2a]k_n(\gamma_{11} - \alpha_\varphi b|\kappa_{12}|)^2 + k_s\gamma_{13}{}^2 + k_\varphi\kappa_{12}{}^2\right\}, \tag{6}$$

where the Cosserat continuum deformation measures – strains γ_{ji} and curvatures κ_{ji} (NOWACKI, 1974) are extracted

$$\gamma_{11} = \frac{\partial u_1}{\partial x_1}, \quad \gamma_{13} = \frac{\partial u_3}{\partial x_1} + \varphi_2, \quad \kappa_{12} = \frac{\partial \varphi_2}{\partial x_1}. \tag{7}$$

Variation of the homogenised energy density (6) with respect to the deformation measures, i.e., strains γ_{11}, γ_{13} and curvature κ_{12}, yields the following constitutive equations:

$$\sigma_{11} = \frac{a}{2bc}k_n\Theta[-(\gamma_{11} - \alpha_\varphi b|\kappa_{12}|)2a](\gamma_{11} - \alpha_\varphi b|\kappa_{12}|), \tag{8}$$

$$\sigma_{13} = \frac{a}{2bc}k_s\gamma_{13}, \tag{9}$$

$$\mu_{12} = \frac{a}{2bc}\{-k_n\Theta[-(\gamma_{11} - \alpha_\varphi b|\kappa_{12}|)2a]\alpha_\varphi b\ \mathrm{sgn}\kappa_{12}(\gamma_{11} - \alpha_\varphi b|\kappa_{12}|) + k_\varphi\kappa_{12}\}, \tag{10}$$

or

$$\mu_{12} = -\sigma_{11}\alpha_\varphi b \operatorname{sgn} \kappa_{12} + \frac{a}{2bc} k_\varphi \kappa_{12}, \tag{11}$$

where σ_{ji} is a non-symmetric stress tensor and μ_{ji} is the moment stress tensor. Equations (8)–(10) are constitutive relations of a new type; they are strongly non-linear and contain 'non-isotropic terms', *viz.* curvature κ_{12} contributes to the normal stress σ_{11}, and normal strain γ_{11} contributes to the component μ_{12} of the moment stress tensor. These constitutive relations (8)–(10) differ from the ones developed by MÜHLHAUS (1993), MÜHLHAUS *et al.* (1997) and SULEM and MÜHLHAUS (1997) for blocky and masonry structures. In their models, the contribution of rotation is linear; furthermore, no effects of curvature twists on the normal stresses occur. By contrast, in the constitutive equations (8)–(10), the rotations appear in the constitutive equations through the *absolute value* of the curvature contributing to the normal stress distribution.

The analysis of (11) shows that the moment created by the applied load is balanced by a moment corresponding to normal stresses created by the load and an additional bending moment proportional to curvature—the gradient of the Cosserat rotation φ_2. Also, the absolute value of the moment stress can be estimated as follows:

$$|\mu_{12}| \geq -\sigma_{11}\alpha_\varphi b > 0, \quad \kappa_{12} \neq 0, \tag{12}$$

$$|\mu_{12}| < -\sigma_{11}\alpha_\varphi b, \quad \kappa_{12} \equiv 0. \tag{13}$$

Formulae (12) and (13) show that the violation of condition (12) means that the applied load gives rise to a moment stress lower than the resistance due to the action of the compressive stress (the assembly supports only compression, otherwise it cannot maintain its integrity). The bending moment generated is not sufficient to produce rotations of the blocks in the assembly. This suggests the existence of two zones within the solution: A classic one in which no Cosserat rotation is present and the zone of Cosserat solution with non-zero Cosserat rotations, its size being determined by the boundary conditions and the geometry of the problem.

The evolution equations can be obtained in the usual manner by substituting deformation measures (7) into the constitutive equations (8)–(10) and the result into the governing equations

$$\sigma_{11,1} = 0, \quad \sigma_{13,1} = 0, \quad \mu_{12,1} - \sigma_{13} = 0. \tag{14}$$

We note that in accordance with the Cosserat continuum theory the stress tensor σ_{ji} is non-symmetric; the non-symmetry is compensated for by the presence of the moment stress tensor μ_{ji} whose gradient balances out the antisymmetric part of the stress tensor. Also, mathematically, equations (7)–(11) and (14) represent a 2-D orthotropic Cosserat continuum.

The solution of the governing equations (14) together with (7)–(10) reads:

1. For $|x_1| < x_1^*$ where no rotation gradient exists

$$\gamma_{11} = -P/(2ak_n), \quad \gamma_{13} = 2bc(ak_s)^{-1}\sigma_{13}^0, \tag{15}$$

$$\varphi_2 = \varphi_2^*, \quad \kappa_{12} = 0, \quad \partial u_3/\partial x_1 = 2bc(ak_s)^{-1}\sigma_{13}^0. \tag{16}$$

2. For $|x_1| > x_1^*$ where a non-zero rotation gradient exists

$$\gamma_{11} = \alpha_\varphi b|\kappa_{12}| - P/(2ak_n), \quad \gamma_{13} = 2bc(ak_s)^{-1}\sigma_{13}^0, \tag{17}$$

$$\partial u_3/\partial x_1 = 2bc(ak_s)^{-1}\sigma_{13}^0 - bc(ak_\varphi)^{-1}\sigma_{13}^0(x_1^2 - L^2) + P(2ak_\varphi)^{-1}\alpha_\varphi b(|x_1| - L)\ \mathrm{sgn}\ \sigma_{13}^0, \tag{18}$$

$$\varphi_2 = bc(ak_\varphi)^{-1}\sigma_{13}^0(x_1^2 - L^2) - P(2ak_\varphi)^{-1}\alpha_\varphi b(|x_1| - L)\mathrm{sgn}\ \sigma_{13}^0. \tag{19}$$

These two zones are separated by a boundary x_1^* which, according to condition (12), is governed by

$$|\sigma_{13}^0|x_1^* = P(4bc)^{-1}\alpha_\varphi b. \tag{20}$$

Figure 3 shows the schematics of the classical zone (no Cosserat rotation) and the Cosserat zone solutions. Figures 4 and 5 show distributions of the normalised rotations and the displacement gradient $u_{3,1}$ (both plotted for the positive x_1 values only, since the solution is symmetric) in the Cosserat ($x_1 > x_1^*$) and the classical ($x_1 < x_1^*$) zones for various values of a dimensionless parameter $s = 4cL\sigma_{13}^0/\alpha_\varphi P$. This parameter expresses the ratio of the moment stress, $L\sigma_{13}{}^0$ created by the applied shear stress to the critical moment stress $\alpha_\varphi P/4c$ associated with resistance to block rotations generated by the applied compression. For the contact stiffness parameters the following relatios were used: $k_s = L^{-2}k_\varphi = k_n$. Both plots show the uniformity of the fields in the classical zone and a non-uniformity in the intervals where the Cosserat solution applies.

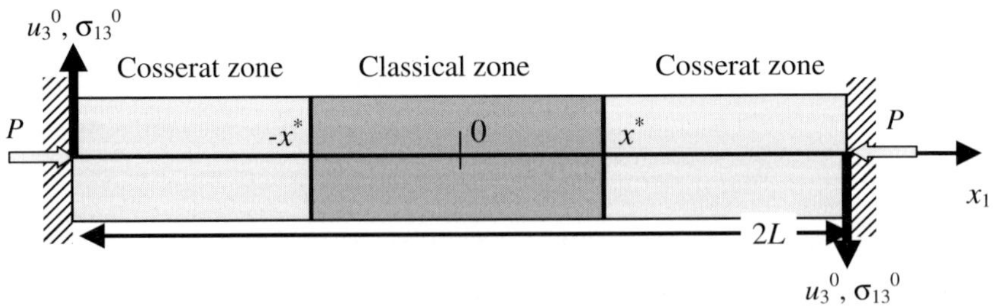

Figure 3
The schematics of the classical (no Cosserat rotation) and the Cosserat zones solution for the cross section.

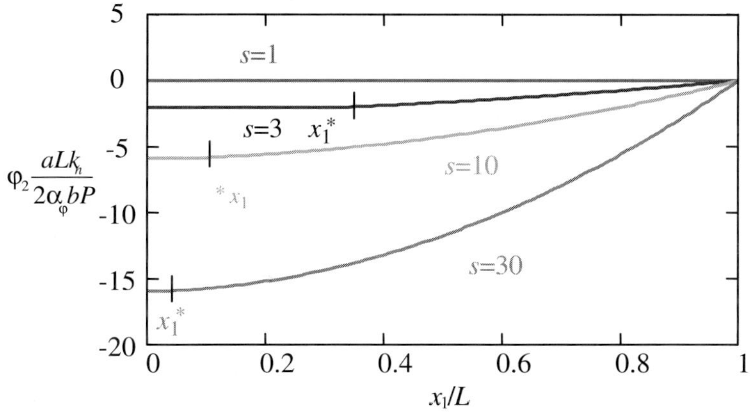

Figure 4

Distribution of normalised rotations plotted for the positive values of x_1 in the Cosserat zone $x_1 > x_1^*$ and the classical zone $x_1 < x_1^*$ for various values of the dimensionless parameter $s = 4cL\sigma_{13}^0/\alpha_\varphi P$, which expresses the ratio of the applied shear stress and the applied compression and for $k_s = L^{-2}k_\varphi = k_n$.

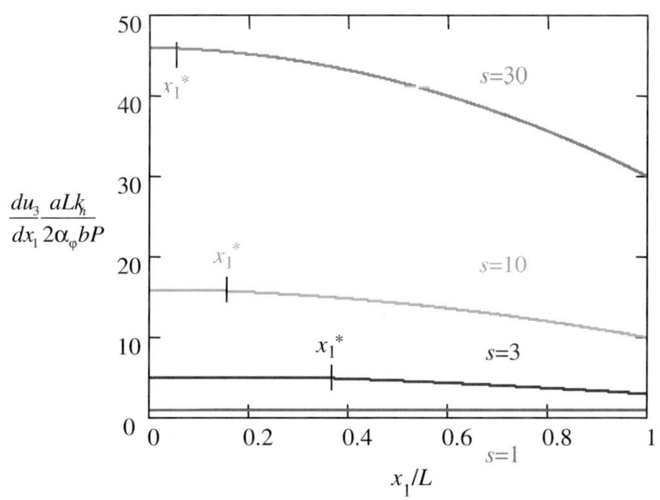

Figure 5

Distribution of normalised gradient of the vertical displacement $u_{3,1}$ plotted for the positive values in the Cosserat $x_1 > x_1^*$ and classical zone $x_1 < x_1^*$ for various values of a dimensionless parameter $s = 4cL\sigma_{13}^0/\alpha_\varphi P$, which expresses the ratio of applied shear stress over applied compression and for $k_s = L^{-2}k_\varphi = k_n$.

As pointed out above, the dilatation is the characteristic feature of the fragmented solids due to the 'elbowing' of the blocks under loading. The non-zero displacement gradient $u_{1,1}$ serves as an expression for the dilatation. Figure 6 shows distributions of the displacement gradient $u_{1,1}$ (plotted for the positive x_1-values only) in the Cosserat $x_1 > x_1^*$ and classical $x_1 < x_1^*$ zones for the same values of the dimensionless parameter s and the same ratios of contact stiffnesses $k_s = L^{-2}k_\varphi = k_n$. The plots show

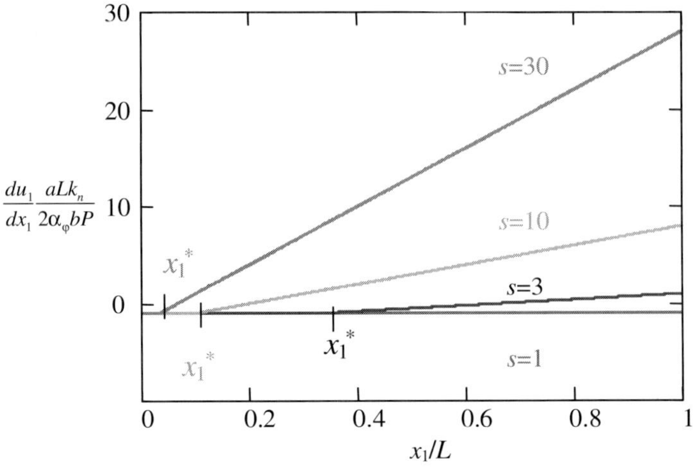

Figure 6
Dilatation in the assembly—distribution of the normalised gradient of the horizontal displacement $u_{1,1}$ plotted for the positive values in the Cosserat zone $x_1 > x_1^*$ and classical zone $x_1 < x_1^*$ for various values of the dimensionless parameter $s = \frac{4cL\sigma_{13}^0}{\alpha_\varphi P}$ and $k_s = L^{-2}k_\varphi = k_n$.

the presence of dilatation zone concentrated at the boundaries of the assembly. The size of the zone $L - x_1^*$ in which dilatation is observed is governed by the boundary conditions and the blocks geometry. According to (20) the higher the applied shear stress σ_{13}^0, the smaller x_1^*, all other parameters being fixed. In other words, the closer x_1^* to the origin, the larger $L - x_1^*$, i.e., the bigger the localisation area.

Figure 7 shows the loading curves as the applied shear load σ_{13}^0 versus the shear displacement u_3^0. The loading curves are non-linear and exhibit apparent plasticity with work hardening for loading past the apparent yield point.

General Formulation of Constitutive Equations

Now we consider a strip of width $2L$ in the x_1 direction, infinite in the x_3 direction: $-L < x_1 < L, -\infty < x_3 < +\infty$ comprised of the cross sections considered in the previous section. As before, each cross section comprises blocks that are held in place by confining pressure applied along the vertical (x_3) direction. We assume that the cross sections are identical, but not decoupled such that the interaction between the cross sections is transmitted through the components of the stress and moment stress tensors (contrary to the case considered above, when the cross sections were decoupled), Figures 8 and 9. The same boundary conditions are specified at the strip edges:

$$u_3(\pm L, x_2, x_3) = \pm u_3^0, \quad \varphi_2(\pm L, x_2, x_3) = 0, \quad \sigma_{11} = -P/(4bc). \tag{21}$$

For an (i,j)-th block at the contact with its $(i+1,j)$ neighbour (Fig. 9) its kinematics is given by the relative normal displacement $\Delta_n u_{1ij}$ and shear displacement

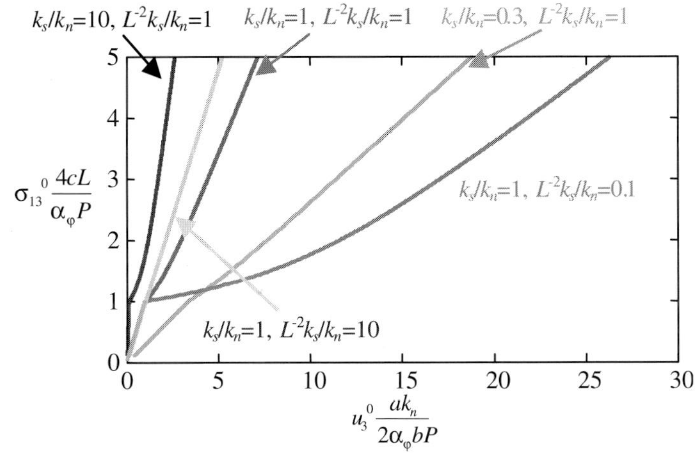

Figure 7
Different loading curves: Applied shear load σ_{13}^0 versus shear displacement u_3^0 at the edges of the layer.

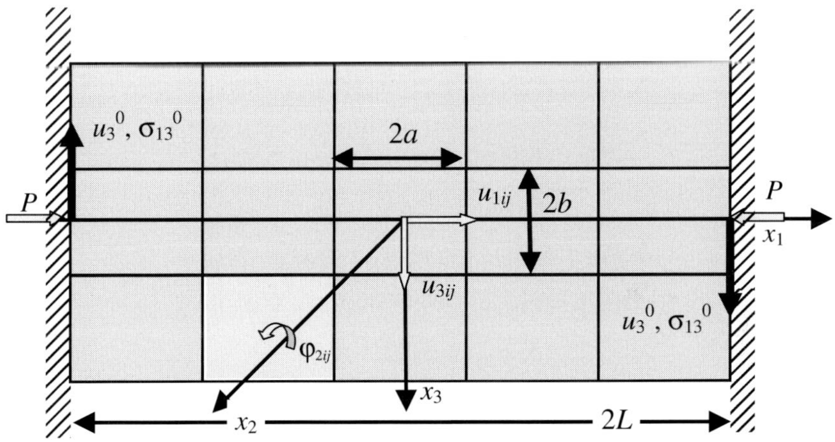

Figure 8
Simple shear in compression of a 2-D block arrangement, $2a$ and $2b$ denoting block length and height respectively, $2L$ is the length of assembly infinite in the x_3-direction, u_3^0 is the applied shear displacement at the boundary, σ_{13}^0 is corresponding shear stress at the boundary, P is the applied compression load at the boundary, u_{1i}, u_{3i}, φ_{2i} are displacements and rotation in the directions x_1, x_3 and x_2, respectively.

$\Delta_s u_{3ij}$ in directions x_1 and x_3, respectively, and the relative rotation $\Delta_1\varphi_{2ij}$ (the subscript 1 here referring to the direction of the upward normal at the contact) along x_2 axis. Furthermore, for the (i,j)-th block at the contact with its $(i,j+1)$ neighbour (Fig. 9) its kinematics is given by the relative normal displacement $\Delta_n u_{3ij}$ and shear displacement $\Delta_s u_{1ij}$ in the directions x_3 and x_1, respectively, and the relative rotation

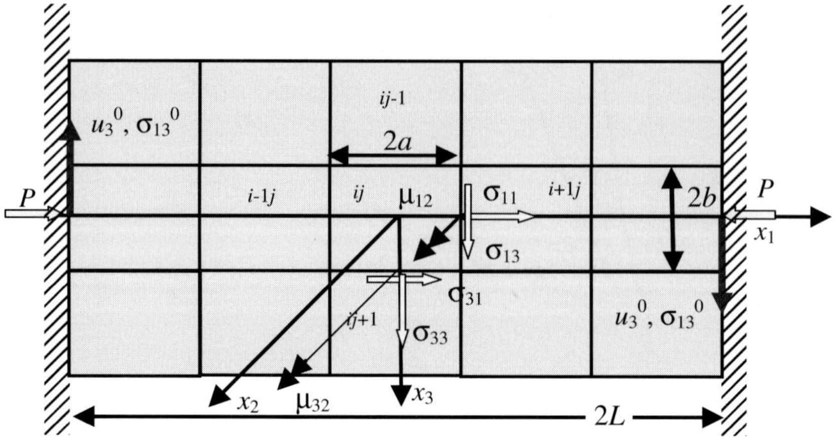

Figure 9
Stresses and moment stresses in the assembly.

$\Delta_3\varphi_{2ij}$ (the subscript 3 here referring to the direction of the upward normal at the contact) along x_2 axis:

$$\Delta_n u_{1ij} = u_{1i+1j} - u_{1ij} - \alpha_\varphi b\left|\varphi_{2i+1j} - \varphi_{2ij}\right|, \tag{22}$$

$$\Delta_s u_{3ij} = u_{3i+1j} - u_{3ij} + a\left(\varphi_{2i+1j} + \varphi_{2ij}\right), \tag{23}$$

$$\Delta_1\varphi_{2ij} = \varphi_{2i+1j} - \varphi_{2ij}, \tag{24}$$

$$\Delta_n u_{3ij} = u_{3ij+1} - u_{3ij} - \beta_\varphi a\left|\varphi_{2ij+1} - \varphi_{2ij}\right|, \tag{25}$$

$$\Delta_s u_{1ij} = u_{1ij+1} - u_{1ij} - b\left(\varphi_{2ij+1} + \varphi_{2ij}\right), \tag{26}$$

$$\Delta_3\varphi_{2ij} = \varphi_{2ij+1} - \varphi_{2ij}, \tag{27}$$

where α_φ and β_φ are the coefficients of 'elbowing' in x_1 and x_3 directions, respectively. These coefficients account for the fact that blocks are not necessarily of regular rectangular shape (see the explanation in the previous section). These relative displacements and rotations are the generalisations of the respective displacements and rotations considered in the previous section.

The discrete energy density associated with the (i,j)-th block can be written as follows:

$$\begin{aligned} W_{ij} = 1/2\Big\{&\Theta(-\Delta_n u_{1ij})k_n\Delta_n u_{1ij}^2 + \Theta(-\Delta_n u_{3ij})k_n\Delta_n u_{3ij}^2 \\ &+ k_s\left(\Delta_s u_{1ij}^2 + \Delta_s u_{3ij}^2\right) + k_\varphi\left(\Delta_1\varphi_{2ij}^2 + \Delta_3\varphi_{2ij}^2\right)\Big\}, \end{aligned} \tag{28}$$

where k_n, k_s, k_φ are generalised normal, shear and rotational stiffnesses of the isotropic contacts.

Now, similar to the procedure used in the foregoing section, we replace the discrete coordinates with the continuous ones $ia \to x_1$, $jb \to x_3$ and apply homogenisation by differential expansions, replacing the finite differences with corresponding Taylor series truncated after the quadratic terms. The result reads:

$$W(x_1, x_3) = a(4bc)^{-1}\Big\{\Theta\big[-(\gamma_{11} - \alpha_\varphi b|\kappa_{12}|)2a\big]k_n\big(\gamma_{11} - \alpha_\varphi b|\kappa_{12}|\big)^2$$
$$+ \Theta\big[-(\gamma_{33} - \beta_\varphi a|\kappa_{32}|)2b\big]k_n\big(\gamma_{33} - \beta_\varphi a|\kappa_{32}|\big)^2$$
$$+ k_s\big(\gamma_{13}^2 + \gamma_{31}^2\big) + k_\varphi\big(\kappa_{12}^2 + \kappa_{32}^2\big)\Big\}, \tag{29}$$

where the Cosserat continuum measures (NOWACKI, 1974) are defined as

$$\gamma_{11} = \frac{\partial u_1}{\partial x_1}, \quad \gamma_{13} = \frac{\partial u_3}{\partial x_1} + \varphi_2, \quad \gamma_{31} = \frac{\partial u_1}{\partial x_3} - \varphi_2, \quad \gamma_{33} = \frac{\partial u_3}{\partial x_3},$$
$$\kappa_{12} = \frac{\partial \varphi_2}{\partial x_1}, \quad \kappa_{32} = \frac{\partial \varphi_2}{\partial x_3}. \tag{30}$$

Variation of the homogenised energy density (29) with respect to the deformation measures, i.e., strains γ_{11}, γ_{13}, γ_{31}, γ_{33} and curvatures κ_{12}, κ_{32}, yields the following constitutive equations:

$$\sigma_{11} = \frac{a}{2bc}k_n\Theta[-(\gamma_{11} - \alpha_\varphi b|\kappa_{12}|)2a](\gamma_{11} - \alpha_\varphi b|\kappa_{12}|), \tag{31}$$

$$\sigma_{33} = \frac{b}{2ac}k_n\Theta[-(\gamma_{33} - \beta_\varphi a|\kappa_{32}|)2b](\gamma_{33} - \beta_\varphi a|\kappa_{32}|), \tag{32}$$

$$\sigma_{13} = \frac{a}{2bc}k_s\gamma_{13}, \tag{33}$$

$$\sigma_{31} = \frac{b}{2ac}k_s\gamma_{31}, \tag{34}$$

$$\mu_{12} = \frac{a}{2bc}\{-k_n\Theta[-(\gamma_{11} - \alpha_\varphi b|\kappa_{12}|)2a]\alpha_\varphi b \operatorname{sgn}\kappa_{12}(\gamma_{11} - \alpha_\varphi b|\kappa_{12}|) + k_\varphi\kappa_{12}\}, \tag{35}$$

$$\mu_{32} = \frac{b}{2ac}\{-k_n\Theta[-(\gamma_{33} - \beta_\varphi a|\kappa_{32}|)2b]b_\varphi a \operatorname{sgn}\kappa_{32}(\gamma_{33} - \beta_\varphi a|\kappa_{32}|) + k_\varphi\kappa_{32}\}, \tag{36}$$

or

$$\mu_{12} = -\sigma_{11}\alpha_\varphi b \operatorname{sgn}\kappa_{12} + \frac{a}{2bc}k_\varphi\kappa_{12}, \tag{37}$$

$$\mu_{32} = -\sigma_{33}\beta_\varphi a \operatorname{sgn}\kappa_{32} + \frac{b}{2ac}k_\varphi\kappa_{32}, \tag{38}$$

where σ_{ji} is a non-symmetric stress tensor, μ_{ji} is the moment stress tensor. The constitutive relations (31)–(38) are again strongly non-linear and contain 'non-isotropic terms', *viz.* curvature κ_{12} contributing to the normal stress σ_{11}, and normal strain γ_{11} contributing to the μ_{12} component of the moment stress tensor, also curvature κ_{32} contributing to the normal stress σ_{33}, and normal strain γ_{33} contributing to the μ_{32} component of the moment stress tensor. We note that, even if the effective lengths of the projections of the blocks $\alpha_\varphi b$ and $\beta_\varphi a$ are equal, the constitutive equations (31)–(38) do not necessarily correspond to the isotropic Cosserat continuum.

Similar to equations (12) and (13), we obtain the condition for the bending moment produced to be sufficient to cause the rotations of the blocks in the assembly:

$$|\mu_{12}| \geq -\sigma_{11}\alpha_\varphi b > 0, \quad \kappa_{12} \neq 0, \tag{39}$$

or

$$|\mu_{12}| < -\sigma_{11}\alpha_\varphi b, \quad \kappa_{12} \equiv 0. \tag{40}$$

Similarly,

$$|\mu_{32}| \geq -\sigma_{33}\beta_\varphi a > 0, \quad \kappa_{32} \neq 0, \tag{41}$$

or

$$|\mu_{32}| < -\sigma_{33}\beta_\varphi a, \quad \kappa_{32} \equiv 0. \tag{42}$$

We now note that the boundary conditions are homogeneous along x_3. Subsequently, we seek the solution homogeneous along x_3 as well, i.e., $\partial()/\partial x_3 \equiv 0$. In this case the Cosserat equations of equilibrium, $\sigma_{ji,j} = 0$ and $\mu_{ji,j} + \varepsilon_{ijk}\sigma_{jk} = 0$ (where ε_{ijk} is the Levi-Civita symbol) assume the form:

$$\sigma_{11,1} = 0, \quad \sigma_{13,1} = 0, \quad \mu_{12,1} - \sigma_{13} + \sigma_{31} = 0, \tag{43}$$

along with the deformation measures (30)

$$\begin{aligned} &\gamma_{11} = \frac{\partial u_1}{\partial x_1}, \quad \gamma_{13} = \frac{\partial u_3}{\partial x_1} + \varphi_2, \quad \gamma_{31} = -\varphi_2, \quad \gamma_{33} = \frac{\partial u_3}{\partial x_3} = 0, \\ &\kappa_{12} = \frac{\partial \varphi_2}{\partial x_1}, \quad \kappa_{32} = \frac{\partial \varphi_2}{\partial x_3} = 0. \end{aligned} \tag{44}$$

The full set of equations for this case can be obtained in the usual way by substituting the deformation measures (44) into the constitutive equations (31)–(36) ($\mu_{32} = 0$, $\sigma_{33} = 0$) and the result in (43).

For the sake of simplifying further analysis, it is convenient to introduce the following normalisation of parameters:

$$\frac{P}{4bc} \mapsto p, \quad \frac{ak_n}{2bc} \mapsto k_n, \quad \frac{ak_s}{2bc} \mapsto k_s, \quad \frac{ak_\varphi}{2bc} \mapsto k_\varphi, \quad b\alpha_\varphi \mapsto \alpha, \quad a\beta_\varphi \mapsto \beta, \quad \frac{x_1}{L} \mapsto x_1. \tag{45}$$

The constitutive equation (37) can be rewritten in the equivalent form by using the normalisation (45) as follows:

$$k_\varphi \kappa_{12} = \Theta(|\mu_{12}| + \alpha\sigma_{11})(|\mu_{12}| + \alpha\sigma_{11})\,\mathrm{sgn}\,\mu_{12}, \tag{46}$$

from which it follows that the condition $\kappa_{12} \neq 0$ is fulfilled when the argument of the Heaviside step function is positive, i.e.,

$$|\mu_{12}| + \alpha\sigma_{11} > 0. \tag{47}$$

On the other hand, the solution of the first equation of (43) gives the uniform value of $\sigma_{11} = -P/(4bc)$ or, after the normalisation (45), $\sigma_{11} = -p$ along the length of the cross section x_1. Then inequality (47) takes the form:

$$|\mu_{12}| - \alpha p > 0. \tag{48}$$

Hence, the condition which defines the transition from the classical solution zone (no rotations), where $|\mu_{12}| - \alpha p < 0$ holds, to Cosserat continuum solution zone (non-zero Cosserat rotation field), where $|\mu_{12}| - \alpha p > 0$ holds, reads:

$$|\mu_{12}| = \alpha p. \tag{49}$$

From the constitutive equation (37) after the normalization (45), one finds that:

$$|\mu_{12}| = -\sigma_{11}\alpha + k_\varphi|\kappa_{12}| \tag{50}$$

or taking into account that $\sigma_{11} = -p$, one obtains

$$|\mu_{12}| = \alpha p + k_\varphi|\kappa_{12}|. \tag{51}$$

Comparing (49) and (51) we find that $\kappa_{12} = 0$ is the condition of transition between the regions of classical and Cosserat solutions; the boundary between the regions being x_1^*.

Substituting the constitutive equations (33)–(35) into the second and third equilibrium equations (43) one obtains the differential equation for the unknown rotations:

$$k_\varphi \frac{\partial^2 \varphi_2}{\partial x_1^2} - k_s\left(\frac{b}{a}\right)^2 \varphi_2 - \sigma_{13}^0 = 0, \tag{52}$$

where σ_{13}^0 is the uniform shear stress that corresponds to the solution of the second equilibrium equation (43).

The solution satisfying boundary condition (21) and the symmetry considerations $\varphi_2(-x_1) = \varphi_2(x_1)$ has the form

$$\varphi_2 = \frac{\sigma_{13}^0}{k_s}\left(\frac{a}{b}\right)^2\left(\frac{\cosh(b_s x_1)}{\cosh(b_s)} - 1\right), \quad x_1^* = 0, \quad b_s = L\sqrt{\frac{k_s}{k_\varphi}\frac{b}{a}}. \tag{53}$$

It is remarkable that opposite to the previous case of decoupled cross sections, here no classical region (with zero rotation gradient) exists, $x_1^* = 0$. The obvious reason for this is the presence of another shear stress component σ_{31}, which allows the blocks to overcome resistance to rotation produced by compression σ_{11}; this stress component is absent in the case of decoupled cross sections.

Using (53) one can find the curvature

$$\kappa_{12} = \frac{\sigma_{13}^0}{k_s}\left(\frac{a}{b}\right)^2 b_s \frac{\sinh(b_s x_1)}{\cosh(b_s)}. \tag{54}$$

From here, using equations of equilibrium and the constitutive equations we obtain the displacement gradient

$$u_{3,1} = \frac{\sigma_{13}^0}{k_s}\left[1 + \left(\frac{a}{b}\right)^2 - \left(\frac{a}{b}\right)^2 \frac{\cosh(b_s x_1)}{\cosh(b_s)}\right]. \tag{55}$$

Using the first two of boundary conditions (21) the displacement u_3 is obtained as follows

$$u_3 = \frac{\sigma_{13}^0}{k_s}\left[\left(1 + \left(\frac{a}{b}\right)^2\right)x_1 - \left(\frac{a}{b}\right)^2 \frac{\sinh(b_s x_1)}{b_s \cosh(b_s)}\right], \tag{56}$$

$$u_3^0 = \frac{\sigma_{13}^0}{k_s}\left[1 + \left(\frac{a}{b}\right)^2 - \left(\frac{a}{b}\right)^2 \frac{\tanh(b_s)}{b_s}\right]. \tag{57}$$

It is clear that in this case the relation between the imposed shear displacement and the produced shear stress in the fault is linear, quite opposite to the case of the fault with decoupled cross sections.

Solving the first equation of equilibrium (43) accompanied by constitutive equation (31) and the obtained expression for curvature (54) one has the expression for dilatation

$$u_{1,1} = -\frac{p}{k_n} + \alpha_\varphi \frac{\sigma_{13}^0}{k_s}\left(\frac{a}{b}\right)^2 b_s \frac{\sinh(b_s|x_1|)}{\cosh(b_s)}. \tag{58}$$

Subsequently, the expression for displacement reads

$$u_1 = -\frac{p}{k_n}x_1 + \alpha_\varphi \frac{\sigma_{13}^0}{k_s}\left(\frac{a}{b}\right)^2 \frac{\cosh(b_s x_1) - 1}{\cosh(b_s)} \operatorname{sgn} x_1. \tag{59}$$

Figures 10–12 show the dependencies (53), (55) and (58). Figure 11 indicates considerable non-uniformity in the distribution of displacement gradient, which being measured in real fault could be passed for strain localisation. The dilatation, Figure 12, strongly depends upon the ratio s between the shear stress σ_{13}^0 and the compression p: when the compression is high the normal strain $\gamma_{11} = u_{1,1}$ is negative; only when s is high enough can dilatation emerge.

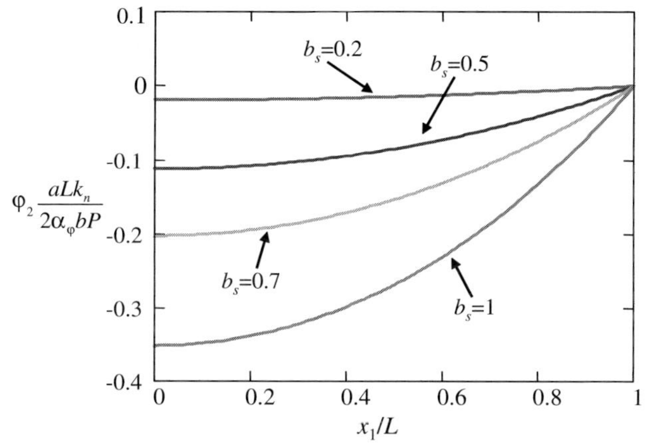

Figure 10
Distribution of normalised rotations plotted for the positive values of x_1 for various effective aspect ratios b_s of blocks. It is assumed that $\alpha_\varphi = 1$ and $k_s = L^{-2}k_\varphi = k_n$.

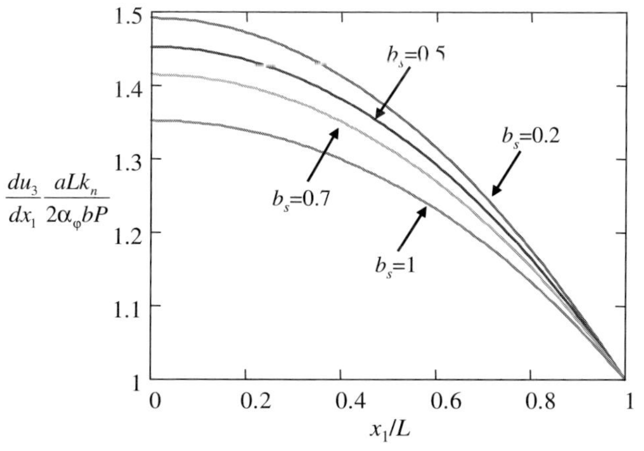

Figure 11
Distribution of normalised displacement gradients plotted for the positive values of x_1 for various effective aspect ratios b_s of blocks. It is assumed that $\alpha_\varphi = 1$ and $k_s = L^{-2}k_\varphi = k_n$.

Conclusions

We have introduced a continuum theory of deformation of fragmented solids which takes into account the block angularity. The absence of the binder phase or any adhesion between the blocks makes it possible for them to rotate independently. The block rotations create normal stresses (referred to as 'elbowing'), this effect being independent of the direction of rotations. Consequently, the continuum description

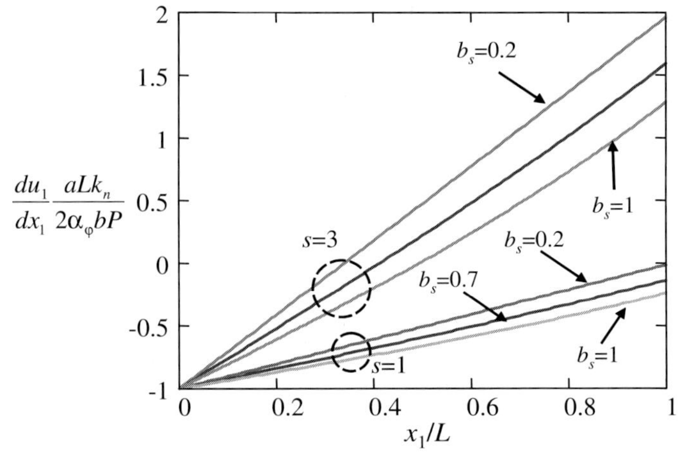

Figure 12
Distribution of normalised dilatation plotted for the positive values of x_1 for various effective aspect ratios b_s of blocks and $s = \frac{4cL\sigma_{13}^0}{\alpha_\varphi P}$. It is assumed that $\alpha_\varphi = 1$ and $k_s = L^{-2}k_\varphi = k_n$.

of the motion of a solid consisting of many fragments involves a modified Cosserat theory with coupling between the curvature-twists and normal stresses as well as between normal strains and moment stresses. A remarkable feature of this theory is the fact that the curvature twists enter with their absolute values which creates a potential for strong non-linear behaviour.

We apply this theory to modelling deformation of transform faults. We consider two somewhat limiting cases. The first case is deformation of a transform fault produced by spreading. In this case, tensile stresses acting in the plate along the fault make the cross sections of the fault deform nearly independently. The second case is deformation of a transform fault in the presence of compression along the fault.

The first case of independent sections proves to be the most interesting one. In simple shear under compression, there exists a displacement threshold: When the displacement is below the threshold, the deformation pattern coincides with the conventional one for a classical elastic isotropic layer, i.e., the shear strain is uniform across the layer. When the imposed displacement exceeds the threshold, boundary zones emerge exhibiting non-uniform rotations, non-uniform displacement gradients and dilatation. These zones grow with the increase of the imposed shear displacement. The displacement gradient present in these zones exhibits strong non-uniformity. The loading curves (shear stress vs. displacement at the layer edges) are non-linear and are characterised by apparent plasticity with work hardening. Thus in a situation which involves neither energy dissipation (for instance through friction) nor plastic behaviour of the material constituents two features emerge: non-uniform displacement gradient and non-linear deformation curves which could be mistaken for non-elastic strain localization and plastic behaviour.

In the other extreme case, the block rotations affect the fault deformation throughout its entire thickness—no zone of classical (rotation-independent) solution exists. This case is characterised by linear stress-strain dependence and pronounced non-uniformity in the displacement gradient.

Thus, the ability of blocks to independently rotate and elbow each other can have a profound effect on the overall deformation of transform faults leading to apparent non-elastic behaviour and localization, even in the absence of energy dissipation.

Acknowledgment

E. Pasternak acknowledges the financial support from the ARC Postdoctoral Fellowship (2003–2006) and Discovery Grant DP0346148. A.V. Dyskin acknowledges the financial support from the ARC Discovery Grant DP0210574. We also acknowledge the support by the Australian Computational Earth Systems Simulator (ACcESS) — Major National Research Facility. The authors also acknowledge support from ARC through a Linkage International Grant (LX0347195).

References

Burkart, B. and Self, S. (1985), *Extension and rotation of crustal blocks in northern Central America and effect on the volcanic arc*, Geology *13*, 22–26.

Earle, S. (2004), *A simple paper model of a transform fault at a spreading-ridge*, J. Geosc. Educ. *52*, 391–392.

Estrin, Y., Dyskin, A.V., Kanel-Belov, A.J., and Pasternak, E., *Materials with novel architectonics: Assemblies of interlocked elements*. In *IUTAM Symposium on Analytical and Computational Fracture Mechanics of Non-homogeneous Materials* (ed. Karihaloo. B.) (Kluwer Academic Press 2002) pp. 51–56.

Estrin, Y., Dyskin, A.V., Pasternak, E., Schaare, S., Stanchits, S., and Kanel-Belov, A.J. (2003), *Negative stiffness of a layer with topologically interlocked elements*, Scripta Materialia *50*, 291–294.

Jones, D.L., Cox, A., Coney, P., and Beck, M., *The growth of Western North America*. In *Shaping the Earth. Tectonics of Oceans and Continents: Readings from Scientific American Magazine* (ed. Moores. E.) (W.H. Freeman and Company, New York 1990) pp. 156–176.

Kasahara, K., *Earthquake Mechanics* (Cambridge University Press 1981).

Kishino, Y., *Characteristic deformation analysis of granular materials*. In *Mechanics of Granular Materials. An Introduction* (eds. Oda, M. and Iwashita, K.) (Balkema, Rotterdam 1999) pp. 106–201.

Lynendyk, B.P. (1991), *A model for Neogene Crystal rotations, transtension and transgression in southern California*, Geol. Soc. Am. Bull. *103*, 1528–1536.

Molnar, P., Molnar, P., and Gipson, J.M. (1994), *Very long baseline interferometry and active rotations of crustal blocks in the Western Transverse Ranges, California*, Geolog. Soc. Am. Bull. *106*, 594–606.

Mühlhaus, H.-B., *Continuum models for layered and blocky rock*. In *Comprehensive Rock Engineering*. (Pergamon Press 1993) pp. 209–230.

Mühlhaus, H.-B. and Oka, F. (1996), *Dispersion and wave propagation in discrete and continuous models for granular materials*, Int. J. So. Struct. *33*, 2841–2858.

Mühlhaus, H.-B., Sulem, J., and Unterreiner, P. (1997), *Discrete and continuous models for dry masonry columns*, J. Eng. Mech., April, 399–403.

Nowacki, W., *The linear theory of micropolar elasticity*. In *Micropolar Elasticity* (eds. Nowacki, W. and Olszak, W.) (Springer–Verlag, Wien, New York 1974) pp. 1–43.

PASTERNAK, E., and MÜHLHAUS, H.-B. (2005), *Generalised homogenisation procedures for granular materials*, Eng Math. *52*, 199–229.

PASTERNAK, E., MÜHLHAUS, H.-B., and DYSKIN, A.V. (2004), *On the possibility of elastic strain localisation in a fault*, Pure Appl. Geophys. *161*, 2309–2326.

SCHELLE, H. and GRÜNTHAL, G. (1996), *Modeling of neogene crustal block rotation: Case study of southeastern California*, Tectonics *15*, 700–710.

SULEM, J. and MÜHLHAUS, H.-B. (1997), *A continuum model for periodic two-dimensional block structures*, Mechanics of Cohesive-Frictional Mat. *2*, 31–46.

TAMURA, T., *Application of rigid plasticity to granular materials based on upper bound theory*, In *Mechanics of Granular Materials. An Introduction* (eds. Oda, M. and Iwashita, K.) (Balkema, Rotterdam 1999) pp. 201–207.

WESNOUSKY, S.G. (2005), *Active faulting in the Walker Lane*, Tectonics *24*, 1645.

(Received December 30, 2004, revised November 11, 2005, accepted December 30, 2005)

Pure appl. geophys. 163 (2006) 2031–2047
0033–4553/06/092031–17
DOI 10.1007/s00024-006-0111-5

Pure and Applied Geophysics

Elasticity, Yielding and Episodicity in Simple Models of Mantle Convection

Hans-Bernd Mühlhaus,[1,2] Matt Davies,[1] and Louis Moresi[3]

Abstract—We explore the implications of refinements in the mechanical description of planetary constituents on the convection modes predicted by finite-element simulations. The refinements consist in the inclusion of incremental elasticity, plasticity (yielding) and multiple simultaneous creep mechanisms in addition to the usual visco-plastic models employed in the context of unified plate-mantle models. The main emphasis of this paper rests on the constitutive and computational formulation of the model. We apply a consistent incremental formulation of the non-linear governing equations avoiding the computationally expensive iterations that are otherwise necessary to handle the onset of plastic yield. In connection with episodic convection simulations, we point out the strong dependency of the results on the choice of the initial temperature distribution. Our results also indicate that the inclusion of elasticity in the constitutive relationships lowers the mechanical energy associated with subduction events.

Key words: Mantle convection, constitutive formulation, elasticity, plastic yielding, episodicity, simulation.

Introduction

The way a planet deforms in response to thermal or gravitational driving forces, depends on the material properties of its constituents. The Earth's behavior is unique in that its outermost layer consists of a small number of continuously moving plates. Venus, another planet of similar size and bulk composition to the Earth, displays signs of active volcanism but there is no evidence of plate movements or plate tectonics.

It is generally accepted that plate tectonics is a manifestation of mantle convection, a natural solid state convection process driven by the thermal gradients of a cooling planet with radiogenic heating. In this context, the pattern of surface motion is determined by the rheology of the cool thermal boundary layer. Purely

[1] ESSCC, The University of Queensland, St Lucia, QLD 4072, Australia.
E-mail: {muhlhaus, matt}@esscu.uq.edu.au
[2] CSIRO Division of Exploration and Mining, 26 Dick Perry Ave, Kensington, WA 6051, Australia.
[3] School of Mathematical Sciences, Building 28, Monash University, VIC 3800, Australia.
E-mail: louis.moresi@sci.monash.edu.au

viscous models of the lithosphere are not capable of producing the narrow plate boundary deformation zones and low strain-rate plate interiors which characterize plate tectonics. Research over the past ten years in the area of computational unified plate-mantle models has demonstrated that additional model ingredients are required including temperature dependence of the viscosity combined with a finite material strength to enable the fracture processes necessary for plate formation (MORESI and SOLOMATOV, 1998; TACKLEY, 1998). In square or cubic convection cells temperature dependence and a finite yield strength leads to three distinct modes of convection: If the yield strength is high compared to the thermal driving force, then convection is confined to a domain underlying a high viscosity, quasi-rigid plate or lid. This mode of convection is usually designated stagnant lid convection. If the yield strength is low compared to the driving forces then yielding of the material at the cold boundary takes place and plate like regions form; this mode is designated mobile lid convection. In between the two extreme modes exists an episodic convection mode with alternating mobile and stagnant lid behavior.

In the following, we study the effect of physical and computational model refinements such as large strain elasticity, stress advection and an incremental decomposition (or tangent form) of the constitutive relationships. The model proposed here is an extension of the model proposed by MORESI *et al.* (2002) in the context of a Lagrangian particle method. While the three distinct convection modes are clearly observed in square or cubic convection domains or in more general domains with periodic initial conditions, the sensitivity of the convection modes with respect to more general initial conditions is not clear and will be examined here as well. The paper is structured as follows: The model equations, constitutive relationships, and computational aspects are outlined in the following two sections. Simulation results are presented subsequently.

Model Formulation

The governing model equations consist of the stress equilibrium conditions

$$\sigma'_{ij,j} - p_{,i} = \rho_0(1 - \alpha_p(T - T_0))g_i \tag{1}$$

and the heat equation

$$\rho_0 c_p(T_{,t} + v_j T_{,j}) = (\kappa T_{,j})_{,j} + \sigma'_{ij} D_{ij} + \rho_0 h. \tag{2}$$

Here we adopt the usual subscripted Cartesian tensor notation; postfixing subscripted differential indices after a comma and adopting the summation convention. In equation (1), σ'_{ij} is the deviatoric stress, D_{ij} is the stretching, p is the pressure, g_i, is parallel and opposite to the direction of gravity (i.e., such that $g = |g_i|$ is the gravitational constant) and α_p is the thermal expansion coefficient. In equation (2), ρ_0 is a reference density at the surface, c_p is the heat capacity, κ is the thermal conductivity and h is a specific energy source term (e.g., radiogenic heating). In the following examples we assume $h = 0$ for simplicity.

In the formulation of the constitutive model we make the standard assumption that the stretching is the sum of an elastic and a visco-plastic part:

$$D_{ij} = D^e_{ij} + D^{vp}_{ij}. \tag{3}$$

We assume incompressibility such that $D_{jj} = v_{j,j} = 0$, where v_j is the velocity. The visco-plastic stretching is then defined as $D^{vp}_{ij} = 1/2\eta_{\text{eff}}\sigma'_{ij}$, where η_{eff} is the effective viscosity, and for the elastic part we assume

$$D^e_{ij} = \frac{1}{2\mu}(\dot{\sigma}^J_{ij})', \tag{4}$$

where μ is the elastic shear modulus and $\dot{\sigma}^J_{ij}$ the Jaumann stress rate. The reader is referred to KOLYMBAS and HERLE (2003) and MÜHLHAUS and REGENAUER-LIEB (2004) for recent discussions and comparisons between the various objective stress rates. The Jaumann stress rate is related to the material stress rate by

$$\dot{\sigma}^J_{ij} = \frac{\partial}{\partial t}\sigma_{ij} + v_k\sigma_{ij,k} - W_{ik}\sigma_{kj} + \sigma_{ik}W_{kj}, \tag{5}$$

where W_{ij} is the non-symmetric part of the velocity gradient.

As a further refinement to this approach, the visco-plastic stretching is adopted in the form of a combined Newtonian and composite power-law creep model (H.-B. MÜHLHAUS and REGENAUER-LIEB, 2005). The composite power-law viscosity includes a contribution from both dislocation glide, with a typical power law exponent ($n \approx 3$), and plastic deformation, with temperature independent co-efficients and a large exponent ($n_{pl} \approx 15$). The effective viscosity is then given by:

$$\eta_{\text{eff}} = \left(\frac{1}{\eta_N} + \frac{1}{\eta_N\left(\frac{\tau}{\tau_0}\right)^{1-n}} + \frac{1}{\eta_Y\left(\frac{\tau}{\tau_Y}\right)^{1-n_{pl}}}\right)^{-1}, \tag{6}$$

where η_Y is a reference viscosity for the plastic yield, τ_0 is the dislocation transition stress, τ_Y is the plastic yield stress and $\tau = \sqrt{1/2\sigma'_{ij}\sigma'_{ij}}$ and η_N is the temperature-dependent Newtonian viscosity for which we adopt an Arrhenius relationship. This relationship is of the form

$$\eta_N = \eta_{\text{N0}}\exp(AT_M/T), \tag{7}$$

where η_{N0} is a reference Newtonian viscosity, A is the activation energy and T_M is the melting temperature.

The yield stress, τ_Y, may depend on the accumulated visco-plastic strain, other state variables including damage variables (not considered here) and the pressure. For completeness, although not used to obtain the results of this paper, we state an expression for the pressure dependence in the form

$$\tau_Y = \tau_{Y0} + \beta p, \tag{8}$$

where β is the pressure sensitivity and τ_{Y0} is the yield stress at zero pressure. BYERLEE (1968) proposes a simplified criterion in which the pressure, p, of (8) is replaced by the overburden pressure $\rho g z$ where z is the coordinate in the direction of gravity with $z = 0$ on the cold boundary. The yield criterion (6) and Byerlee's simplified model are mathematically quite different however: The fourth-order tensor relating the stress to the stretching is non-symmetric for the yield criterion (8), whereas the corresponding fourth-order tensor is symmetric in the case of Byerlee's model. Non-symmetry is preferred (RUDNICKI and RICE, 1975), as non-symmetric models have a stronger tendency for strain localization.

The governing equations are highly non-linear with respect to the introduced constitutive relationships. In the numerical solution of the model equations the non-linearities are usually considered iteratively in that the effective viscosity (6) is updated in each step of the iteration on the basis of the results from the previous iteration step (e.g., MORESI and SOLOMATOV, 1998). This approach is commonly referred to as a secant method. As an alternative, the convergence of this procedure can be improved drastically if the predictor step of the iteration is based on the incremental (= tangent) form of the constitutive relationship instead of the secant form as described above. In this paper we shall apply the tangent approach. The tangent or incremental form of the constitutive relations (3–8) have been derived in (MÜHLHAUS and REGENAUER-LIEB, 2005) and are outlined here in the appendix for easy reference.

Computational Aspects

In the next section we present the results of finite-element simulations of plane strain, natural convection problems in a rectangular L by H domain, where H is the dimension in the direction of gravity. Typical values are $H = 700$ km for upper mantle convection and $H = 3000$ km for whole mantle convection. We make the standard assumptions that the shear stress and normal velocity vanish on all boundaries of the domain; the temperatures are fixed on the top and the bottom and the normal gradient of the temperature vanishes on the sides.

The non-dimensional forms of the governing equations are respectively given by

$$\sigma'_{ij,j} - p^{th}_{,i} + Ra^c/gTg_i = 0, \tag{9}$$

$$T_{,t} + v_j T_{,j} = T_{,jj} + \frac{Di^c}{Ra^c}\sigma'_{ij} D_{ij}. \tag{10}$$

Equation (9) is obtained from equation (1) by decomposing the pressure as $p = \rho_0 g(H - x_z) + p^{th}$, where p^{th} is the pressure due to thermal expansion, x_z is the coordinate opposite to the direction of gravity and Ra^c and Di^c designate the computational Rayleigh number and the dissipation number, respectively. The stresses in (9) and (10) are normalized with respect to $\eta^* t_D$ where η^* is a viscosity parameter and t_D is the characteristic thermal diffusion time. The significance of the viscosity parameter η^* follows from the transformation (for numerical purposes) of the Arrhenius law (see below). The thermal diffusion time, t_D, is given by the expression

$$t_D = \frac{\rho_0 c_p H^2}{\kappa}. \tag{11}$$

With $\kappa/(\rho_0 c_p) = 10^{-6}\mathrm{m}^2\mathrm{s}^{-1}$ we obtain $10^{17} < t_D < 10^{19} s$. The Ra^c and Di^c, consistent with the way the stresses are non-dimensionalised, are defined as

$$Ra^c = \frac{\rho_0^2 c_p g \alpha \Delta T H^3}{\kappa \eta *} \quad \text{and} \quad Di^c = \frac{\alpha \rho_0 g H}{\rho_0 c_p}. \tag{12}$$

In (12) ΔT is the temperature difference between the hot and the cold boundary of the domain under consideration. In the present paper we assume $Di^c = 0$ since our main focus here is on the influence of temperature dependence of the viscosity, plasticity and elasticity. The temperature and the velocities are non-dimensionalised with respect to ΔT and H/t_D, respectively.

The governing equations have been implemented using the computational software toolkit *Escript* (DAVIES *et al.*, 2004). Through adoption of the *Python* scripting language, *Escript* is designed to provide an extensible interface to various computational kernel modules such as equation solvers and matrix assembly routines. These kernels in turn provide implicit parallelism, permitting an expert user to focus on scripting the computational scheme for a model as opposed to the low-level implementation detail of parallelization primitives. In this way it is possible to script high performance parallel software without scripting a single line of parallel code. All of the parallelism is delegated to (and encapsulated within) the computational kernel modules selected for the task at hand.

An important computational kernel module interfaced to *Escript* is the *Finley* FEM library (DAVIES *et al.*, 2004). To use *Finley*, an unsteady initial boundary value problem (IBVP) must first be transformed into a sequence of steady BVPs by means of a suitable time discretisation procedure applied at each time step. The steady, linear BVP can then be provided to *Finley* which will then assemble a stiffness matrix associated with a given unstructured mesh using a discretisation based on the standard variational formulation. For an unknown function u, the system of partial differential equations (PDEs) associated with a BVP can be defined through the specification of the coefficients of the following templated system:

$$-\left(A_{ijkl} u_{k,l}\right)_{,j} - \left(B_{ijk} u_k\right)_{,j} + C_{ikl} u_{k,l} + D_{ik} u_k = -X_{ij,j} + Y_i, \quad x_i \in \Gamma_i. \tag{13}$$

The tensorial coefficients A, B, C, D, X, and Y are functions of their location in the physical domain. In a similar manner, *Finley* also accepts a system of implicit natural and Dirichlet boundary conditions for which template parameters can be specified.

In the implementation we solve sequentially the stress equilibrium equation and the heat equation. The incompressibility constraint is satisfied iteratively by means of the algorithm

$$p^{\alpha+1} = p^{\alpha} - \lambda v_{j,j}^{\alpha+1}, \quad \alpha = 1, 2, 3, \ldots, \tag{14}$$

where α is the iteration counter $\lambda = \lambda_0 \mu_{\text{eff}} \delta t$ is a penalty function (see appendix for definition of the effective shear modulus μ_{eff}). A typical value for the constant λ_0 is 100. In connection with direct solvers, convergence is fastest with a larger value for λ_0. However, there are usually limits to the value of λ_0 in connection with iterative solvers.

After the pressure iteration the stresses are calculated by solution of matrix problems for each column of the stress tensor. For consistency the order of interpolation is one order less than the one for the velocities. For the velocities and the temperature we use eight noded rectangular elements with bi-quadratic shape functions. For the stresses we use bilinear rectangular elements and evaluate the values of the stresses at the mid-side nodes after the solution of the stress equations. Subsequently the heat equation is solved using backward Euler time differencing.

The stress equations and the heat equation involve convective derivatives of the stresses and temperature, respectively. We use a basic upwinding scheme (ZIENKIEWICZ and TAYLOR, 2000, p. 30) to avoid spurious oscillations of the fields in advection dominated regimes. The common procedure to implement the standard streamline upwind Petrov-Galerkin (SUPG) formulation is to modify the test or weight functions used in the formulation of the finite-element method. However, in the present case, it is convenient to modify the differential equation. In our formulation we replace the material time derivative of a function a as follows:

$$a_{,t} + v_j a_{,j} \leftarrow a_{,t} + v_j a_{,j} - \left((a_{,t} + v_j a_{,j}) \frac{h}{2\sqrt{v_k v_k}} v_i\right)_{,i}. \tag{15}$$

The function a can be either a scalar (e.g., temperature), a vector (e.g., velocity) or a tensor (e.g., stress). For pure advection problems, this approach is equivalent to the SUPG method.

For unstructured meshes, the (mid) time step δt^+ is determined from a 1st order implicit extrapolation of the Courant condition $\delta t = Ch/v_{\max}$ where C is an appropriate Courant number, h is the discretization scale and $v_{\max}(t) \in \mathbf{C}^1$ is the maximum magnitude of the nodal point velocities. For the results of this paper the following condition is used:

$$\delta t^+ = \frac{2Ch}{v_{\max}^{(t+\delta t)} + v_{\max}^{(t)}} = \frac{\sqrt{\delta t^- \left(\delta t^- v_{\max}^{(t)^2} + 2(v_{\max}^{(t)} - v_{\max}^{(t-\delta t)})Ch\right)} - \delta t^- v_{\max}^{(t)}}{v_{\max}^{(t)} - v_{\max}^{(t-\delta t)}}. \tag{16}$$

Due to the adaptive nature of the 1st order implicit extrapolation scheme, larger values of C can be trialled in a guarded manner, potentially yielding larger time steps.

In the case of an explicit algorithm, C is set to ½ for stability of the numerical solution in connection with regular grids. As a fully implicit integration scheme is used, the Courant (-like) condition is not needed for numerical stability. However, we retain the condition to achieve accuracy of the transient solution of the non-linear equations. In general, higher order approaches such as the condition (16) can be used if $v_{\max}$ can be shown to be continuously differentiable to the required order. Furthermore, a corresponding initial predictor of the velocity field for the next time-step can then be calculated if required and may in turn be used as a time step corrector.

In the convection study in the next section we ignore the pressure dependence of T_M in the Arrhenius relation. The main emphasis in the study will be on the role of elasticity, power-law creep and plasticity on the emergence of different modes of convection. In the dimensionless formulation we write the Arrhenius relationship as follows (see TACKLEY, 1998):

$$\eta_{N0} \exp(AT_M/T) \rightarrow \eta_{N0} \exp\left(\frac{2\hat{A}}{3}\right) \exp\left(\hat{A}\left(\frac{1}{1+T} - \frac{2}{3}\right)\right) = \eta * \exp\left(\hat{A}\left(\frac{1}{1+T} - \frac{2}{3}\right)\right) \tag{17}$$

The exponent of (15) varies between 0, for $T = 0.5$, and $-\hat{A}/6$ or $\hat{A}/3$, for $T = 1.0$ or $T = 0.0$, respectively. For $\hat{A} = 23$, this corresponds to a Newtonian viscosity contrast of about 10^5 across the convection cell. In the absence of convection, the Newtonian viscosity varies slowly due to temperature change in the lower half of the cell, from 1 in the middle to 0.022 on the bottom and rapidly in the upper half from 1.0 to 2087.0 on the top.

Simulations

In the first subsection we consider the effect of refinements such as elasticity and stress advection for the case of episodic convection. Episodic convection is suited for this study as it involves phases of rapid change of the flow pattern within which we expect elasticity to be important and stagnant lid phases where elasticity is expected to be unimportant. In the second subsection we consider the influence of non-periodic initial conditions in connection with two-dimensional 4 by 1 convection cells on the persistence of the different convection modes.

The temperature dependence of the viscosity was given by (17) with $\hat{A} = 23$ which gives a viscosity ratio from the cold to the hot boundary due to temperature alone of 10^5. More extreme viscosity contrasts can easily be considered in the present formulation because the upper limit for the effective dimensionless viscosity is set by the dimensionless elastic shear modulus and the time increment $(\mu t_D/\eta*)\delta t$. We assign $Ra^c = 10^4$, $\tau_0 = 0.866\ 10^{2.5}$, $\tau_Y = 3\tau_0$ (i.e., a constant for ideal plasticity) and $(\mu/\eta*)t_D = 10^4$. In the simulations we used the power-law plasticity model with $n_Y = 15$ and $\eta_Y = \eta_{N0}$ (see also the appendix for further detail). The initial temperature distribution was

$$T = \frac{1}{10}\sin(x_2\pi)\cos(x_1\pi) + (1 - x_2). \tag{18}$$

The corresponding Nusselt number *vs.* time plot is shown in Figure 1. After initial, rapidly decaying, primarily elastic oscillations (not shown), the system settles temporarily into a stagnant lid mode of convection similar to the temperature and velocity profile shown in Figure 2a. During this initial period, stress accumulates until the yield stress is reached within the cold boundary layer. The locally-increased mobility was accompanied by thermal advection, a narrow plume formed, hot material was advected underneath toward a narrow, cold boundary layer. This process continued until the cold layer plunged back into the lower mantle along the boundary opposite to the plume (Fig. 2b). This process repeated itself at approximately regular intervals. The initial time intervals between the first peaks of the episodic mode in Figure 1 are 0.071, 0.044, 0.045, 0.044, with constant intervals

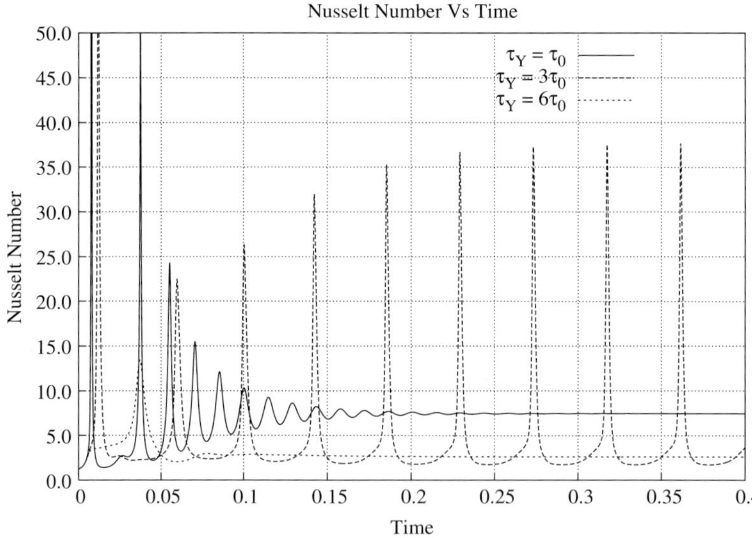

Figure 1

A comparison of Nusselt numbers for stagnant-lid (lowest at steady state), episodic and mobile lid convection modes arising from harmonic initial conditions (18). The plastic yield stress, τ_Y, is 1,3, or 6 times the Newtonian-power-law transition stress $\tau_0 = 0.866\ 10^{2.5}$. An Arrhenius relation describes the temperature dependence of creep with viscosity contrast across the layer of 10^5. The composite power-law exponents are $n = 3$ and $n_{pl} = 15$ for dislocation glide and plastic deformation, respectively. The dimensionless shear modulus is $\mu = 10^4$.

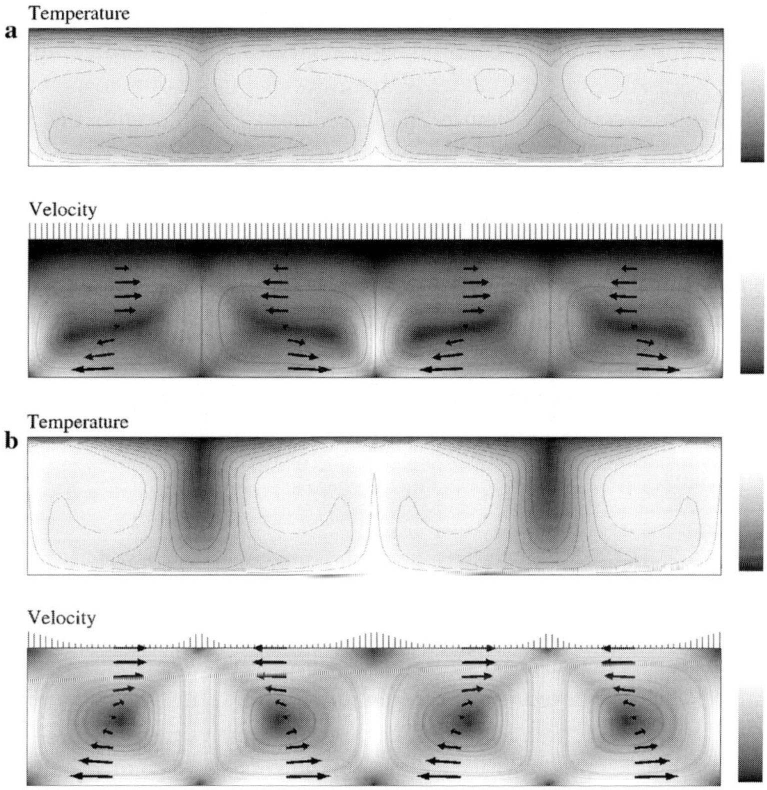

Figure 2
(a) Typical temperature and velocity distributions for an episodic convection mode at a minimum of the Nusselt number (refer to Fig. 1). The length of the vertical lines on top of the velocity plot are inversely proportional to the cold boundary velocities. (b) Typical temperature and velocity distributions for an episodic convection mode at a maximum of the Nusselt number (refer to Fig. 1). The length of the vertical lines on top of the velocity plot are inversely proportional to the cold boundary velocities.

afterward. Also shown in Figure 1 are the Nusselt numbers for mobile lid convection (obtained for $\tau_Y = \tau_0$) and stagnant lid convection ($\tau_Y \geq 6\tau_0$).

The length y of the vertical lines plotted along the cold (top) boundary of the velocity plots in Figures 2, 3 and 6 are given by:

$$y = \frac{\max|v_1| - |v_1|_{x_2=1}}{\max|v_1|}, \tag{19}$$

where $\max|v_1|$ is the magnitude of the largest horizontal velocity anywhere in the convection cell and $|v_1|_{x_2=1}$ is the local horizontal velocity on the top boundary surface. Within the framework of our model, $|v_1|_{x_2=1}$ is representative of the plate velocity. If $y = 0$ at a point on the boundary, then this point is mobile with a velocity of $\max|v_1|$. A comparison of the mobile lid convection mode (Fig. 3) with the stagnant lid phase of the episodic convection mode (Fig. 2a) indicates that the horizontal velocities on the cold boundary are largest for the mobile lid convection mode. Large segments of the top layer with values of y close to zero as in Figure 3 or during episodic convection in Figure 2b are indicative of plate like behaviour.

The isotherms and velocity arrows displayed on Figures 2a and 2b are representative for the stagnant lid phases (minima of Nusselt plots in Figure 1) and the subduction events (maxima of Nusselt plots) for an episodic convection mode. The isotherms and velocity arrows corresponding to the steady state of mobile lid convection is shown in Figure 3.

Next we investigate the influence of the choice of the initial temperature distribution. Will the three distinct modes of convection be stable with respect to a change of the initial conditions? We adopt a nonharmonic condition at t = 0 such that

$$T = \frac{1}{10}\sin(x_2\pi)\cos(x_1^2\pi/L) + (1 - x_2). \tag{20}$$

Figure 4 shows Nusselt number *vs.* time plots corresponding to the three cases considered before. The mobile lid mode settles at a slightly lower Nusselt number at steady state but the features of the isotherms and velocity arrows are qualitatively the same as for the case with harmonic initial conditions. Instead of the expected episodic convection mode we observed for harmonic initial conditions, we observe a tendency toward mobile lid convection, although the initial oscillation decays at a slower rate than for the mobile lid mode with $\tau_Y = \tau_0$. For non-harmonic initial conditions, the stagnant lid mode of Figure 5 becomes comparable to a dampened episodic convection mode. In this case, we observe an extended oscillatory phase but again ultimately the system settles into a steady state (observable in Figs. 4 and 5).

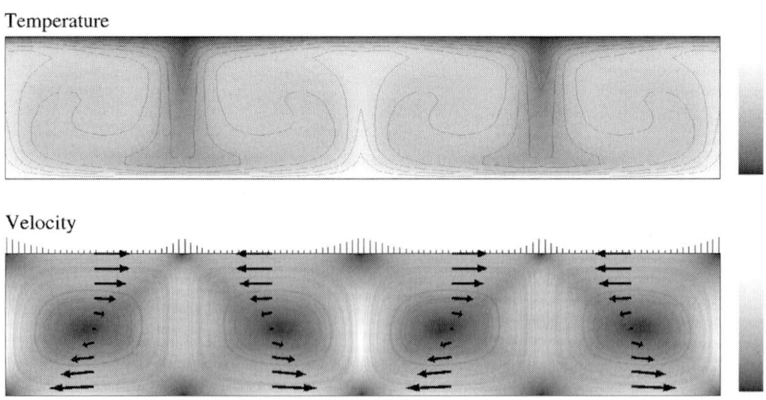

Figure 3
Typical temperature and velocity distributions at steady state for the mobile lid convection mode. The length of the vertical lines on top of the velocity plot are inversely proportional to the cold boundary velocities. For mobile lid convection, significant parts of the top layer move like rigid bodies.

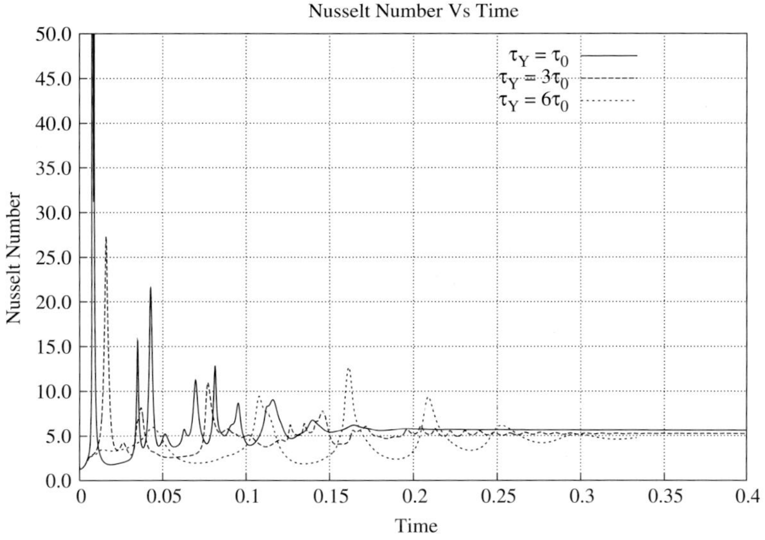

Figure 4
A comparison of Nusselt numbers for convection modes arising from nonharmonic initial conditions (20). The parameters are the same as those described in Figure 1.

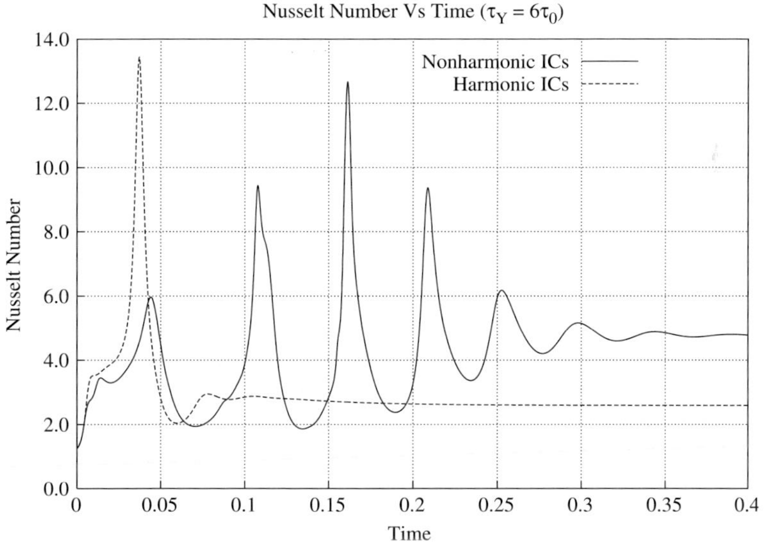

Figure 5
A comparison between Nusselt numbers for stagnant lid convection arising from a harmonic initial temperature distribution (18) and a nonharmonic initial temperature distribution (20).

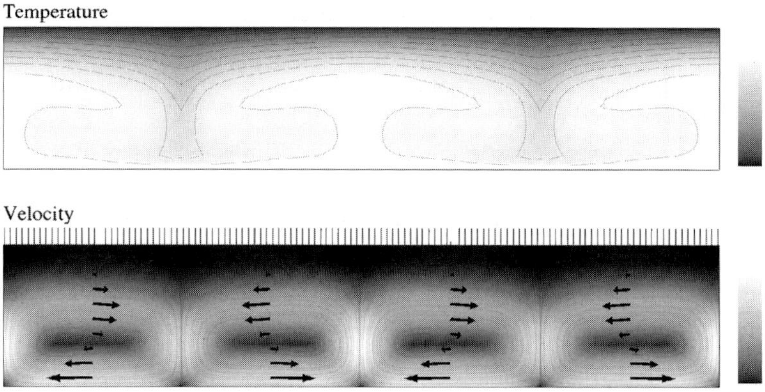

Figure 6
Temperature and velocity distributions at steady state for the stagnant lid convection mode arising from a nonharmonic initial temperature distribution (20). The length of the vertical lines on top of the velocity plot are inversely proportional to the cold boundary velocities. Here the horizontal velocities on top are only a small fraction of the horizontal velocities closer to the hot boundary (compare with Fig. 3).

A comparison of the distribution and length of vertical lines (see equation (19)) on top of the velocity plots of Figures 5 and 6, indicates that, in contrast to the mobile lid convection mode, the horizontal velocities on the cold boundary are much smaller than the horizontal velocities closer to the hot boundary.

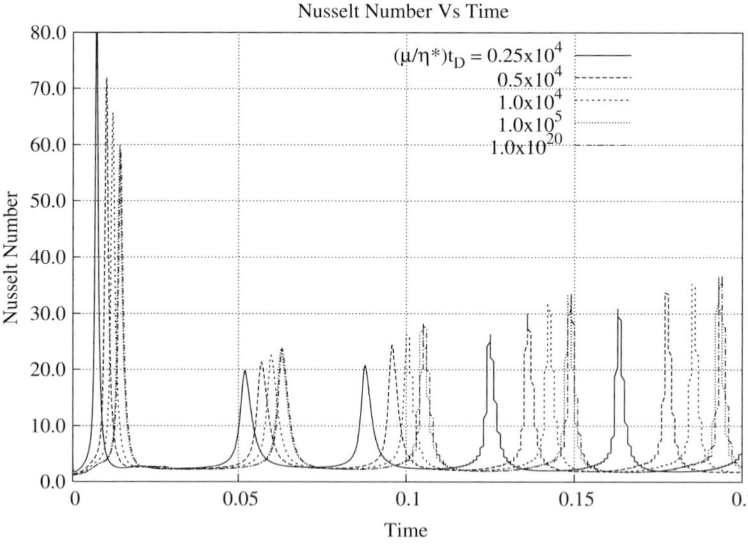

Figure 7
Influence of the elastic shear modulus: $(\mu/\eta^*)t_D = 0.25 \times 10^4, 0.5 \times 10^4, 10^5, 10^{20}$; $n = 3$, $n_{pl} = 15$, $Ra^c = 10^4$, $\tau_0 = 0.866\ 10^{2.5}$, $\tau_Y = 3\tau_0$.

The influence of the elastic shear modulus becomes important if $(\mu/\eta^*)t_D < 10^5$. A comparion of the results reveals that there is little difference between the $(\mu/\eta^*)t_D = 10^5$ and the $(\mu/\eta^*)t_D = 10^{20}$ cases. Smaller shear moduli cause a shortening of the period between the subduction events.

By multiplying (9) with v_i, integration of the resulting expression over the domain V and application of Gauss theorem we obtain the total mechanical power, $\dot{W}$, as $\dot{W} = RaV(Nu - 1)$. The reader is referred to PARMENTIER *et al.* (1976) for details of the derivation. This equation relates the area under the graph of a Nusselt plot to the mechanical work associated with the convection process. In particular, the area under a spike of the Nusselt plot is proportional to the work associated with a subduction event. This area decreases with the decreasing value of the elastic shear modulus. It is remarkable that this effect is evident in a global energy measure such as the Nusselt number.

Conclusions

We have outlined a formulation for visco-elastic convection based on a combined Newtonian and composite power-law rheology where the effect of plastic yielding is considered as a power-law term with a high ($n_{pl} \approx 15$) power-law-coefficient (6). The model is suitable for studying the geodynamics of mantle convection amongst other problems in geophysical flow. The non-linear equations of motion are solved incrementally based on a consistent tangent formulation producing second-order accurate results and avoiding the computationally expensive iterations that are otherwise necessary to handle the onset of plastic yield in most cases. In MORESI and SOLOMATOV (1998) and TACKLEY (1998), plastic yielding is considered by introducing an upper limit to the viscosity given by the ratio of the yield stress and the equivalent viscous strain rate. Since the strain rate at the current time is unknown, an initial estimate has to be based on the strain rate from the last time step producing a result of first-order accuracy. For this reason, a time consuming, iterative approach is necessary. The iterative approach is often more time-consuming than the present incremental approach including the occasional iterative reduction of residuals. With the iterative approach, the constitutive operator is sparser than in the consistent incremental approach, which can be used to advantage.

The convection model with a strongly temperature-dependent viscosity has some unique characteristics: The produced strains can be very large, necessitating a fluid-dynamics formulation, yet the relaxation time in the cool thermal boundary layer is significant compared to the characteristic time associated with fluid flow. In the bulk of the fluid, the relaxation time is small compared to the time taken for convective features to evolve due to the much lower viscosity of the warm fluid.

As elastic stresses in the strongly convecting part of the system relax rapidly, the introduction of elasticity does not produce a qualitative change to the stagnant lid convection regime (see SOLOMATOV, 1995). In the episodic and mobile lid modes, there is competition between the build-up of stresses in the cool lid and the stress-limiting effect of the yield criterion. Our results show that the introduction of elastic deformation does not quantitatively influence this balance either, although we do expect a difference in the distribution of stresses in the lid, which explains the variation in the onset of overturns and their increasing frequency which we observed as the elastic shear modulus was reduced. It should be noted that a global flow characteristic such as the Nusselt number may not be suitable to investigate the effect of elasticity. This is because the significant elastic effect is limited to the plate (or top boundary layer) which occupies only a small fraction of the total volume of the convection cell. In a future study we plan to differentiate between the energy rates associated with elastic, Newtonian creep, power-law creep and plastic deformation, evaluated in those parts of the convection cell in which plate like behavior is expected (e.g., for $T < 1400K$, see SCHUBERT *et al.*, 2001, p.26).

The presence of an elastic deformation mechanism also permits significant deformation of the highly viscous lid with lower energy rates as the plate is able to bend without necessarily having to form a plastic hinge. This is reflected in lower energy levels during episodic overturns which we observed by integrating the system Nusselt number. In the Earth, this effect may be important in subduction zones where the prediction of dissipation rates due to slab bending is larger than what seems physically plausible in this context (CONRAD and HAGER, 1999).

We observed a breakdown in the highly regular boundary layer turnover time when transitioning from harmonic initial condition to a non-harmonic initial condition.

This is similar to the results of MORESI and SOLOMATOV (1998), who noted for the purely viscous case that the regularity of the episodic mode was an artifact of the small convection domains. Once we break the perfect symmetry of the convection pattern, the evolution also becomes significantly more time-dependent.

Acknowledgment

Support is gratefully acknowledged by the Australian Computational Earth Systems Simulator Major National Research Facility (ACcESS MNRF), the Queensland State Government, The University of Queensland, and SGI. The Australian Commonwealth Government, participating institutions, and the Victorian State Government fund the ACcESS MNRF.

Appendix

The constitutive relation (3) can be rewritten as:

$$D_{ij} = \frac{1}{2\mu}\dot{\sigma}'^{J}_{ij} + \left(\frac{1}{2\eta_N} + \frac{1}{2\eta_N(\frac{\tau}{\tau_0})^{1-n}}\right)\sigma'_{ij} + \dot{\gamma}^p \frac{\sigma'_{ij}}{2\tau}, \tag{A1}$$

where we have adopted a power-law expression for the equivalent plastic strain rate, $\dot{\gamma}^p$, of the form:

$$\dot{\gamma}^p = \frac{\tau_Y}{\eta_Y}\left(\frac{\tau}{\tau_Y}\right)^{n_{pl}}. \tag{A2}$$

The coefficient of $\sigma'_{ij}/2$ in (A1) obtained by inserting (A2) into (A1) is equal to the inverse of the effective viscosity (6). The incremental form of our constitutive relationship is obtained by inserting the expansions $\sigma'^{t+\delta t}_{ij} = \sigma^t_{ij} + \delta\sigma_{ij}$, $T^{t+\delta t} = T^t + \delta T$, $p^{t+\delta t} = p^t + \delta p$ into (A1). Retaining only the linear terms in the stress, temperature and pressure increments yields:

$$D_{ij} - \frac{\delta t}{2\eta_{\text{eff}}}(W_{ik}\sigma_{kj} - \sigma_{ik}W) - \frac{1}{2\mu}\sigma'_{ij,k}v_k = \frac{1}{2\mu_{\text{eff}}\delta t}\delta\sigma'^{J}_{ij} + \underbrace{\left(\frac{1}{h_{vis}} + \frac{1}{h_Y}\right)}_{1/h}\frac{\sigma'_{ij}}{2\tau}\frac{\sigma'_{kl}}{2\tau}\delta\sigma'^{J}_{kl}$$
$$+ \frac{1}{2\eta_{\text{eff}}}\sigma'_{ij} + \frac{1}{2\eta_{\text{eff}}}\sigma'_{ij}\left(\frac{\delta T}{h_T} + \frac{\delta p}{h_p}\right) \tag{A3}$$

where

$$\dot{\sigma}'_{ij} = \sigma'_{ij,t} + \sigma'_{ij,k}v_k, \tag{A4}$$

$$\delta\sigma^J_{ij} = (\sigma_{ij,t} - W_{ik}\sigma_{kj} + \sigma_{ik}W)\delta t, \quad \delta T = T_{,t}dt, \quad \delta p = p_{,t}dt, \tag{A5}$$

$$h_{vis} = \frac{1}{(n-1)}\eta_N\left(\frac{\tau}{\tau_0}\right)^{1-n}, \quad h_Y = \frac{1}{(n_{pl}-1)}\eta_Y\left(\frac{\tau}{\tau_Y}\right)^{1-n_{pl}}, \tag{A6}$$

$$h_T = \frac{T^2}{AT_M}, \quad h_p = -\frac{T}{A\frac{\partial T_M}{\partial p}}, \tag{A7}$$

$$\mu_{\text{eff}} = \left(\frac{1}{\mu} + \frac{\delta t}{\eta_{\text{eff}}}\right)^{-1}. \tag{A8}$$

Here, η_{eff} and η_{N} are defined by (6) and (7) respectively. Inversion of (A3) yields:

$$\delta\sigma_{ij}^{\prime J} = (\mu_{\text{eff}}\delta t(\delta_{ik}\delta_{jl} + \delta_{jk}\delta_{il}) - \left(\frac{(\mu_{\text{eff}}\delta t)^2}{h + \mu_{\text{eff}}\delta t}\frac{\sigma'_{ij}}{\tau}\frac{\sigma'_{kl}}{\tau}\right)\left(D_{kl} - \frac{\sigma'_{kl,m}v_m}{2\mu}\right)$$

$$-\frac{\mu_{\text{eff}}\delta t}{\eta_{\text{eff}}}\left(1 - \frac{\mu_{\text{eff}}\delta t}{h + \mu_{\text{eff}}\delta t}\right)\left(1 + \frac{\delta T}{h_T} + \frac{\delta p}{h_p}\right)\sigma'_{ij} - \frac{\mu_{\text{eff}}\delta t^2}{\eta_{\text{eff}}}(W_{ik}\sigma_{kj} - \sigma_{ik}W), \qquad \text{(A9)}$$

where

$$h = \frac{h_{vis}h_Y}{h_{vis} + h_Y}. \qquad \text{(A10)}$$

The function h_p in (A3) and (A9) considers the pressure dependence of T_M. The slope of the $T_M - p$ curve is usually assumed as constant; a typical order of magnitude value for the slope is 10^{-7} K Pa^{-1}. For $n = 1$ and $\mu \to \infty$ we obtain Newtonian flow with the viscosity ½ η_N .

In many practical applications the velocity-pressure incompressibility and the heat equation are solved and advanced sequentially. In this case, the terms associated with δT and δp in equation (A9) are not needed. In linear instability analyses, however, the full incremental form (A9) is required.

In the viscous limit, $\mu \to \infty$, (A9) reduces to

$$\delta\sigma'_{ij} = \left(\eta_{\text{eff}}(\delta_{ik}\delta_{jl} + \delta_{jk}\delta_{il}) - \frac{\eta_{\text{eff}}^2}{h + \eta_{\text{eff}}}\frac{\sigma'_{ij}}{\tau}\frac{\sigma'_{kl}}{\tau}\right)D_{kl} - \left(1 - \frac{\eta_{\text{eff}}}{h + \eta_{\text{eff}}}\right)\left(1 + \frac{\delta T}{h_T} + \frac{\delta p}{h_p}\right)\sigma'_{ij}. \qquad \text{(A11)}$$

In steady states, the stress, temperature and pressure increments vanish so that the remaining terms in (A11) cancel. This cancellation is indeed the case as can be shown by insertion of $D_{ij} = 1/2\eta\sigma'_{ij}$.

The limit $\eta_N \to \infty$ does not yield a simpler expression for the incremental relationship (A9) since the effective moduli still depend on the viscosity $\eta_Y(\tau_Y/\tau)^{n_{pl}-1}$. However, in the rate-independent limit as $n_{pl} \to \infty$, this viscosity tends to infinity so that the expression (A14) reduces to

$$\partial\sigma'_{ij}/\partial t = \left(\mu(\delta_{ik}\delta_{jl} + \delta_{jk}\delta_{il}) - \mu\frac{\sigma'_{ij}}{\tau}\frac{\sigma'_{kl}}{\tau}\right)\left(D_{kl} - \frac{1}{2\mu}\sigma'_{ij,k}v_k\right) + (W_{ik}\sigma_{kj} - \sigma_{ik}W_{kj}). \qquad \text{(A12)}$$

In the above derivation it was assumed that the yield stress is constant. A large deformation model with power-law plasticity and state variable dependence of the yield stress will be presented in a forthcoming paper.

References

Byerlee, J. (1965), *The brittle ductile transition in rocks*, J. Geophys. Res. *73*, 4741–4750.

Conrad, C.P. and Hager, B.H. (1999), *Effects of plate bending and fault strength at subduction zones on plate dynamics*, J. Geophys. Res. *104* (B8), 17551–17571.

Davies, M., Gross, L., and Mühlhaus, H.-B. (2004), *Scripting high performance Earth systems simulations on the SGI Altix 3700*, Proc. 7th Int. Conf. HPC Grid Asia Pac. Reg., pp. 244–251.

Kolymbas, D. and Herle, I. (2003), *Shear and objective stress rates in hypoplasticity*, Int. J. Numer. Anal. Meth. Geomech. *27*, 733–744.

Moresi, L., Dufour, F., and Mühlhaus, H.-B. (2002), *Mantle convection models with viscoelastic/brittle lithosphere: Numerical methodology and plate tectonic modeling,* Pure Appl. Geophys. *159* (10), 2335.

Moresi, L. and Solomatov, V.S. (1998), *Mantle convection with a brittle lithosphere: thoughts on the global tectonic styles of the Earth and Venus*, Geophys. J. Int. *133* (6), 669–682.

Mühlhaus, H.-B. and Regenauer-Lieb, K. (2005), *A Self-consistent plate mantle model that includes elasticity: Computational aspects and application to basic modes of convection,* Geophys. J. Int., still subm.

Parmentier, E.M., Turcotte, D.L., and Torrance, K.E. (1976), *Studies of finite amplitude non-Newtonian thermal convection with application to convection in the Earth mantle*, J. Geophys. Res. *81*, 1839–1846.

Rudnicki, J.W. and Rice, J.R. (1975), *Conditions for the localisation of deformation in pressure-sensitive dilatant materials*, J. Mech. Phys. Sol. *23*, 371–394.

Schubert, G., Turcotte, D.L., and Olson, P., *Mantle Convection in the Earth and Planets* (Cambridge University Press, Cambridge 2001).

Solomatov, V.S. (1995), *Scaling of temperature and stress dependent viscosity convection*, Phys. Fluids *7*, 266–274.

Tackley, P. (1998), *Self-consistent generation of tectonic plates in three-dimensional mantle convection,* Earth and Planet. Sci. Lett. *157*, 9–22.

Zienkiewicz, O.C. and Taylor R.L., *The Finite Element Method*, vol. 3, 5th Ed., (Butterworth/Heinemann 2000).

(Received November 2, 2004, revised September 29, 2005, accepted December 30, 2005)
Published Online First: September 5, 2006

Notes to Authors

PAGEOPH welcomes original contributions in English (and occasionally in French and German) on all aspects of geophysics. All manuscripts should be submitted to the Regular Issues Editor-in-Chief, in triplicate, formatted with double spacing and wide margins. For further details see the following paragraphs.

Format of Manuscripts

Length and Page Charges: A paper should not exceed 16 printed pages including tables and figures. For articles exceeding 16 printed pages the authors will be charged sFr. 80.00 for each additional page. No page charges, except those for color prints, are required for contributors to special issues.

Title Page: This should include the the complete title, full names and addresses of all authors. In addition, corresponding authors should provide their fax number and e-mail address if they are available.

Abbreviated Title: It is necessary to indicate an abbreviated title, which will be used as a running head (no more than 50 characters including spaces).

Abstract: The abstract should be in English, and in the language of the text, if different. It has to be of no more than 10 sentences and should be concise and self-contained.

Keywords: Up to 6 keywords should be listed, suitable for incorporation into information-retrieval systems.

Text: The text must include a citation for each item listed under References; the approximate position of each figure should be indicated in the text. The metric system should be used throughout the text, figures, and tables.

Tables: Tables are to be presented on separate pages, with a brief title for each.

Figures: Figure captions and legends are to be typed on a separate page or pages as the last element in the manuscript. Make sure that line thickness and lettering allow an adequate size reduction. Heliographic or photocopies are not suitable for reproduction. Highquality, glossy, photographic prints must be submitted. Color prints are permitted but authors will be charged for them.

References: They are to be listed in alphabetical order in the following style:
Journal article: Haurwitz, B., and Cowley, A.D. (1973), The Diurnal and Semidurnal Barometric Oscillations, Global Distribution and Annual Variation, Pure Appl. Geophys. 102, 193-222.

Whole book: Bath, M., Introduction to Seismology (Birkhäuser, Basel 1973).

Article in a book: Haurwitz, B., and Cowley, A.D., Barometric oscillations, In Introduction to Seismology (ed. Bath. M.) (Birkhäuser, Basel 1973) pp. 193-222.

Submission of Manuscripts

Manuscripts must be submitted in triplicate, formatted with double line spacing and wide margins. Copies of the figures should be attached at the end of the manuscript. High-quality, prints may be submitted later. All manuscript pages, including references, tables, and captions, should be numbered consecutively, starting with the title page as page one.

All manuscripts should be submitted to Regular Issues Editor-in-Chief
Brian Mitchell
Department of Earth & Atmospheric Sciences
Saint Louis University
3507 Laclede Avenue
St. Louis, MO 63103, USA
e-mail: mitchbj@eas.slu.edu

The final version of a manuscript accepted for publication in PAGEOPH should be sent as both a hard copy and on diskette to the Editor-in-Chief. Delivering manuscripts in electronic form may substantially facilitate the publication process provided certain points are taken into consideration:

- Texts should be delivered in Ms Word or Wordperfect
- Mathematics in Tex, LaTex and any Tex formats
- Figures in TIFF (high resolution) or EPS

The electronic and printed version must be absolutely identical. All pictorial and graphic illustrations should be delivered as hard copy originals. Do not fail to include a hard copy for ready vicwing. Back-up copies of the diskettes must be kept. Diskettes must be adequately protectedfor transport.

Galley Proofs
Unless indicated otherwise, galley proofs will be sent to the first-named author directly from Birkhäuser Verlag AG and should be returned with the least possible delay. Textual alterations made in the galley proof stage will be charged to the author. One copy of the corrected proof is to be returned immediately.
The editorial office assumes no responsibility for delayed proofs, errors in the original manuscript, or major alterations in proofs for any reason.

Reprints
The authors will receive 50 reprints of each article without charge. Additional reprints may be ordered in lots of 50 when the final corrected page proofs are returned. Orders submitted thereafter are subject to considerably higher rates.